Studies in Systems, Decision and Control

Volume 261

Series Editor

Janusz Kacprzyk, Systems Research Institute, Polish Academy of Sciences, Warsaw, Poland

The series “Studies in Systems, Decision and Control” (SSDC) covers both new developments and advances, as well as the state of the art, in the various areas of broadly perceived systems, decision making and control–quickly, up to date and with a high quality. The intent is to cover the theory, applications, and perspectives on the state of the art and future developments relevant to systems, decision making, control, complex processes and related areas, as embedded in the fields of engineering, computer science, physics, economics, social and life sciences, as well as the paradigms and methodologies behind them. The series contains monographs, textbooks, lecture notes and edited volumes in systems, decision making and control spanning the areas of Cyber-Physical Systems, Autonomous Systems, Sensor Networks, Control Systems, Energy Systems, Automotive Systems, Biological Systems, Vehicular Networking and Connected Vehicles, Aerospace Systems, Automation, Manufacturing, Smart Grids, Nonlinear Systems, Power Systems, Robotics, Social Systems, Economic Systems and other. Of particular value to both the contributors and the readership are the short publication timeframe and the world-wide distribution and exposure which enable both a wide and rapid dissemination of research output.

** Indexing: The books of this series are submitted to ISI, SCOPUS, DBLP, Ulrichs, MathSciNet, Current Mathematical Publications, Mathematical Reviews, Zentralblatt Math: MetaPress and Springerlink.

More information about this series at http://www.springer.com/series/13304

Andrey E. Gorodetskiy · Irina L. Tarasova
Editors

Smart Electromechanical Systems

Situational Control

With 132 Figures and 15 Tables

Editors
Andrey E. Gorodetskiy
Institute for Problems in Mechanical
Engineering of the Russian
Academy of Sciences
Saint Petersburg, Russia

Irina L. Tarasova
Institute for Problems in Mechanical
Engineering of the Russian
Academy of Sciences
Saint Petersburg, Russia

ISSN 2198-4182 ISSN 2198-4190 (electronic)
Studies in Systems, Decision and Control
ISBN 978-3-030-32712-5 ISBN 978-3-030-32710-1 (eBook)
https://doi.org/10.1007/978-3-030-32710-1

© Springer Nature Switzerland AG 2020
This work is subject to copyright. All rights are reserved by the Publisher, whether the whole or part of the material is concerned, specifically the rights of translation, reprinting, reuse of illustrations, recitation, broadcasting, reproduction on microfilms or in any other physical way, and transmission or information storage and retrieval, electronic adaptation, computer software, or by similar or dissimilar methodology now known or hereafter developed.
The use of general descriptive names, registered names, trademarks, service marks, etc. in this publication does not imply, even in the absence of a specific statement, that such names are exempt from the relevant protective laws and regulations and therefore free for general use.
The publisher, the authors and the editors are safe to assume that the advice and information in this book are believed to be true and accurate at the date of publication. Neither the publisher nor the authors or the editors give a warranty, expressed or implied, with respect to the material contained herein or for any errors or omissions that may have been made. The publisher remains neutral with regard to jurisdictional claims in published maps and institutional affiliations.

This Springer imprint is published by the registered company Springer Nature Switzerland AG
The registered company address is: Gewerbestrasse 11, 6330 Cham, Switzerland

Preface

Smart electromechanical systems (SEMS) are used in cyber-physical systems (CPhS). Cyber-physical systems are the ability to integrate computing, communication and storage of information, monitoring and control of the physical world objects. The main tasks in the field of theory and practice CPhS are to ensure the efficiency, reliability and safety of functioning in real time. It is important to bear in mind that recently, the task of ensuring the functioning of robots as a part of collective performing joint work is set. In this case, control consists in making control decisions as problems arise in accordance with the dynamically changing environment of choice. This requires another level of control, which would be the interface between the group and the operator, setting targets, and which can be attributed to the optimization problems of situational control. Moreover, the operator or decision maker can be not only a person but also a computer program that makes decisions.

The purposes of the publication is to introduce the latest achievements of scientists of the Russian Academy of Sciences in the theory and practice of SEMS Situational Control and development of methods for design and simulation of SEMS Situational Control based on the principles of safety, flexibility and adaptability in behavior and intelligence and parallelism in information processing, computation and control.

Topics of primary interest include, but are not limited to the following:

Problems and principles of situation control
Methods and algorithms of situational control
Information and measuring support of situational control systems
Simulation of situation control.

This book is intended for students, scientists and engineers specializing in the field of smart electromechanical systems and robotics and includes many scientific domains such as receipt, transfer and pre-treatment measurement information, decision making theory, control theory, working bodies of robots that imitate the complexity and adaptability of biological systems.

We are grateful to many people for the support received during the writing of this book. A list of their names cannot be represented here, but all of them we are deeply grateful.

Saint Petersburg, Russia
August 2019

Andrey E. Gorodetskiy
Irina L. Tarasova

Contents

Contributors

A. K. Akhmerov ITMO University, Saint Petersburg, Russia

Pavel P. Belonozhko Bauman Moscow State Technical University, Moscow, Russia

Daria A. Drozdova ITMO University, Saint Petersburg, Russia

Ivan L. Ermolov Ishlinsky Institute for Problems in Mechanics of the Russian Academy of Sciences, Moscow, Russia

Alexander Ya. Fridman Institute for Informatics and Mathematical Modelling, Kola Science Centre of RAS, Apatity, Russia

Alexey A. Gorbachev ITMO University, Saint Petersburg, Russia

Andrey E. Gorodetskiy Institute of Problems of Mechanical Engineering, Russian Academy of Sciences, Saint Petersburg, Russia

Valery G. Gradetsky Ishlinsky Institute for Problems in Mechanics of the Russian Academy of Sciences, Moscow, Russia

Jianwen Huo Bauman Moscow State Technical University, Moscow, Russia

Anatoliy P. Karpenko Bauman Moscow State Technical University, Moscow, Russia

Maxim M. Knyazkov Ishlinsky Institute for Problems in Mechanics of the Russian Academy of Sciences, Moscow, Russia

Igor A. Konyakhin ITMO University, Saint Petersburg, Russia

Valery V. Korotaev ITMO University, Saint Petersburg, Russia

Dmitry A. Kozlov MIREA—Russian Technological University, Moscow, Russia

Andrey Yu. Kuchmin Institute of Problems of Mechanical Engineering, Russian Academy of Sciences, Saint Petersburg, Russia

Boris A. Kulik Institute of Problems in Mechanical Engineering, Russian Academy of Sciences (RAS), Saint Petersburg, Russia

Vugar G. Kurbanov Institute of Problems of Mechanical Engineering, Russian Academy of Sciences, Saint Petersburg, Russia;
Saint-Petersburg State University of Aerospace Instrumentation, Saint Petersburg, Russia

Ilia A. Leshchev Bauman Moscow State Technical University, Moscow, Russia

Anastasiya Y. Lobanova ITMO University, Saint Petersburg, Russia

Tong Minh Hoa ITMO University, Saint Petersburg, Russia

Anaid V. Nazarova Bauman Moscow State Technical University, Moscow, Russia

Ivan S. Nekrylov ITMO University, Saint Petersburg, Russia

Anton A. Nogin ITMO University, Saint Petersburg, Russia

Hoang Anh Phuong ITMO University, Saint Petersburg, Russia

Victoria A. Ryzhova ITMO University, Saint Petersburg, Russia

Sergey N. Sayapin Blagonravov Mechanical Engineering Research Institute of the Russian Academy of Sciences, Moscow, Russia;
Bauman Moscow State Technical University, Moscow, Russia

Eugeny A. Semenov Ishlinsky Institute for Problems in Mechanics of the Russian Academy of Sciences, Moscow, Russia

Vladimir A. Serov MIREA—Russian Technological University, Moscow, Russia

Leonid V. Smirnov ITMO University, Saint Petersburg, Russia

Artem N. Sukhanov Ishlinsky Institute for Problems in Mechanics of the Russian Academy of Sciences, Moscow, Russia

Irina L. Tarasova Institute of Problems of Mechanical Engineering, Russian Academy of Sciences, Saint Petersburg, Russia

Alexander N. Timofeev ITMO University, Saint Petersburg, Russia

A. S. Vasilev ITMO University, Saint Petersburg, Russia

A. V. Vasileva ITMO University, Saint Petersburg, Russia

Evgeny M. Voronov Bauman Moscow State Technical University, Moscow, Russia

Stanislav L. Zenkevich (Deceased)

Problems and Principles of Situational Control

The Principles of Situational Control SEMS Group

Andrey E. Gorodetskiy

Abstract *Problem statement*: Solving the problems of situational control of SEMS modules in complex robotic systems (CRS) play an important role in the intellectualization of the CRS. The article describes and analyzes the principles of situational control SEMS group in CRS. *Purpose*: Statement of the problem of situational control group SEMS and analysis of the principles of situational control in terms of stochasticity and uncertainty of the environment of choice. *Results*: The concept of situational control is considered, the generalized mathematical description of a problem of situational control of group SEMS is reccived, the methodology and various approaches to the organization of situational control of group SEMS are analyzed. *Practical significance*: The possibility of realization of the proposed mathematical formulation of the problem of situational control of SEMS by means of computer technology with parallel organization of calculations is shown.

Keywords Situation control · SEMS groups · Selection environment · Deterministic · Stochastic and not fully defined constraints · Fuzzy mathematical models · Operations and algorithms for finding optimal solutions

1 Introduction

Situational control (from lat situatio-position) is the operational control of a group of interacting dynamic objects with appropriate behavior [1]. Such control consists in making control decisions as problems arise in accordance with the dynamically changing environment of choice [2]. This requires another level of control, which would be the interface between the group and the operator, setting targets [3, 4], and which can be attributed to the optimization problems of situational control [5, 6]. Moreover, the operator or decision-maker (DM) can be not only a person, but also a computer program that makes decisions. In the works [7–9] on situational management of enterprises and other socio-economic objects under the environment

A. E. Gorodetskiy (✉)
Institute of Problems of Mechanical Engineering, Russian Academy of Sciences, St. Petersburg, Russia
e-mail: g27764@yandex.ru

© Springer Nature Switzerland AG 2020

A. E. Gorodetskiy and I. L. Tarasova (eds.), *Smart Electromechanical Systems*, Studies in Systems, Decision and Control 261, https://doi.org/10.1007/978-3-030-32710-1_1

of choice understand the management situation as a specific set of circumstances that have a significant impact on the work of the organization at the moment. Pospelov [5] expands the concept of the situation, adding to it information about the relationships between objects and says that "the current situation is a set of all information about the structure of the object and its functioning at a given time." All information also includes cause-and-effect relationships, which can be expressed in a variety of sequential events or processes. In this sense, the situation is fundamentally different from the state and events that can only correspond to one point in time.

In case of situational control of the SEMS group, similarly to the group of robots, the environment of choice or the environment of the control problem can be understood as a subjective assessment of specific characteristics of SEMS and the external environment (situational variables) and the relationships between them that take place at the present time, but depend on the events that occurred and developing in time and space [10–14].

The creation and development of situational control systems of the SEMS group requires a lot of resources to collect information about the objects and the control environment, their dynamics, control methods, as well as to systematize this information within the semiotic model. Therefore, it is considered that the method of situational control is appropriate to apply only in cases where other methods of formalization lead to the problem of too large (for practical implementation) dimension [15].

2 The Concept of Situational Management

The concept of situational control of a group of interacting dynamic objects can be reduced to the following system of basic provisions:

- there is no universal approach to control, as it depends on the specific problem situation in the environment of choice, requiring different approaches to their resolution;
- situational probabilistic and not fully defined factors are taken into account in the strategies, structures and control methods, thereby achieving effective decision-making;
- there is more than one way to achieve the control goal and it is necessary to find the optimal way;
- the results of decision-making on control in the same environment of choice due to its stochasticity and incomplete certainty may differ significantly from each other;
- any decision on control should be considered only in close connection with other problems of interacting dynamic objects;
- decision-makers (computer programs) can adapt the characteristics of interacting dynamic objects to the situation (the environment of choice) or change the situation according to the requirement of the control goal;

- decision-makers should have psychologically comfortable conditions for interaction with controlled dynamic objects.

Thus, situational control is primarily the art of the decision-maker, to correctly identify and assess the situation and to choose the most effective control methods that best meet the situation and control goals. In this case, the control process should consist of the following mandatory steps that must be implemented by the control system to achieve effective control in each specific situation:

- creating the necessary conditions for changes in the dynamic configuration space of managed interacting dynamic objects;
- creation of a database and knowledge base on the environment and management facilities;
- identification and analysis of the situation in the environment of choice;
- selection of approaches and methods of optimal control in the current situation;
- development of control actions for the conduct required to achieve control objectives change in the dynamic objects;
- assessment of the likely consequences of situational control.

By formalizing the environment of choice or environment tasks of situational control you must meet the following conditions:

- the selection environment must contain a finite number of factors and describe their state and relationships;
- the environment of choice should contain factors that significantly affect the objects of control, since it is impossible to take into account the influence of absolutely all factors when making a decision;
- the selection environment must contain factors that affect the interaction of control objects in the group;
- in the environment of choice, it is necessary to take into account the causes and consequences of possible control situations.

Compliance with the latter condition leads to the need to classify control situations. This is due to the fact that their recognition is the first stage of the process of resolving situational control tasks. To date, a large number of classifications of control situations with different classification characteristics and depth of decomposition have been developed. As a basis for the analysis and resolution of control situations can be used a model based on the account of a number of sources of control situations in the environment of the choice of a group of robots, considering their content characteristics and the use of known strategies for resolving situations to control a group of SEMS. The following main types of restrictions can be distinguished:

- technological, which are determined by the type and flexibility of interacting SEMS;
- intellectual, reflecting the levels of competence of the Mat. providing control computing systems that make decisions considering changes in the environment of choice;

– limitations in the formulation of the problem due to the actual nature of the work performed by the group of interacting SEMS.

3 The Principles of Situational Control SEMS Groups

In any approach to the organization of situational management of the SEMS group, it is necessary to collect information about the environmental parameters, the current state of individual SEMS from the group, the planned actions of the group members, etc. Generally after collecting the information, a model of the selection environment is created $O(t_k)$. Then the planning situation control group SEMS will be:

– in the division of group tasks into subtasks:

$$O(t_0) \underset{U(t_1)}{\Rightarrow} O(t_1) \ldots O(t_0) \underset{U(t_f)}{\Rightarrow} O(t_f) \tag{1}$$

where $U(t_k) = \{u_{a_1}(t_k), u_{a_2}(t_k), \ldots, u_{a_n}(t_k)\}$, $u_{a_i}(t_k)$—control action applied to a_i SEMS at a time t_k, $k = 0, 1, \ldots f$, in distribution between SEMS of group of solutions of subtasks so that the solution of a group task was carried out in the minimum time taking into account the available restrictions, including on information interaction.

In general, the solution of the group problem of situational control will be the synthesis of the search algorithm, i.e. the ordered set $\varpi \subset \Omega$ of the many alternative combinations of controls $U(t_k)$. This should be the best combination of management laws for each member of the SEMS group $u_{a_i}(t_k) \in U(t_k)$, obtained on the basis of quality estimates Q, constructed taking into account the system of preferences E and selection environments $O(t_k)$.

$$\varpi \subset 2^U x\, Q^U, \tag{2}$$

where 2^U—denotes the set of all subsets U, and Q^U—set of all quality ratings (tuples of length from 2 to $|U|$), x—sign of Cartesian product.

In order to make the most effective decisions in a given situation in an $O(t_k)$ selection environment and to make the appropriate changes in the group of interacting SEMS in the best possible way, the decision makers must have certain principles or rules containing fundamental requirements for effective control, the most important of which are the following:

– competence, i.e. the correct use of the mathematical apparatus in the formalization of the environment of choice and completeness of information about the current state of control objects and the environment;
– the ability to make decisions in the absence of precedents, when no control situation, no matter how standard it may seem, can not be absolutely similar to any situation that has taken place in the past;

- the presence of the relationship of situational variables, when all the factors of the situation constitute a single whole, a system and therefore somehow affect each other;
- the presence of dual influence of factors when situational factors have different, sometimes even contradictory characteristics;
- continuity of changes, when changes in control objects and their external environment, one way or another, occur constantly;
- irreversibility of changes when any control action puts the SEMS group on a new stage of configuration space change;
- the ability to react quickly when the constant change of situational variables requires continuous development of control decisions aimed at adapting the SEMS group to these changes;
- the presence of prerequisites for changes, when along with the constant monitoring of changes, it is necessary to continuously monitor the presence of prerequisites and conditions necessary to bring the parameters of the SEMS group in line with the changed situation;
- compliance with the optimal ratio of results and costs in the formation of the optimality criteria with the greatest approximation of the SEMS group to the goals;
- the possibility of a priori decision, which allows not only to correctly assess the situation and respond to its change in a timely manner, but also to anticipate possible changes in this situation;
- the ability to make decisions not only the formation of changes in the SEMS group when the situation changes, but also the partial adaptation of the environments of choice to the control goals;
- compliance with the psychological comfort of the interaction of the decision-maker with the control SEMS.

It is desirable that all the requirements and principles are implemented in the creation of situational management systems in conjunction. Their combination depends on the specific tasks solved by the SEMS group, the state of the environment of choice, the hardware and software of the situational control system and some other factors.

To implement these principles, certain methods of situational management of dynamic objects have been developed [16]. Most often in situational control methods used system and situational analysis, factor and cross-factor analysis, genetic analysis, diagnostic method, expert-analytical method, methods of analogy, morphological analysis and decomposition, simulation methods, game theory, etc. However, the greatest effect and quality of control are achieved when the system of methods is used in the complex, which allows you to see the control object from all sides and avoid miscalculations.

It is possible to distinguish the following main tasks solved at creation of systems of situational control of SEMS group:

- development of a system for monitoring the occurrence of critical situations requiring situational analysis;

- selection, adaptation and development of methods for collecting, analyzing and synthesizing information about the environment of choice;
- selection of methods of statistical data analysis;
- definition and formalization of reference situations for each environment of choice;
- formation and updating of the database of situations;
- preparation of instrumentation, including the Mat. apparatus for determining the factors characterizing the development of the situation and indices for assessing their condition;
- definition and actualization of the factors characterizing the state of the situation, assessment of their comparative importance, development of indices of the state of the situation;
- selection and adaptation of methods of formation of evaluation systems;
- development of scenarios for the possible development of situations, including a meaningful description and definition of a list of the most likely scenarios (options) of situations, the formation of a list of the main factors affecting the development of the situation, the identification of the most likely factors that affect the development of the situation, and discarding those factors which can not have a significant impact on the change in the situation;
- analysis of options for the development of situations to identify the main dangers, threats, risks, strengths, prospects in the development of the situation;
- development of expert forecast of changes in factors and indices characterizing the situation, presented in the form of the most likely scenarios of the situation with the assessment of the stability of situations for the developed alternative scenarios of their development;
- assessment of the development of the situation in terms of the possibility of achieving the goals of the SEMS group;
- synthesis of alternative management solutions and control actions to achieve the goals of the SEMS group;
- development of recommendations for strategic and tactical decision-making in the analyzed situation; on the mechanisms of their implementation; on control over the implementation of decisions; on support of the implementation of decisions; on the analysis of results, including an assessment of the effectiveness of decisions taken and the effectiveness of their implementation;
- development of methods for the selection of the most effective solutions for the situational control of the SEMS group under conditions of incomplete definiteness;
- development of interfaces between the decision maker and the SEMS group, providing psychologically comfortable interaction conditions.

The development of interfaces between decision-makers and SEMS one of the main objectives is the establishment of psychologically comfortable conditions of interaction of the DM with the group control SEMS. This issue has recently received a lot of attention. In particular, in [17] analyzes the ways of assessing the success of the dialogue, as well as the means of registration and analysis of preferences of DM, necessary to configure the interfaces of DM and intelligent robots, and concludes the need to take into account the psychological aspects of their interaction. In [18] there is a need to equip robots with tools that would help him to respond to the rapid "decision-making", to exclude the choice of a bad decision and to act in such a way as to maximize the safety of people, thereby increasing human confidence in the robotic system. In [19] an approach to control the behavior of robots based on the mechanism of emotions and temperament is proposed. It is demonstrated that these psychological features can be modeled in quite simple ways. The proposed emotional architecture of the robot control system is based on the V. P. Simonov's information theory of emotions, and the features of temperament are reduced to a two—parameter model of the "excitation-inhibition" type. A number of experiments based on mobile robots are described. These experiments demonstrate a set of different types of robot behavior: melancholic, choleric, sanguine, and phlegmatic. All these types were implemented using the so-called temperament regulator, which determines the balance between the values of the excitation and inhibition parameters of the robot control system. The paper also proposes an automatic model of temperament, which allows describing the behavior of an individual. On the basis of this model, it is shown that in solving some problems of collective behavior it is advisable to have individuals with different behavior in a group. And this behavior is also determined by the individual emotions and temperament of the robot.

4 The Methodology of Situational Control SEMS Group

Situational control of the SEMS group uses the method of control of complex technical and organizational systems, based on the ideas of the theory of artificial intelligence and the representation of knowledge about the objects of control, ways to control them and the environment of choice at the level of logical-interval, logical-probabilistic and logical-linguistic models, the use of training as the main procedures in the construction of control procedures for current situations and the use of deductive systems to build multi-step solutions [20].

The solution to the problem of situational control of the SEMS group can be represented as the following sequence of operations (see Fig. 1):

- formalization on the basis of collected by the sensors of the Central nervous system SEMS in the database and knowledge of information about the current situation in the current environment of the choice of dynamic control objects creates its Mat. description that is supplied to the Analyzer input.
- analysis, that is, evaluation of received messages and determination of the need for intervention of the control system in the process occurring in the control objects. If the current situation does not require such intervention, the Analyzer does not send it for further processing. Otherwise, the description of the current situation enters the Classifier.
- classification using the information stored in the Analyzer and classification of the situation to one or more classes, which correspond to one-step solutions. The analyzer passes this information to the Correlator.

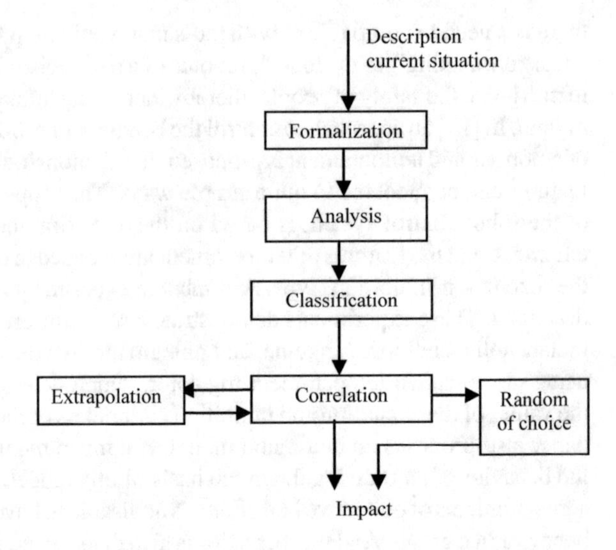

Fig. 1 Scheme of solving the problem of situational control

- define the SEMS group transformation rules to be used. All rules of transformation of the SEMS group in different situations are stored in the Correlator. These rules are called logical-transformational rules (LTR) or correlation rules. The full list of LTR determines the capabilities of the situational control system to influence the control objects to solve the task. The correlator determines the LTR, which should be used. If such rule is unique, it is issued for execution.
- if there are several such rules, the choice of the best of them is made after processing the preliminary decisions in the Extrapolator, after which the Correlator gives a decision on the impact on the object.
- if the Correlator or Classifier cannot make a decision on the received description of the current situation, the random Selection block is triggered and one of the influences that have not too much influence on the objects is selected, or the system refuses any impact on the objects. This suggests that the control system does not have the necessary information about its behavior in this situation.

In solving the problems of situational management is often used to build situational models that simulate the processes occurring in the control objects and the control system. They are constructed on the basis of the following basic operations:

1. creation of databases and knowledge about the environment, management objects and control system in the computer memory;
2. construction of models of control objects, the environment of the choice and the control system;
3. description of states of objects and environment of choice in the class of semiotic models;
4. formation of a hierarchical system of generalized descriptions of the States of control objects and the environment of choice;
5. classification of states (situations) for possible solutions;

6. forecasting the consequences of decisions;
7. training and self-study.

The necessity of operation 1 is determined by the need to turn on the computer into the control loop at the earliest possible stages of assessment and control search in order to increase the efficiency of the decision maker (DM). Typically, a decision maker in situational control systems uses semantics, for example, semantic networks in the form of a graph, whose nodes correspond to concepts and objects, and arcs to relations between objects.

The content of operation 3, complementing the second, is that the representation of all the necessary models is carried out with the help of elements of the language in which the DM describes the management system and its functioning. Usually DM in situational control systems uses semantics, such as semantic networks in the form of a graph, the nodes of which correspond to the concepts and objects, and the arc relationships between objects.

The differences between semiotic and formal systems are as follows:

- semiotic systems have the set of signs absent in formal systems, possessing, in particular, plans of expression (syntax) and content (semantics);
- unlike formal systems, semiotic systems can independently change their syntax and semantics;
- semiotic systems are open, not closed, as formal.

The main stages of the system based on situational models include:

- obtaining a description of the current situation available on the analyzed control object and the environment of choice;
- replenishment of micro description of the situation;
- classification of the situation and identification of classes of possible solutions to the used estimates of system;
- output of permissible values of estimates (in this case there is a reverse movement on hierarchical levels of knowledge representation of the situational model);
- predicting the impact of acceptable decisions as final estimates;
- decision making on estimates.

The work of such systems under conditions of incomplete certainty is based on the analysis of logical-probabilistic and logical-linguistic rules, logical inference and optimization procedures of mathematical programming in ordinal scales and generalized mathematical programming [20].

5 Conclusion

The problem of finding the optimal algorithm of situational control always arises when it is necessary that some group of SEMS together to perform some work.

The natural restriction on the time of optimal decision-making on situational control of the SEMS group in real time imposes restrictions on the number of members of the controlled group and the distances between them associated with the dynamics of the environment of choice and the dynamics of the controllability of the SEMS themselves.

The classification of situations, which consists in assigning the current situation to one or several classes corresponding to some control, allows to simplify and accelerate the planning of situational control of the SEMS group. If the resulting solution to the classification problem is unique and the selected class of situations requires some certain impact on the objects, then the objects are served associated with this control class. At the same time, one of the prerequisites for the effective application of the situational approach should be observed: the number of possible control decisions is significantly less than the number of possible situations [5]. Otherwise, it is necessary to solve the problem of analysis of the current situation in the environment of choice $O(t_k)$, which turns into the task of assessing the previous control in order to make a decision on changing the plan of situational control. Adjustment of the plan of situational control at each subsequent step does not necessarily lead to the construction of an optimal plan, since in the aggregate the selected step-by-step path to the goal will not be guaranteed to be optimal (the control process may not be Markov). It is therefore necessary before making a decision about the plan situational control to simulate the control group SEMS. For example, on the basis of fuzzy mathematical modeling of poorly formalized processes and systems [20], which allows for step-by-step construction of the path to go back and discard non-effective sections of the path. Naturally, such a search for the optimal plan of situational control based on modeling requires additional computing power and time from the control system. Significant progress in this direction can provide parallelization of calculations, i.e. simultaneous passage of all possible ways of the plan of situational control with the subsequent decision on optimality.

Acknowledgements This work was financially supported by Russian Foundation for Basic Research, Grants 16-29-04424, 18-01-00076 and 19-08-00079.

References

1. Gorodetskiy, A.E., Tarasova, I.L.: Situational control a group of robots based on SEMS. In: Gorodetskiy, A.E., Tarasova, I.L. (eds.) Smart Electromechanical Systems: Group Interaction. Studies in Systems, Decision and Control, vol. 174, pp. 9–18. Springer International Publishing (2019). https://doi.org/10.1007/978-3-319-99759-9-2
2. Gorodetskiy, A.E., Kurbanov, V.G., Tarasova, I.L.: Decision-making in central nervous system of a robot. Inf. Control Syst. **1**, 21–30 (2018). (in Russia). https://doi.org/10.15217/issnl684-8853.2018.1.21
3. Vorob'ev, V.V.: Logical inference and action planning elements in robot groups. In: Proceedings of 16th National Conference on Artificial Intelligence KII-2018, vol. 1, pp. 88–96, Moscow (2018). (in Russia)
4. Ivanov, D.YA., Shabanov, I.B.: Model of application of coalitions of intelligent mobile robots with limited communications. In: Proceedings of 16th National Conference on Artificial Intelligence KII-2018, vol. 1, pp. 97–105, Moscow (2018). (in Russia)

5. Pospelov, D.A.: Situation Management: Theory and Practice [Situacionnoe upravlenie: Teoriya i praktika], 286 p. Nauka, Moscow (1986). (in Russia)
6. Kunc, G., O Donnel, S.: Management: System and Situation Analysis of Control Functions [Upravlenie: sistemnyj i situacionnyj analiz upravlencheskih funkcij], 588 p. Progress, Moscow (2002). (in Russia)
7. Sokolov, B., Ivanov, D., Fridman, A.: Situational modelling for structural dynamics control of industry-business processes and supply chains. In: Sgurev, V., Hadjiski, M., Kacprzyk, J. (eds.) Intelligent Systems: From Theory to Practice, pp. 279–308. Springer-Verlag Berlin Heidelberg, London (2010)
8. Friedman, A.Ya.: Situational Control of the Structure of Industrial and Natural Systems. Methods and Models. LAP, Saarbrucken, Germany (2015)
9. Mishin, S.P.: Optimal Control Hierarchies in Economic Systems [Optimal'ny'e ierarxii upravleniya v e'konomicheskix sistemax]. PMSOFT, Moscow (2004). (in Russia)
10. Kalyaev, I.A., Kapustyan, S.G., Gaiduk, A.R.: Self-organizing distributed control systems for groups of intelligent robots built on the basis of the network model [Samoorganizuyushhiesya raspredelenny'e sistemy' upravleniya gruppami intellektual'ny'x robotov, postroenny'e na osnove setevoj modeli]. UBS **30**(1), 605–639 (2010). (in Russia)
11. Kalyaev, I.A., Gaiduk, A.R., Kapustian, S.G.: Control of a team of intellectual objects based on schooling principles [Upravlenie kollektivom intellektual'ny'x ob"ektov na osnove stajny'x principov]. Bull. Sci. Cent. Russ. Acad. Sci. **1**(2), 20–27 (2005). (in Russia)
12. Kapustian, S.G.: Decentralized method of collective distribution of goals in the group of robots [Decentralizovanny'j metod kollektivnogo raspredeleniya celej v gruppe robotov]. In: Kapustian, S.G. (ed.) Proceedings of the Higher Educational Institutions, Electronics, no. 2, pp. 84–91 (2006). (in Russia)
13. Kalyaev, I.A.: Principles of collective decision making and control in the group interaction of robots [Principy' kollektivnogo prinyatiya resheniya i upravleniya pri gruppovom vzaimodejstvii robotov]. In: Mobile Robots and Mechatronic Systems: Material Scientific Schools Conference, pp. 204–221. Publishing House of Moscow State University, Moscow (2000). (in Russia)
14. Kapustian, S.G.: The method of organizing multi-agent interaction in distributed control systems of a group of robots when solving the area coverage problem [Metod organizacii mul'tiagentnogo vzaimodejstviya v raspredelenny'x sistemax upravleniya gruppoj robotov pri reshenii zadachi pokry'tiya ploshhadi]. Artif. Intell. **3**, 715–727 (2004). (in Russia)
15. Fridman, A.Ya.: SEMS-based control in locally organized hierarchical structures of robots collectives. In: Gorodetskiy, A.E., Kurbanov, V.G. (eds.) Smart Electromechanical Systems: The Central Nervous System. Studies in Systems, Decision and Control, vol. 95, pp. 31–47. Springer International Publishing Switzerland (2017)
16. Vasiliev, S.N., et al.: Intellectual Control of Dynamic Systems [Intellektual'noe upravlenie dinamicheskimi sistemami], 352 p. FIZMATLIT, Moscow (2000). (in Russia)
17. Prishchepa, M.V.: Development of a user profile with account of the psychological aspects of human interaction with an information mobile robot [Razrabotka profilya pol'zovatelya s uchetom psixologicheskix aspektov vzaimodejstviya cheloveka s informacionny'm mobil'ny'm robotom]. Tr. SPIIRAN **21**, 56–70 (2012). (in Russia)
18. Ladygina, V.: Social and ethical problems of robotics [Social no-eticheskie problemy robototechniki]. Vyatka State Univ. Bull. **7**, 27–31 (2017). (in Russia)
19. Karpov, V.E.: Emotions and temperament of robots: behavioral aspects. J. Comput. Syst. Sci. Int. **5**, 126–145 (2016). (in Russia)
20. Gorodetskiy, A.E., Tarasova, I.L.: Fuzzy Mathematical Modeling of Poorly Formalized Processes and Systems [Nechetkoe matematicheskoe modelirovanie ploxo formalizuemy'x processov i sistem], 336 p. SPb Publishing House Polytechnic, Un-ta (2010). (in Russia)

The Problem of the Choice of the Satellite Orbit in the Formation Adaptive Mirror System of the Space Radio Telescope Antenna

Andrey E. Gorodetskiy, Vugar G. Kurbanov and Irina L. Tarasova

Abstract *Problem statement*: The correct choice of the orbit on which the spacecraft is placed, largely determines its functionality in the distant operation. The article deals with the issues related to the choice of orbits of satellites that form the mirror system of the radio telescope. *Purpose*: Approaches to solving the problem of the choice of the satellite orbit in the formation adaptive mirror system of space radio telescope antenna. *Results*: The factors influencing the motion of the artificial earth satellite are analyzed. The suitability of various orbits of artificial earth satellites for the formation of a mirror system of a space radio telescope is evaluated. *Practical significance*: Recommendations on the choice of the orbit for artificial earth satellites used to form an adaptive mirror system of the space radio telescope antenna are given.

Keywords Space radio telescope · The adaptive mirror system · Control elements · SEMS · Alignment · Focus adjustment · Automatic control system · Satellite orbit · Equation of motion of satellites · Influencing factors

1 Introduction

The reflecting surface of the mirror two modern large radio telescopes with a parabolic shape (the primary mirror) and ellipsoidal (correlator) reflect the boards installed in the managed elements [1]. Alignment and focusing of the mirror system is carried out by correcting signals from the radio telescope control system [2, 3]. To eliminate

A. E. Gorodetskiy · V. G. Kurbanov (✉) · I. L. Tarasova
Institute of Problems of Mechanical Engineering, Russian Academy of Sciences, St. Petersburg, Russia
e-mail: vugar_borchali@yahoo.com

A. E. Gorodetskiy
e-mail: g27764@yandex.ru

I. L. Tarasova
e-mail: g17265@yandex.ru

V. G. Kurbanov
Saint-Petersburg State University of Aerospace Instrumentation, St. Petersburg, Russia

© Springer Nature Switzerland AG 2020

A. E. Gorodetskiy and I. L. Tarasova (eds.), *Smart Electromechanical Systems*, Studies in Systems, Decision and Control 261, https://doi.org/10.1007/978-3-030-32710-1_2

the influence of the earth's astronomical climate, reflecting shields with controlled elements are installed on artificial earth satellites (AES), which are put into earth orbit [4]. In this case, it is possible to achieve high values of the aperture surface utilization factor (SUF) by adjusting the antenna surfaces to the operating frequency range, the absence of weight and wind deformations of the antenna design elements and the elimination of the influence of the earth's astronomical climate. However, in this case, the formation of a mirror system in orbit by artificial satellites is much more complicated and a number of problems arise related to a large temperature drop, solar radiation, periodic adjustment of the position and orientation of satellites that require appropriate energy consumption, etc. In this case, one of the primary problems is the problem of choosing the orbit of the satellite in the formation of an adaptive mirror antenna system of such a space radio telescope.

2 Factors Affecting the Movement of Satellites

The shape of the orbit, which is derived from the final stage of the flight of the launch vehicle, is determined by the amount of kinetic energy reported to the spacecraft by the launch vehicle, i.e. the value of the final speed of the latter. In this case, the value of the kinetic energy reported by the satellite must be in a certain relation to the energy of the field of the Central body, which exists at a given distance r from its center (see Fig. 1). For circular orbits with values r close to the radius of the Earth $R = 6371$ km, the final speed of the launch vehicle to launch the spacecraft into such an orbit will be $V_0 \sim 7900$ m/s. This is the so-called first space velocity. For elliptical orbits, the final velocities will be $V_3 = (7900–11{,}200)$ m/s.

The motion of the AES in orbit can be represented as the motion of a material particle of infinitesimal mass (m) in the gravitational field of a Central body of mass M under the action of forces determined by the potential function U, and a set of forces P that have no potential. Then the differential equations of particle motion in the inertial rectangular coordinate system associated with the Central body M can be represented as [5]:

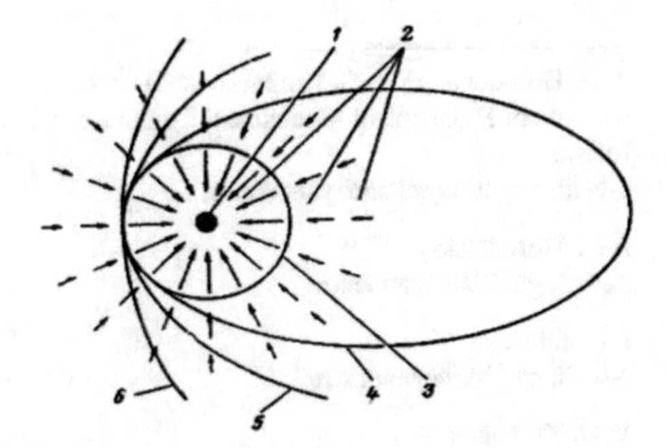

Fig. 1 Orbit of the spacecraft in the field of the Central body. 1—Central body; 2—force field of the Central body; 3—circular orbit; 4—elliptical orbit; 5—parabolic orbit; 6—hyperbolic orbit

$$m\frac{d^2x}{dt^2} = -\frac{\partial U}{\partial x} + P \quad (1)$$

with initial conditions

$$x_0 = x(t_0),\ \ \dot{x}_0 = \dot{x}(t_0) \quad (2)$$

where $U = -\frac{\mu}{r} - R, x = (x_1, x_2, x_3)^T$—the position vector of the satellite; t—physical time; r—the module of the vector position; $\mu = k^2M$, k—constant of gravitation; $U = U(t, x)$, $\partial/\partial x$—gradient. And the first term in U—the potential due to the attraction of the spherical Earth, considered as a material point, and the second term is the potential of disturbing forces. The influence of the non-spherical component of the earth's gravitational field is usually considered as a disturbing force having a potential in the problems of the dynamics of the satellite. All other forces, including the influence of the moon and the Sun, the position of which is given either in a table or in the form of series obtained outside the problem of the movement of the satellite, are forces that have no potential.

When introducing the force due to the influence of the earth's non-sphericity into equations (1), it should be remembered that the equations of motion are written in the inertial coordinate system, and the potential of the earth's gravitational field is related to the rotating coordinate system rigidly connected to the Earth.

The solution of equations (1) in the case of undisturbed motion is obtained, for example, in [6]. Lagrange equations for osculating Kepler elements are also used in the construction of analytical and numerical-analytical algorithms for predicting the motion of satellites. In these equations, the changes of the elements are associated with partial derivatives of the perturbing function of the elements. Newton-Euler equations are used, as a rule, in the problems of numerical prediction of the motion of satellites, as well as in the construction of analytical and semi-analytical theories of motion to account for perturbations from forces without potential. Lagrange equations are used to account for perturbations from potential forces in analytical methods. In addition, canonical variables can be used in the construction of analytical and numerical-analytical theories taking into account all types of perturbations.

A significant contribution to the perturbation of the satellite is influenced by tidal deformations occurring in the body of the planet under the influence of the external body attraction. At present, complete tidal models are developed that take into account the elastic properties of the Earth, for example, the model [5]. In addition, modern models take into account the influence of tidal deformations occurring in the ocean and the earth's atmosphere [7].

The perturbing influence of the moon and the Sun, especially noticeable in the motion of satellites in an elliptical orbit, is considered to be independent of each other. The main difficulty in taking into account the lunar-solar influence will be to represent the coordinates of the disturbing body in the required form [8].

One of the non-gravitational perturbations acting on the satellite is the light pressure. It is usually assumed that the power of the solar radiation flux is constant and the light pressure is always directed along the Earth–Sun line. The main difficulty in calculating the perturbations caused by light pressure is to take into account the effect of the satellite entering the earth's shadow. To remedy this difficulty is possible by the introduction of a perturbing acceleration of the so-called shadow mode to δ [9]. And $\delta = 1$, if the satellite is illuminated by the Sun, $\delta = 0$—otherwise. In the General case, the shadow has a conical shape, but in those cases where high precision is not required, it can be considered that, given the remoteness of the source the shadow has a cylindrical shape.

The main factor determining the lifetime of satellites in circular and elliptical orbits is the height of the first and the height of the perigee of the second, where the main braking occurs. On the satellite, moving at an altitude of 150–1500 km, a noticeable effect is the resistance of the atmosphere. The air resistance force acting on the forward motion of the satellite is directed opposite to the speed of the object relative to the air. The greatest difficulty in determining disturbances from the atmospheric resistance is in calculating the density of the atmosphere [10]. In changing the parameters of the atmosphere there is a periodicity associated with the rotation of the Earth around the Sun, with the rotation of the Sun around its axis, with the change in solar activity during the eleven-year cycle, etc.

Models of the atmosphere, taking into account the dependence of its parameters not only on the height, but also on the above factors, are called dynamic. Since the construction of these models is a laborious process, in practice they use various simplified models. For example, a static model of the atmosphere, which makes it possible to determine the density as a function of height and does not take into account the dependence of density on time. Local models of the atmosphere, suitable only for a given range of heights and for a certain time interval, are also widely used in practice. The simplest example of a local model is the so-called isothermal model.

As a rule, simplified models of the atmosphere are used in the construction of analytical and numerical-analytical algorithms for predicting the motion of satellites. With numerical methods, depending on the conditions of the problem, models of all types can be used: dynamic, statistical, local.

3 Assessment of the Suitability of Various Orbits of Artificial Earth Satellites for the Formation of a Mirror System

The orbits of artificial earth satellites (AES) are determined by the initial conditions set when the satellite is put into orbit, and by the action of natural forces on the passive flight section [11]. You can divide the appearance of the orbit into five groups : $e = 0$—circle; $0 < e < 1$—ellipse; $e = 1$—parabola; $1 < e < \infty$—hyperbole; $e = \infty$—straight (degenerate case). For artificial satellites, the plane of the earth's equator is usually chosen for the reference plane. At the moment, for the placement of the space radio telescope (SRT) perspective circular and elliptical orbits (Fig. 2).

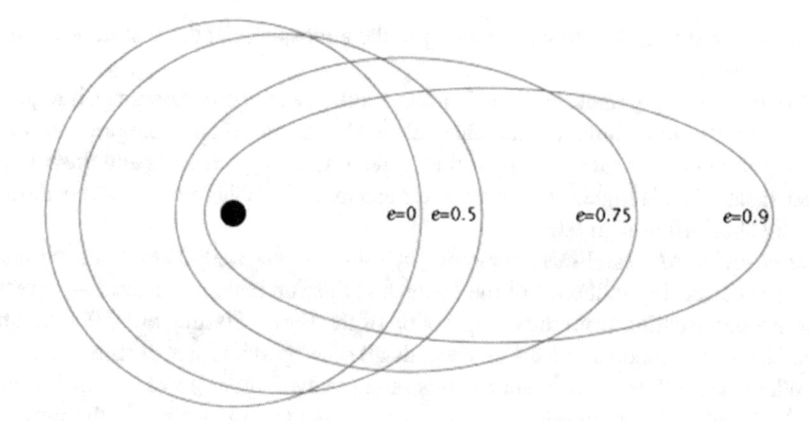

Fig. 2 Changing the appearance of an elliptical orbit with increasing values of eccentricity

3.1 Circular and Near-Circular Orbits

These orbits have eccentricities of 0–0.03. The height above the Earth's surface for such satellites varies little during the flight, which facilitates the description of the movement of AES and their stabilization in orbit, depending on the action of various disturbing forces, which in turn depend on the altitude of the flight of AES.

Depending on the altitude of the satellites of this class of orbits are divided into low-flying, medium-high and high-flying.

Low-flying AES move at altitudes $H = 200$–1500 km. The main sources of disturbances at such distances from the Earth are the non-sphericity of the Earth and the resistance of the atmosphere. The prevailing is the compression of the Earth. Atmospheric resistance plays a significant role up to a height of 500–600 km and significantly affects the movement for altitudes up to 1500 km. Although the perturbations caused by the compression of the Earth are much more perturbations due to the resistance of the atmosphere, they do not significantly change the orbit of the satellite, but only rotate it in space. Braking in the atmosphere even at high altitudes (1000–15,000 km) changes the orbit of the satellite and, in fact, determines the duration of its existence.

When controlling AES in such orbits, the lowest power of control signals and their delay in the communication channels are required. However, their stabilization in orbit requires the greatest energy reserves. Therefore, the life time of such satellites is small and their use in the SRT is impractical.

Medium high AES is the exoatmospheric AES moving at altitudes up to 30,000–35,000 km. The main disturbing factors are the Earth's non-sphericity and the gravitational influence of the moon and the Sun. The prevailing factor is still the compression of the Earth. Starting at an altitude of about 20,000 km, the perturbations from the attraction of the Moon and the Sun become comparable to the influence of anomalies of the earth's gravity. Braking in the atmosphere at these altitudes will be significantly lower.

When controlling satellites in such orbits, more power of control signals is required and their delay in communication channels is also greater than in the previous case. Stabilization of such satellites in orbit requires less energy reserves and the lifetime of such satellites is higher. However, their use in the SRT is impractical because of the still short lifetime in orbit.

High-flying AES satellites are moving at altitudes $H > 30{,}000$ km. For such satellites, the perturbing influence of the Moon and the Sun becomes equal to or greater than the perturbation from the compression of the Earth. Starting at 50,000 km, the attraction of the moon and the Sun exceeds all other gravitational perturbations.

When controlling AES in such orbits requires even more power control signals and their delay in the communication channels as even more than in the previous case. However, to stabilize such satellites in orbit requires less energy and a higher lifetime of such satellites. Therefore, such orbits are most suitable for use in the SRT.

Among the circular orbits of the satellites, it is necessary to distinguish those that have periods of revolution commensurate with the period of rotation of the Earth. These are *geostationary orbits*, which are very popular and are used to accommodate many types of satellites, including satellites for scientific research. However, the phenomenon of resonance, which occurs due to the commensurability of the average motion of the satellite with the earth's rotation frequency, generates additional disturbing forces. They can have a significant impact on their orbits and require additional stabilization with energy costs. In addition, the disadvantages of geostationary orbits are the following:

- the control signal travels a greater distance, and therefore, there are large, in comparison with LEO or MEO, losses;
- the cost of delivery and placement of the satellite into GEO-orbit is higher-due to the higher altitude above the Earth;
- long distance from the Earth to the satellite leads to large signal delays;
- the geostationary satellite orbit can only lie above the equator, and therefore there is no coverage of polar latitudes.

For geostationary AES is considered as satellites with periods from 22 to 26 h, eccentricities "*e*" is not more than 0.3 and the angles of tilt of the orbital plane to the plane of the equator "*i*" to 15°, but in some sources, you can find more detailed classification, and more stringent border.

3.2 Elliptical Orbit

Weakly elliptical orbits are characterized by moderate eccentricity, $0.03 < e < 0.2$. For them, the height difference in the apogee and perigee ranges from several hundred to several thousand kilometers. By the influence of disturbing forces, satellites of this class of orbits are amenable to the classification given for circular orbits depending on the height of H.

High elliptical orbits are characterized by large eccentricities, $e > 0.2$. At the same time, the height of satellites in the apogee can exceed the height of satellites in perigee by tens or hundreds of times. Satellites with such orbits are called *high-altitude*. They do not lend themselves to the convenient classification given by us for circular orbits, since all the main types of disturbances can be essential for them: from the earth's non-sphericity, the resistance of the atmosphere, the attraction of the moon and the Sun.

The lifetime of such AES in orbit is much longer, as they experience less inhibition in the atmosphere on most of the trajectory in orbit. However, effective management with a small delay is possible on a site with a small distance from the Earth (in the perigee region). Nevertheless, these orbits can be considered promising for the placement of the mirror antenna system of the space radio telescope.

The elliptical orbit of the AES is shown in Fig. 3 in the absolute geocentric (Equatorial) coordinate system. The origin of the system is combined with the center of the Earth. The OZ axis is directed along the axis of rotation of the Earth towards the North pole. The axis OX lies in the Equatorial plane and is directed to the point of the spring equinox. The OY axis completes a Cartesian right-handed coordinate system.

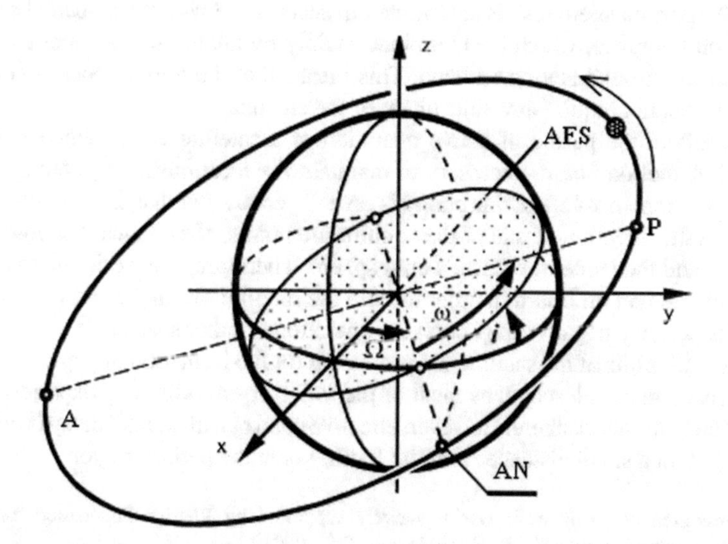

Fig. 3 Elements of the orbit of AES in space. Ω—ascending node longitude (AN), *i*—orbit inclination, *ω*—perigee argument, *P*—perigee, *A*—apogee

4 Conclusion

The advantage of the geostationary orbit is that the satellite located on it is always in the same position. This allows you to send a fixed antenna to the ground station. The laws of geometry tell us that the only option for making one revolution per day, the satellite always remained above one point of the earth's surface, is its circulation in the same direction in which the Earth itself rotates. In addition, the satellite must not move in its orbit either to the North or to the South. All this can be achieved only if the orbit of the satellite passes over the equator.

Even if the satellite is in geostationary orbit, it is affected by some forces that can slowly change its position over time. Such factors as the elliptical shape of the Earth, the attraction of the Sun and the moon, as well as a number of others increase the potential deviation of the satellite from its orbit. In particular, not quite round shape of the Earth near the equator leads to the fact that the satellite attracts to two stable points of equilibrium—one of them is over the Indian ocean, and the second—approximately on the opposite part of the Earth. As a result, there is a phenomenon called East–West libration, or movement back and forth. In order to overcome the consequences of such movement, there must be a certain amount of fuel on board the satellite, which allows it to carry out "supporting maneuvers" that return the satellite clearly to the required orbital position. The necessary time interval between such "support manoeuvres" is determined in accordance with the so-called satellite deflection tolerance, which is determined mainly by taking into account the beam width of the ground station antenna. This means that the normal operation of the satellite does not require any adjustment of the antenna.

Very often, the period of active operation of a satellite is calculated from the amount of fuel on board, necessary to maintain the location of the satellite in one orbital position. Most often, this period is several years. After that, the satellite begins to drift in the direction of one of the equilibrium points, after which it is possible to decrease and then enter the Earth's atmosphere. Therefore, it is desirable to use the last available fuel on Board in order to raise the satellite to a higher orbit in order to avoid its possible negative impact on the operation of other spacecraft.

Elliptical orbits of the satellite are promising for the formation of a mirror system of the space radio telescope, as most of the trajectory in orbit they experience less inhibition in the atmosphere. However, effective control with a small delay is possible on a site with a small distance from the Earth, i.e. in the perigee region.

Acknowledgements This work was financially supported by Russian Foundation for Basic Research Grants 16-29-04424, 18-01-00076 and 19-08-00079.

References

1. Razdorkin, D.Ya., Romanenko, M.V.: Algorithm of optimization of a two-mirror antenna with a reflector from parabolic shields. Antennas **5**(60), 44–47 (2002). (in Russian)
2. Gorodetskiy, A.E., et al.: Radio telescope automatic guidance system. Patent RU No. 2319171. G01S13/66, from July 17, 2006, bulletin No. 7 (2008)
3. Artyomenko, Yu.N., et al.: The method of adaptation of the reflecting surfaces of the antenna. Patent RU No. 2518398, G01S13/66, from Nov 20, 2012, bull. No. 16 (2014)
4. Gorodetskiy, A.E., Tarasova, I.L., Kurbanov, V.G.: The formation of the mirror system of the space radio telescope antenna (in this volume)
5. Bordovitsyna, T.V., Avdyushev, V.A.: The Theory of Motion of Artificial Earth Satellites. Analytical and Numerical Methods: Studies. Allowance, 2nd edn, 254 p. Publishing House of Tomsk State University, Tomsk (2016). (in Russian)
6. Duboshin, G.N.: Heavenly Mechanics. The Main Tasks and Methods, 800 p. Nauka, Moscow (1968). (in Russian)
7. Prokhorenko, V.I.: The long-term evolution of the satellite orbits under the influence of gravitational perturbations caused by the compression of the earth, taking into account perturbations from external bodies. Izv. Universities. Physics. Application. Heavenly Mechanics and Applied Astronomy, vol. 49, no. 2, pp. 63–73. Publishing House of Tomsk State University Press, Tomsk (2006). (in Russian)
8. Kholshevnikov, K.V., Drinking, N.V., Titov, V.B.: The attraction of celestial bodies, p. 104. SPb Publishing House of St. Petersburg University, St. Petersburg (2005). (in Russian)
9. Chernitsov, A.M., Tamarov, V.A.: On the method of constructing an analytical algorithm for calculating the effect of light pressure on the motion of an artificial satellite. In: Astronomy and Geodesy, vol. 16, pp. 239–245. Publishing House of Tomsk State University Press, Tomsk (1998). (in Russian)
10. Tamarov, V.A., Chernitsov, A.M.: Analytical algorithm for calculating disturbances in the motion of an artificial satellite due to atmospheric resistance. In: Astronomy and Geodesy, vol. 16, pp. 134–148. Publishing House of Tomsk State University Press, Tomsk (1998). (in Russian)
11. Schmude Jr., R.: Artificial Satellites and How to Observe Them. Astronomers Observing Guides, 181 p. Springer Science+Business Media, New York (2012). ISBN 978-1-4614-3915-8-4

References

Assessment of Situational Awareness in Groups of Interacting Robots

Alexander Ya. Fridman and Boris A. Kulik

Abstract *Problem statement*: Groups of proactive SEMS-based robots need to be taught for cooperative functioning. Efficiency of learning and teaching algorithms has to be checked by modeling. The concept of situation awareness (SA) provides a promising tool for such checks. *Purpose of research*: Concretization of the concept of SA for the tasks of organizing the teamwork of interacting robots. *Results*: Quantitative assessment of SA and its three main aspects (perception of environmental elements, comprehending of the situation and projecting future statuses) for groups of robots. *Practical significance*: Objective specification of self-assessment functions for equal-ranking robots allows to prevent conflicts among them, and to ensure coordination of robots interactions within hierarchical groups.

Keywords SEMS-based robot · Robots cooperation · Quantitative assessment of situation awareness · Coordination of robots interactions

1 Introduction

Cyber-Physical Systems (CPSs) are considered to be the most promising platform to develop the basic paradigm of adaptive self-learning systems, which integrate control information systems with elements of accumulation and application of knowledge to achieve complex objective functions. Control remains the key question in such models, because the effectiveness of system operation depends on how the chosen behavior corresponds with reality. Besides, it is vitally necessary to create principles of network organization and group control for individual intelligent systems that make up a distributed environment of artificial intelligence. The determining factor

A. Ya. Fridman
Institute for Informatics and Mathematical Modelling, Kola Science Centre of RAS, Apatity, Russia
e-mail: fridman@iimm.ru

B. A. Kulik (✉)
Institute of Problems in Mechanical Engineering, Russian Academy of Sciences (RAS), St. Petersburg, Russia
e-mail: ba-kulik@yandex.ru

© Springer Nature Switzerland AG 2020

A. E. Gorodetskiy and I. L. Tarasova (eds.), *Smart Electromechanical Systems*, Studies in Systems, Decision and Control 261, https://doi.org/10.1007/978-3-030-32710-1_3

in this case is priority of coordination—horizontal interactive links above the vertical "purely competitive" strategies in complex integrated systems. This includes relations between control agents (Vehicle-to-Vehicle, V2V) and between each agent and the outer surrounding infrastructure (Vehicle-to-Infrastructure, V2I) [22].

Relations of cooperation and coordination between individual cognitive units play a key role in formation of multi-level integrated structures with elements of group control strategies due to inefficiency (and in some cases impossibility) of solving complex problems by individual isolated subsystems [2].

Despite all evident advantages, robots are still rarely used in large groups simultaneously. This is caused by the fact that most of modern robots are still not ready for group interaction [11].

The above-introduces citations confirm the authors' opinion that techniques to model and estimate group behavior of individual robots are topical now. With this in mind, below we propose a specification of the concept of the situational awareness (SA, see, for instance, one of the first publications in this area [4]) for the task of comparative analysis of robots' interactions by means of the SA paradigm.

2 SA Basics and Their Features for Robots and Groups

At present, the concept of SA (for example, [3, 4, 18]) describes the most general principles of preparing and processing information for implementing a situational approach in dynamic subject areas. SA is especially important in professional activities, where the flow of information can be quite high, and bad decisions can lead to serious consequences. This is especially evident in high-dynamic subject areas (for instance, piloting an aircraft, military actions, handling seriously ill or wounded patients, etc.).

SA has become a widely used construct within the human factors community, the focus of considerable research over the past 25 years. Patrick and Morgan [20] found some 17,500 articles discussing SA in a Google Scholar search as early as in 2010.

This paradigm is not widespread in Russia yet, there are only few ideas for applying this approach to solve specific problems [1, 21, 26]. That is why we consider SA a promising tool for the task of modeling and testing cooperation within groups of robots. Some researchers criticize this approach for being too general (see, in particular, [3]). It seems that the general principles of SA really become constructive only in relation to a specific model of decision-making in a particular subject area.

Situation awareness includes consciousness of what is happening in the environment in order to understand how information, events and personal actions will affect goals and objectives in the current moment and in near future. Insufficient or incorrect SA is considered to be one of the main factors associated with accidents caused by the "human factor" [18]. Nevertheless, to our minds, it also needs to be taken into account in the considered below problems of modeling robots' interaction [2, 11, 15, 22], where aspects of team SA [10] are especially significant during teaching the robots.

For better perception, main statements of SA are rendered below in *italics*. First we analyze general features of basic SA requirements to robots' interactions and then proceed to our proposals on considering these features.

To achieve SA, it is necessary to ensure correct processing of information at three levels: perception of environmental elements, comprehension of the current situation and projection of future status [4]. In our case, perception of a robot is modeled by a share of essential parameters (cues [4]) of the current situation this robot considers in its internal model; comprehension depends on the degree of correspondence between the robot's actions and the actions desired for fulfillment of the group task; projection is determined by correctness of choosing the robot's local plan compared to the plan desired for the whole group. To combat the "information explosion" at the levels of perception and comprehension, effective means are needed to determine essential factors from the set of available measurements and observations.

In the works by M. R. Endsley, the founder and leader of the SA paradigm, it is repeatedly noted (for example, [7]) that *a high level of SA is most often achieved by experts in solving a particular task, and not by beginners (although she always notes that miracles do happen)*. For a robot, solving of this problem is shifted onto the personnel in charge for modeling and teaching, and eventually onto decision makers (DMs) who ought to be experts in their subject domain; therefore a sufficient level of SA looks quite achievable.

Correct setting of priorities during selection and subsequent analysis of available information is one of the main problems in achieving an acceptable level of SA [7]. To this end, it is necessary to properly form performance criteria systems for individual robots and their groups; our proposals will be introduced below.

The input to the SA attainment procedures is the state of the environment; these procedures are followed by decision making and performing some actions [8, 9]. To our mind, decision making and estimation of robots' behavior will be easier to implement if we model hazardous situations and emergencies as extension of the model of normal functioning [16] by assigning every essential variable with a safe range (SR) [16, 25]. Technologically, this is done by analyzing certain expertly formed conditions and the degree of danger of getting out of an SR. The degree of danger of the current situation must be taken into account when assessing SA. This approach also allows to search for critical processes and objects, problems in which greatly reduce SA, determine moments of transition of a robot from the mode of normal operation to an abnormal mode, calculate the degree of SA loss during such transitions, if the simulation is done in a unified discretized space [13] that allows for presence of both numeric and string variables in the state vector of an object. Thus, the "usual" conceptual space (for example, [17, 24, 27]) expands and becomes applicable also for hazardous operating modes of a group of robots.

Sets of critical cues (parameters) *allow for using a mental scheme (model) to indicate, instantly classify and understand prototypical situations* [7]. In the task under study, this is a significant and fast raising of the general performance criterion, which indicates inefficiency of the current mode of functioning (assuming correct values of input parameters of the task), as well as getting of values of certain essential variables beyond their SRs that displays a possibility of occurring some events initiating emergencies and accidents [25].

Situation awareness is an internal mental model of the state of the environment comprehended by an operator (in our case, a DM). It is built by applying system knowledge, knowledge about interfaces of some software means and the world around it. SA, decision making and effective performance are different steps that interact within a continuous cycle, which can be broken by other factors [7, 8]. Applying SA to the robots' interactions results in choosing an initial situation (this is system knowledge of the most important parameters characterizing the state and behavior of the group under study) and, after simulation, in choosing a new (or conserving the former) criteria system for further functioning of the group or its any part, as well as in selecting among possible alternatives to implement the chosen class of behavior.

When analyzing the temporal characteristics of SA, M. Endsley points out importance of considering the speed of change in the surrounding world, which should correspond to the speed of decision making by the operator [7]. For this purpose, it looks reasonable to use values of gradients (increments) [12] of the quality criteria for the whole group and the means of analyzing sensitivity of supposed decisions to changes in parameters of these criteria. The values of the gradients of the criteria naturally show the possible time interval for predicting (projecting) the behavior of the modeling object: the larger these gradients are, the shorter is the interval of a reliable forecast due to inevitable uncertainties in assessing characteristics of robots and their environment.

3 SA Measurement

In original papers on SA measurements and measures, the general idea is based on test polls of operators working on simulators either in real time or with resuming simulation to fill polling lists. The first technique is implemented by SPAM (Situation Present Assessment Method) [5]; its main drawback is considered as restricted number of test questions. To apply the second approach, SAGAT (SA Global Assessment Technique) [5, 6] was created.

According to the SA authors, SA measures and metrics should [5–7]:

- *measure the construct that is needed, and not a reflection of other processes;*
- *SA depends on states, not on processes;*
- *for SA, relative values are more important than absolute values;*
- *the ideal SA is the exact knowledge of all relevant aspects of the situation.*

Most of these requirements are automatically satisfied if we find quantitative SA measures independent on human factors, the rest ones can be achieved to the extent defined by adequacy and detailedness of the model used for teaching robots. Sure, this model (just as any other model) cannot reveal situations, which were not provided in it, but it gives opportunities to investigate SA for different modes of robots' operation.

As for providing teamwork of intelligent agents (where robots evidently belong to), below we render few ideas from [10].

In a team, the certain role played by a team member determines his/her current subgoal that assists in reaching the general team goal. For peer-ranked groups of robots, such subgoals are set during modeling and teaching procedures; in hierarchical collectives, the subgoals are formed by the coordinator(s) of the collective.

Accumulated team SA can be defined as "the degree to which every member obtains the SA needed for implementing his/her duties" [4]. Hence, we can estimate the total degree of SA for a team according to the "principle of the weakest link" as the minimal value of SA achieved by a member of the team.

Let us now advance to some proposals on measuring and considering SA in robots' group functioning. These measures resulted from rethinking the above-briefed basic ideas of the SA paradigm.

3.1 Peer-Ranked Group of Robots

In a peer-ranked group of proactive robots, each of them ought to be treated as a separate DM. Then it is possible to introduce and calculate the achieved degree of SA for each of the DMs.

Since M. Endsley and her colleagues have repeatedly shown (for example, [7]) that not absolute, but relative values of SA are important, we assume that values of the total degree of SA (SAD) and each of its three components (perception of environmental elements, comprehension of the situation and forecast of future status) are characterized by a non-negative number with the maximal value of 1. Since these components are considered independent on each other [9], the total SAD is reasonable to calculate as their product. Thus, for a quantitative assessment of the degree of SA, currently achieved by each decision maker who has a given area of responsibility (decision area—DA), we introduce the following formula:

$$\mathrm{SAD}_i = \mathrm{PD}_i * \mathrm{CSD}_i * \mathrm{FD}_i, \tag{1}$$

where PD_i is the degree of perception of the environment that depends on the share of the number of input cues for the DA of a DM, which can change without her/his participation, in the total number of cues needed for performance of this DM. Thus, we propose that

$$\mathrm{PD}_i = \frac{n}{n+m}, \tag{2}$$

where n is the number of the input cues for the DA_i, m is the number of cues generated by the DA_i and used by other DMs;

CSD_i is the degree of comprehension of the situation; *it is a synthesis of the elements of the* PD_i *(measure of proximity of the current state of the system to the ideal state)* [7];

FD_i (forecast degree) is determined by the change rate for the current situation in the DA_i.

To calculate CSD_i and FD_i for a robot, it looks sensible to immediately estimate deviations between its current and desirable state. In our opinion, the overall deviation can be calculated by our earlier developed generalized performance quality criterion (PQC) (see, for example, [14]):

$$\Phi ::= \left(\frac{1}{m} \sum_{i=1}^{m} \left(\frac{a_i - a_{i0}}{\Delta\, a_i} \right)^2 \right)^{1/2} ::= \left(\frac{1}{m} \sum_{i=1}^{m} \delta a_i^2 \right)^{1/2}, \tag{3}$$

where a_i are essential output parameters of this object, their total number is m;

a_{i0} and $\Delta a_i > 0$ are adjusting parameters that reflect requirements of the coordinator or a teacher to the nominal value a_i and its allowable deviation Δa_i from this value, respectively;

$\delta\, a_i ::= \frac{a_i - a_{i0}}{\Delta\, a_i}$ is the relative deviation of the actual value of the signal a_i from its nominal value a_{i0}.

If we assume that a_i are current values of scalar PQCs for a robot, and their nominal values equal a_{i0}, then (3) is a generalized criterion with importance coefficients inversely proportional to the allowable deviations of scalar criteria, which corresponds to common sense: the more important this scalar criterion is for the decision maker, the less acceptable are its deviations from the nominal value.

The value of the criterion (3) will be equal to unity if values of all its arguments are on the verge of tolerances:

$$\Phi = 1, \ \text{if}\, |a_i - a_{i0}| = \Delta\, a_i, \quad i = \overline{1, m}. \tag{4}$$

This value does not exceed unity if all arguments are within tolerances.

The specific value of a change in the criterion (3) when one of its arguments changes:

$$\delta\, \Phi_i ::= \frac{\partial \Phi / \partial a_i}{\Delta\, a_i} = m\, \Phi\, \delta\, a_i, \tag{5}$$

characterizes the relative sensitivity of the criterion to a change in this argument.

Considering (3)–(5), we assume that CSD_i depends on the vector of residuals of the own PQC for the DA_i $\delta a_i^{own} = \sqrt{\sum_{j=1}^{n} \delta a_j^2}$ and the vector $\delta a_i^{in} = \sqrt{\sum_{k=1}^{m} \delta a_k^2}$ of residuals for the input cues of the DA_i, while FD_i is determined by temporal increments of the quality criterion (3). The greater is the magnitude of these increments in time, the shorter becomes the interval of reliable forecast due to inevitable uncertainties in assessment of characteristics of the robot's state and the environment.

When developing formulas for estimating components of SA, it is necessary to take into account their desirable asymptotic properties that appear from the semantics of these concepts.

For CSD_i, they are as follows:

when $\delta a_i^{in} \to 0$ and $\delta a_i^{own} \to 0$, $CSD_i \to 1$;

when $\delta a_i^{own} >> 1$, $CSD_i \to 0$;

when $\delta a_i^{in} >> 1$, $CSD_i \to 0$.

Hence, a permissible formula for evaluating CSD is:

$$CSD_i = \frac{2 - \delta a_i^{own} - \delta a_i^{in}}{2 - (\delta a_i^{own})^2 - (\delta a_i^{in})^2}. \quad (6)$$

Asymptotic properties of FD_i are:

with $\Phi_i \to 0$ and $\Delta\Phi_i \to 0$ $FD_i \to 1$, $T_i \to \propto$, where T_i is the interval of a reliable forecast;

at $|\Delta\Phi_i| \to \infty$ $FD_i \to 0$, $T_i \to 0$;

at $\Phi_i >> 1$ $FD_i \to 0$, $T_i \to 0$.

So, a valid formula for FD is:

$$FD_i = 1 - e^{-\tilde{T}_i}, \quad (7)$$

where $\tilde{T}_i$ is the smoothed value of T_i,

$$T_i = \frac{\alpha}{\Phi_i |\Delta\Phi_i|}, \quad (8)$$

and $\alpha > 0$ sets the time scale (dynamics) of operation of the DA_i.

If the used above presupposition regarding equal importance of all involved cues for performance of a robot is inacceptable, this importance can be made different by weighing shares of cues with some expert-estimated coefficients similar to moving totals techniques. Additional options to make formulas more flexible and adjustable appear within the discretized state space [13] where an expert can consider different degree of danger for each value of every cue by assigning weights to distances between those values.

3.2 Hierarchical Groups of Robots

Any hierarchy can be reduced to a two-level system with a Coordinator on the upper level and a set of local DMs on the lower level [15]. There we proposed coordination techniques for such a system with usage specific values (5). All local DMs were there supposed to have peer ranks. Now that we can calculate SAD for every DM (robot in our case), it looks reasonable to consider these SADs for coordination as well. First, we can scale feedbacks from local DMs to the Coordinator by their SADs. Second, the SAD of the Coordinator is equal to the minimal SAD of its subordinated DMs (robots), see Endsley and Jones [10] and the previous section of this paper. This way we can calculate SADs for robots' collectives with arbitrary structure.

4 Conclusion

The proposed correlations (6)–(8) make it possible to objectively evaluate importance of decisions of every DM and take this importance into account when searching for a balance of interests of all DMs who take part in teamwork, in order to coordinate their actions and eliminate conflicts.

These ideas were derived from the earlier developed approach to measuring SA in industry-natural complexes [14] explored by means of the situational conceptual model (SCM). Besides investigation of such complexes, the introduced approach looks prospective for estimating SA in SCM-based supply chain management (for instance, [23]) and development networks of intelligent situational centres [19].

Acknowledgements The authors would like to thank the Russian Foundation for Basic Researches (grants 16-29-04424, 18-29-03022, 18-07-00132, 18-01-00076, and 19-08-0079) for partial funding of this research.

References

1. Afanasyev, A.P., Baturin, Yu.M., Yeremchenko, E.N., Kirillov, I.A., Klimenko, S.V.: Information-analytical system for decision-making on the basis of a network of distributed situational centres. Inf. Technol. Comput. Syst. **2**, 3–14 (2010). (in Russian)
2. Baldassarre, G., Nolfi, S., Parisi, D.: Evolving mobile robots able to display collective behaviors. Artif. Life **9**(3), 255–267 (2003)
3. Banbury, S., Tremblay, S.: A Cognitive Approach to Situation Awareness: Theory and Application, pp. 317–341. Ashgate Publishing, Aldershot, UK (2004)
4. Endsley, M.R.: Toward a theory of situation awareness in dynamic systems. Hum. Factors **37**(1), 32–64 (1995)
5. Endsley, M.R.: Situation awareness measurement in test and evaluation. In: O'Brien, T.G., Charlton, S.G. (eds.) Handbook of Human Factors Testing & Evaluation, pp. 159–180. Lawrence Erlbaum, Mahwah, NJ (1996)
6. Endsley, M.R.: Direct measurement of situation awareness in simulations of dynamic systems: validity and use of SAGAT. In: Garland, D.J., Endsley, M.R. (eds.) Experimental Analysis and Measurement of Situation Awareness, pp. 107–113. Embry-Riddle University, Daytona Beach, FL (2000a)
7. Endsley, M.R.: Theoretical underpinnings of situation awareness: a critical review. In: Endsley, M.R., Garland, D.J. (eds.) Situation Awareness Analysis and Measurement, pp. 3–32. LEA, Mahwah, NJ (2000b)
8. Endsley, M.R.: Final reflections: situation awareness models and measures. J. Cogn. Eng. Decis. Mak. **9**(1), 101–111 (2015a)
9. Endsley, M.R.: Situation awareness misconceptions and misunderstandings. J. Cogn. Eng. Decis. Mak. **9**(1), 4–32 (2015b)

10. Endsley, M.R., Jones, W.M.: A model of inter- and intra team situation awareness: implications for design, training and measurement. In: McNeese, M., Salas, E., Endsley, M. (eds.) New Trends in Cooperative Activities: Understanding System Dynamics in Complex Environments, pp. 46–67. Human Factors and Ergonomics Society, Santa Monica, CA (2001)
11. Ermolov, I.L.: Emerging issues of robots to be used in groups. In: Gorodetskiy, A.E., Kurbanov, V.G. (eds.) Smart Electromechanical Systems: Group Interaction. Studies in Systems, Decision and Control, vol. 174, pp. 3–8. Springer Nature Switzerland AG, Basel, Switzerland (2019)
12. Fridman, A., Fridman, O.: Gradient coordination technique for controlling hierarchical and network systems. Syst. Res. Forum **4**(2), 121–136 (2010)
13. Fridman, A.Ya.: Expert space for situational modelling of industrial-natural systems. Herald of the Moscow University Named after S.Y. Witte **1**(4), 233–245 (2014). (in Russian)
14. Fridman, A.Ya.: Situational Control of the Structure of Industrial-Natural Systems. Methods and Models. LAP, Saarbrucken, Germany (2015). (in Russian)
15. Fridman, A.Ya.: SEMS-based control in locally organized hierarchical structures of robots collectives. In: Gorodetskiy, A.E., Kurbanov, V.G. (eds.) Smart Electromechanical Systems: The Central Nervous System. Studies in Systems, Decision and Control, vol. 95, chap. 3, pp. 31–50. Springer, Switzerland (2017)
16. Fridman, A.Ya., Kurbanov, V.G.: Situational modelling of reliability and safety in industrial-natural systems. Inf. Manag. Syst. **4**(71), 1–10 (2014). (in Russia)
17. Gärdenfors, P.: Conceptual Spaces: The Geometry of Thought. A Bradford Book. MIT Press, Cambridge, MA (2000)
18. Lundberg, J.: Situation awareness systems, states and processes: a holistic framework. Theor. Issues Ergon. Sci. (2015)
19. Oleynik, A., Fridman, A., Masloboev, A.: Informational and analytical support of a network of intelligent situational centers in the Russian Arctic. In: IT&MathAZ 2018 Information Technologies and Mathematical Modeling for Efficient Development of Arctic Zone. Proceedings of the International Research Workshop on Information Technologies and Mathematical Modeling for Efficient Development of Arctic Zone, pp. 57–64, Yekaterinburg, Russia, 19–21 Apr 2018
20. Patrick, J., Morgan, P.L.: Approaches to understanding, analysis and developing situation awareness. Theor. Issues Ergon. Sci. **11**(1–2), 41–57 (2010)
21. Popovich, V.V., Prokaev, A.N., Sorokin, P.P., Smirnova, O.V.: On recognizing the situation based on the technology of artificial intelligence. In: SPIIRAS Proceedings, issue 7, pp. 93–104. SPb Science (2008). (in Russian)
22. Shkodyrev, V.P.: Technical systems control: from mechatronics to cyber-physical systems. In: Gorodetskiy, A.E. (ed.) Smart Electromechanical Systems. Studies in Systems, Decision and Control, vol. 49, pp. 3–6. Springer International Publishing, Switzerland (2016)
23. Sokolov, B., Ivanov, D., Fridman, A.: Situational modelling for structural dynamics control of industry-business processes and supply chains. Intelligent Systems: From Theory to Practice. Studies in Computational Intelligence, pp. 279–308. Springer-Verlag Berlin Heidelberg, New York (2010)
24. Sowa, J.F.: Conceptual Structures—Information Processing in Mind and Machines. Addison-Wesley Publ. Comp, Reading, MA (1984)
25. Yakovlev, S.Yu., Isakevich, N.V., Ryzhenko, A.A., Fridman, A.Ya.: Risk assessment and control: implementation of information technologies for safety of enterprises in the Murmansk region. In: Barents Newsletter on Occupational Health and Safety, vol. 11, no. 3, pp. 84–86, Helsinki (2008)
26. Yampolsky, S.M., Kostenko, A.N.: Situational approach to management of organizational and technical systems in operations planning. Sci. Intens. Technol. Space Res. Earth **8**(2), 62–69 (2016). (in Russian)
27. Zenker, F., Gärdenfors, P.: Applications of Conceptual Spaces. The Case for Geometric Knowledge Representation. Synthese Library, vol. 359. Springer, New York (2015)

Features of Individual and Collective Operation of Mobile SEMS Modular Type on Basis of Octahedral Dodekapod in Conditions of Incomplete Certainty

Sergey N. Sayapin

Abstract *Problem statement*: the article presents the problem of situational monitoring and manipulation operations using mobile robotic systems for individual and collective (Swarm Systems) operations under conditions of incomplete certainty, for example, in an environment opaque to optical, radio, ultrasonic and other physical control methods. *Purpose of research:* the aim of the research is to solve the problem with the help of an original mobile intelligent adaptive spatial parallel robot of modular type with 12 d.o.f., called the Octahedral dodekapod (OD). *Results:* it is shown that the OD can be operated individually and as part of Swarm Systems, as well as combined with other similar robots in reconfigurable mobile robotic structures. The ends of the rods of the adjacent edges of the octahedral structure are interconnected by spherical hinges. This provides the spatial structure with geometric immutability when the linear rod drives are switched off and all the rods work only on tension/compression. As a result, the OD has a higher specific carrying capacity compared to other types of spatial mobile robots. The description of the OD and its functionality are given. *Practical significance:* One of the functionality of this robot is the ability through mechanical contact to build 3D-maps of the surrounding space with reference to the base (inertial) coordinate system. This makes it possible to move the OD and to monitor the environment, as well as to make manipulation operations "blindly" in conditions of incomplete certainty, for example, in an environment opaque to optical, radio, ultrasonic and other physical methods of control. This gives undeniable advantages over other types of mobile robotic systems when operating under conditions of incomplete certainty. Examples of possible application of OD under extreme conditions with incomplete certainty in different environments are shown.

Keywords Mobile parallel robot · SEMS · Octahedral dodekapod (OD) · 3D maps of the surrounding space · Monitoring under conditions of incomplete certainty

---

S. N. Sayapin (✉)
Blagonravov Mechanical Engineering Research Institute of the Russian Academy of Sciences, Moscow, Russia
e-mail: S.Sayapin@rambler.ru

Bauman Moscow State Technical University, Moscow, Russia

© Springer Nature Switzerland AG 2020

A. E. Gorodetskiy and I. L. Tarasova (eds.), *Smart Electromechanical Systems*, Studies in Systems, Decision and Control 261, https://doi.org/10.1007/978-3-030-32710-1_4

1 Introduction

Currently, in a number of areas of national economies there is a problem of conducting situational monitoring and manipulation operations with the help of mobile Smart Electromechanical Systems (SEMS) of modular type in conditions of incomplete certainty, for example, in an opaque environment to optical, radio, ultrasonic and other physical methods of control. At the same time, such robotic systems should be able to be operated both individually and collectively as part of swarm systems. One of the areas of the economy requiring situational monitoring and manipulation operations is the global network of pipelines for various purposes. It is well known that pipelines are pipes which are used for carrying oil, gas, water, and etc. over long distances. Therefore, in today's world pipelines are one of the most important values, both in the economy and directly in our daily lives. The pipeline can consist of underground, overhead, subsea pipeline sections, or a combination of these. The pipeline includes horizontal and vertical pipe sections with different pipe bends (Fig. 1). The pipeline can contains some pipes with variable cross-section also.

The pipeline damage may lead to a leak of dangerous oil, gas, etc. A water leak can also be a problem. Therefore, pipeline inspection is an essential activity to better understand the condition of any pipeline, ultimately to ensure that all assets are fully operational with a long life cycle which can be well over 25 years since initial installation. It should be noted that some pipeline repairs and modification are inevitable and necessary during the course of an asset lifecycle. Mobile SEMS modular type for pipe inspection and repair as necessary are needed where people cannot go, or where the hazards of human presence are great (smaller size, longer range, increased

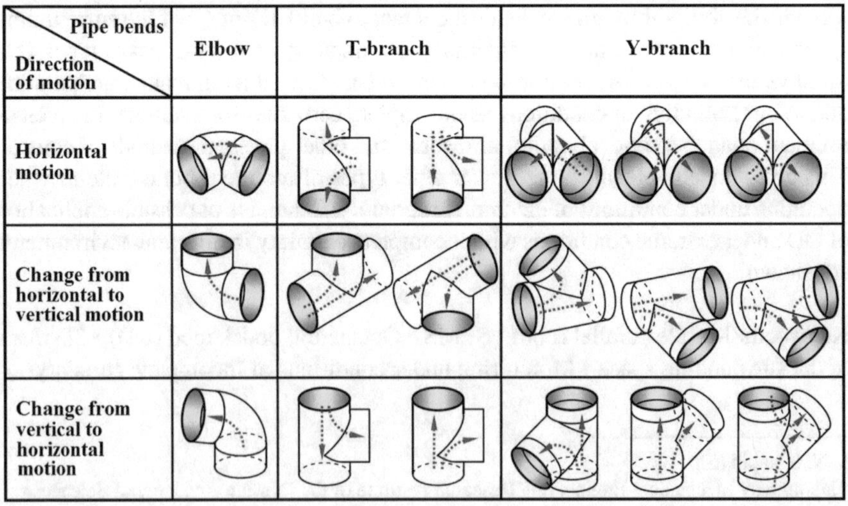

Pipe bends / Direction of motion	Elbow	T-branch	Y-branch
Horizontal motion			
Change from horizontal to vertical motion			
Change from vertical to horizontal motion			

Fig. 1 The main types of pipe connections in branched pipelines and the planned directions of movement of in-pipe robots

maneuverability, high level of irradiation, and etc.). Therefore, the main use of such robotic vehicles is nuclear power plants, conventional power plants, refineries, chemical and petrochemical plant, offshore rigs, long distance city heating pipelines, food and drinks industries, communal waste water pipe systems, gas pipelines, etc. This type of the robotic vehicle is one of the fastest growing the fields of intelligent robotic vehicle of today. The types of inspection tasks are very different. As a result there are a large number of different robotic vehicles for pipe inspection which can be divided into three groups: inside [1–3], outside [4–7], and dual-purpose [8–11]. According to [3], existing pipe inspection robots can be classified by their different movement pattern into one or more of the seven main types, as shown in Fig. 2: pig type (a), wheel type (b), caterpillar type (c), wall press type (d), walking type (e), inchworm type (f), and screw type (g). These robots can also be categorized by their different locomotion methods: wheeled, inchworm, snake, and legged, as in [2].

However, there are a number of limitations that make it difficult or impossible to use them for in-pipe inspection and repair operations. Such limitations may include: inability to operate in active and vertically located pipelines, including pipelines with a steep slope to the horizon, inability to pass through curved sections of pipes and sections with variable cross-section, including T-shaped and Y-shaped pipe connections and bends (Fig. 1), and the inability to move in the intertube space of coaxial pipes.

At the present stage, mobile robots for in-line inspection in General should be SEMS modular type built on mechatronic principles and high-performance information technologies, capable of operating effectively and autonomously inside active pipelines in changing conditions and in real time [1, 3, 12]. Therefore, there is the desire of developers to create a universal in-pipe self-propelled robot modular type, able to fully solve complex problems in the conduct of diagnostic and repair work.

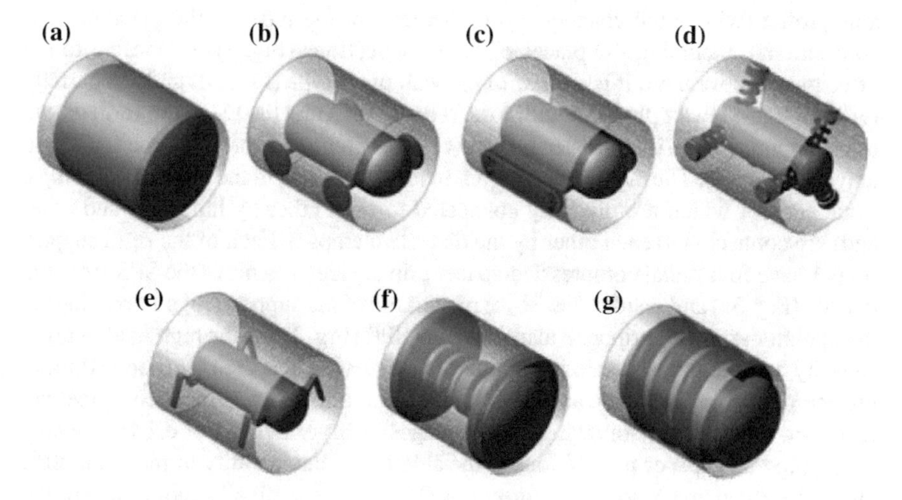

Fig. 2 Classification of in-pipe robots

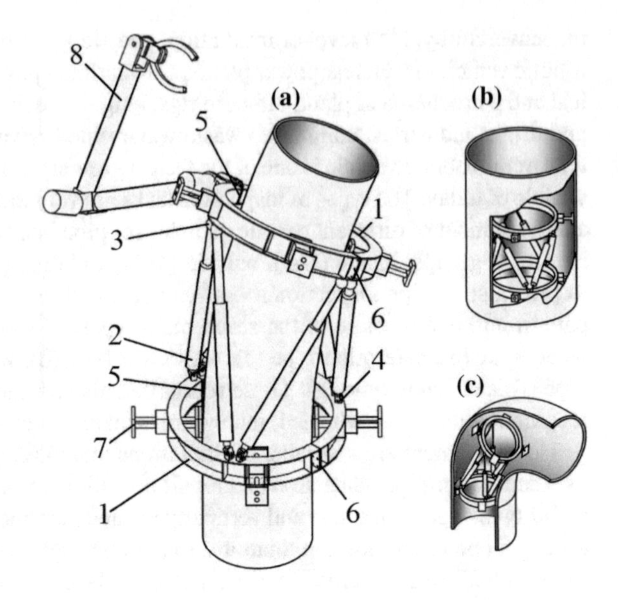

Fig. 3 Design of the climbing parallel robot

In recent years, there has been an interest in considering the possibility of using spatial parallel robots (SPR) based on SEMS for these purposes. Compared with serial robots (SR) SPR have higher stiffness with low weight, and accuracy of the output level, as well as the ability to manage high loads with high speeds of their movements. Due to these qualities, SPR in recent decades are widely used in a wide variety of objects of modern robotics, including pipeline diagnostics [13]. For example, the paper [4] presents an original development of SPR with a manipulator capable of self-displacement not only inside, but also outside the pipes of constant and variable profile (with small changes in the diameter of the pipes in the place of their connections), including the places of their connections (Fig. 3). According to the principle of movement, it is similar to the wall press type (Fig. 2d) and structurally is the Gough-Stewart platform with 6 degrees of freedom [13–15] in the form of two rigid support rings 1, interconnected by six linear actuators 2 through spherical 3 and universal 4 joints. The support rings 1 are made detachable in the form of half-rings, some ends of which are pivotally connected to each other by hinges 5, and other ends are connected to each other by the discrete clamps 6. Each of the rigid support rings 1 have four radial compressive-spacer grip devices 7 to move the SPR from the outside (Fig. 3a) and inside (Fig. 3b, c) pipes. One of the support rings 1 contains the manipulator arm with gripper 8 also. Such the SPR (Fig. 3) has the high load-bearing capacity and the free space in central part. As a result it has the ability to self moving and performs diagnostic and manipulation operations inside the active pipeline. However, there are the following disadvantages: a larger number of drives, the lack of gripping and spacer possibilities at the side faces, the inability to move the SPR on the T-shaped and Y-shaped connections (Fig. 1), the small allowable value of the difference in pipe diameters at the junction, the lack possibility of spatial positioning, measurement of trajectory of in-pipe movement as well as of geometry pipe and its physical and mechanical properties. Also, this SPR is not able to move across open surfaces and to unite with similar structures.

The new universal mobile SEMS modular type on basis of the Octahedral dodekapod (from the Greek words *dodeka* meaning twelve and *pod* meaning foot or equivalent support rod) with 12° of freedom which devoid of these disadvantages is presented below.

2 SEMS on Basis of Octahedral Dodekapod and Its Individual and Collective Applications in Conditions of Incomplete Certainty

2.1 Description of Octahedral Dodekapod

Octahedral dodekapod (OD) was developed at the A.A. Blagonravov Mechanical Engineering Institute of the Russian Academy of Sciences and was first reported at the XXII International Congress on theoretical and applied mechanics (ICTAM 2008, 25–29 August 2008, Adelaide, Australia) [16]. The results of kinematic and dynamic studies of the OD are presented in [17, 18]. The OD is the parallel robotic mechanism with twelve degrees of freedom. Every one of its triangular faces is formed by linear drives that are connected with vertices of octahedron by spherical joints. As a result, all faces of the OD can clamp/unclamp a thing with a closed loop surface of various forms as well as put pressure on environmental surface of contact. This feature opens new functionality of the OD versus a hexapod and others parallel spatial robots. The schematic view of the single-module OD (a), the simplified structural scheme of the control system (b), and the pneumo-hydraulic prototype of OD are shown in Fig. 4 [10, 11, 16–19].

The structural scheme includes maximal number of sensors, radial stops and grippers. This number is dependent on the applications and it may be decreased. OD is executed as the octahedral module (OM) 1. All ribs of the octahedron are executed as the rods with the linear drives 2 each of which have the axial force sensor 3, the medial force sensor 4, the relative displacement sensor 5, and the relative velocity sensor 6. The ends of the adjacent ribs are connected by the spherical joints in the points 7 (A, B, C, D, E, and F) of OM 1. The points 7 of OM 1 contain the radial stops and the middles of the rods contain the grippers (on Fig. 4 were not shown) each of which have the temperature sensor 8. OM 1 has 12 d.o.f., which is a spatial farm as soon as all linear drives 2 are turned off. All points 7 of OM 1 have the spatial position sensors 9 which are integrated with the three-axial acceleration sensors 10. The control system (CS) 11 includes: the neural computer 12, the software 13 and the digital-analog converter (DAC) 14. The inputs of CS 11 are connected to outputs of the analog-digital converter (ADC) 15 of sensors 3 and 4, ADC 16 of sensors 5, ADC 17 of sensors 9 and 10, ADC 18 of sensors 6, and ADC 19 of sensors 8. Outputs of CS 11 are connected to inputs of the software 13 and DAC 14. The outputs DAC

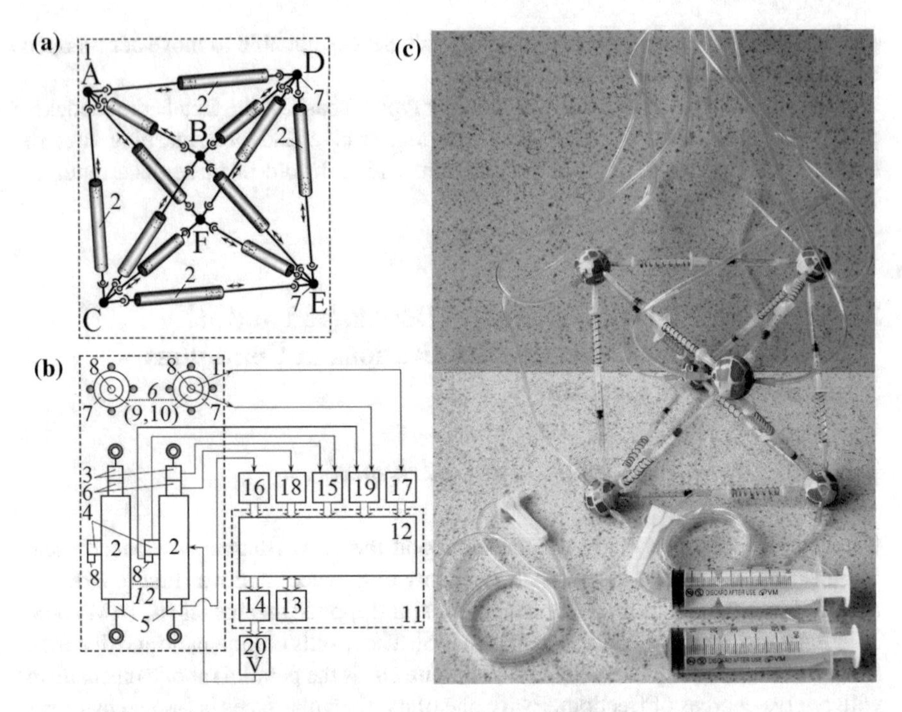

Fig. 4 **a** Schematic view of the single-module OD, **b** Simplified structural scheme of the control system, **c** pneumo-hydraulic prototype of OD

14 are connected to the power amplifier 20 which is connected to each of the linear drives 2. The radial stops and the grippers (on Fig. 4 were not shown) provide the transmission of the efforts from linear drives toward the internal and external contact surfaces. The force sensors 3, 4 and temperature sensors 8 provide the operative control of these efforts and temperature in the contact places. The spatial position sensors 9 with three-axial acceleration sensors 10 provide the operative control of the spatial position of points of octahedral module 7 and of vibration along each of axes of rods with linear drives 2. The relative displacement sensors 5 and the relative velocity sensors 6 of the linear drives 2 register their relative movements and velocities. The conditions and the algorithms for movements of OD in straight and coaxial pipes of constant and variable cross-sections with different bends as well as along outside lengthy objects are represented below [10, 11, 16–19].

2.2 Applications of OD to Use in-Pipe Inspection and Repair Operations in Active Branched Pipelines with Variable Cross-Sections and Outside Operations on Lengthy Objects

2.2.1 Basic Parameters of OD and Its Possible Movements in Different Kinds of Pipes and Along Column

It is obvious that the operation of OD in active pipelines with an opaque transported product should be carried out under conditions of uncertainty and requires the development of new approaches to diagnostic and repair operations. The situation is significantly complicated in case of clogging of the internal space of the pipeline. Possibilities of OD movement in different kinds of pipes and along column is shown in Fig. 5.

The OD, as shown in Fig. 6a contains 6 spherical joints (vertices *A, B, C, D, E, F*) and 12 rods (ribs *AB, BC, AC, DE, EF, DF, AD, AE, BD, BF, CE, CF*) with linear drives.

The vertices *A, B, C* and *D, E, F* form two parallel triangular faces (*ABC* and *DEF*). The vertices *A, B, C* and *D, E, F* are the OD's stops for a moving in the pipe. When one of the faces in the pipe is fixed, the other face moves by simultaneous

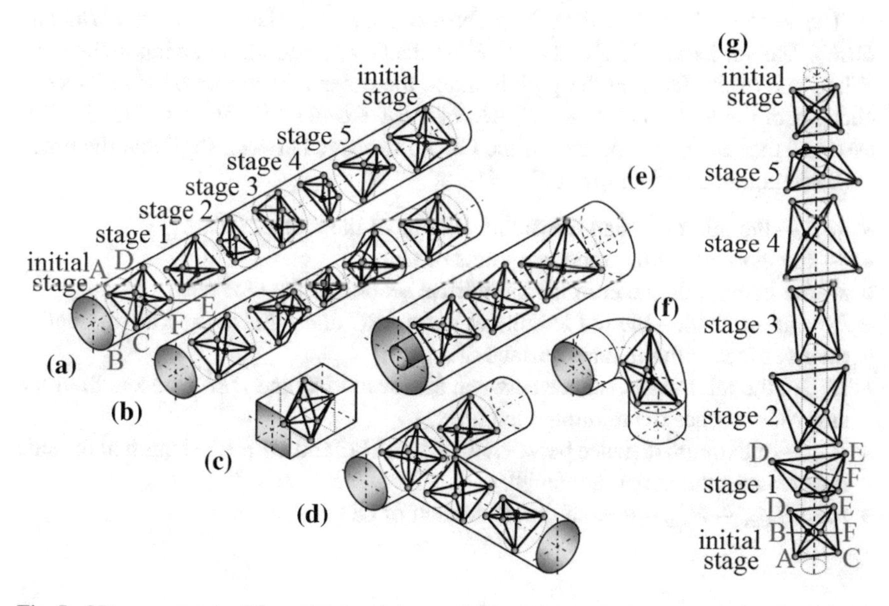

Fig. 5 OD movement in different kinds of pipes and along column: **a** terete or oval pipe, **b** pipe with constant and variable cross-sections, **c** quadrate pipe, **d** T- or Y-branch, **e** coaxial pipes, **f** elbow, **g** column

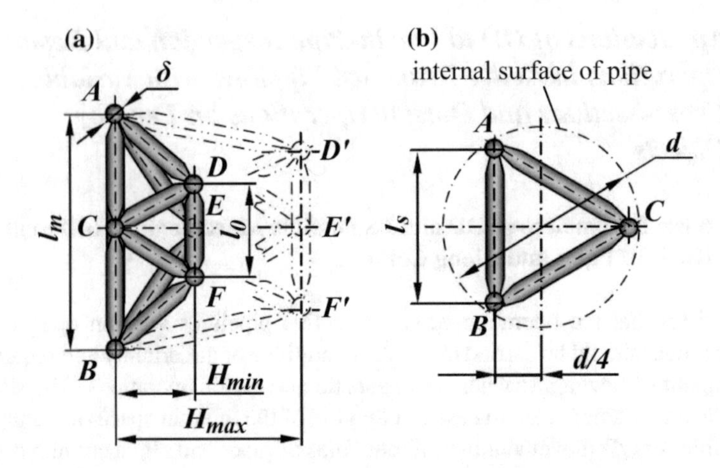

Fig. 6 The main parameters of OD when it is moved in a pipe of constant cross-sections

changes in the lengths of rods *AD, AE, BD, BF, CE, CF*, as shown in Fig. 6b. It is assumed that during the motion of the OD no slippage between stops and the pipe.

The OD, as shown in Fig. 6a contains 6 spherical joints (vertices *A, B, C, D, E, F*) and 12 rods (ribs *AB, BC, AC, DE, EF, DF, AD, AE, BD, BF, CE, CF*) with linear drives.

The vertices *A, B, C* and *D, E, F* form two parallel triangular faces (*ABC* and *DEF*). The vertices *A, B, C* and *D, E, F* are the OD's stops for a moving in the pipe. When one of the faces in the pipe is fixed, the other face moves by simultaneous changes in the lengths of rods *AD, AE, BD, BF, CE, CF*, as shown in Fig. 5a. It is assumed that during the motion of the OD no slippage between stops and the pipe.

Basic notations, as shown in Fig. 6:

- l_0, l_m—the minimum and maximum lengths of the rods of OD;
- δ—the diameter of the spherical joints of OD;
- d—the internal diameter of the cylindrical section of the pipe length L
- $l_S = (d{-}\delta)3^{0.5}/2$—the rod lengths of faces *ABC* and *DEF* at the time of contact their vertices with the inner surface of a pipe;
- $H_{\min}$—the minimum distance between the faces *ABC* and *DEF* (the length of the side rods reaches the minimum length);
- $H_{\max}$—maximum distance between the faces *ABC* and *DEF* (the length of the side rods reaches the maximum length);
- $h_s = H_{\max} - H_{\min}$—step of the movement of OD.

2.2.2 The Algorithm for the Movement of OD in Pipe of Constant Cross-Section

The algorithm for the movement of the OD in pipe of constant cross-section, as shown in Fig. 5a:

- *Initial stage.* The initial lengths of rods and initial position of the spherical joints are assigned from conditions: $l_{AB} = l_{BC} = l_{CA} = l_{DE} =$ $l_{EF} = l_{DF} = l_S = 3^{0.5}d/2$; $l_{AF} = l_{BF} = l_{BD} = l_{CD} = l_{CE} = l_{AE} = l_m$.
- *Stage* 1. The reduced length of rods *AB, BC, CA* from l_s to l_0.
- *Stage* 2. The reduced length of rods *AD, AE, BD. BF, CE, CF* from l_m to l_0.
- *Stage* 3. The increased length of rods *AB, BC, CA* from l_0 to l_s.
- *Stage* 4. The reduced length of rods *DE, EF, DF* from l_s to l_0.
- *Stage* 5. The increased length of rods *AD, AE, BD. BF, CE, CF* from l_0. to l_m.
- *Initial stage.* The increased length of rods *DE, EF, DF* from l_0 to l_s.
  A repeat of all stages from stage 1 to initial stage.

2.2.3 The Algorithm for the Movement of OD in Pipe of Variable Cross-Section

The main parameters when moving OD in a cylindrical pipe of variable cross section are shown in Fig. 7. The pipe consists of two extreme cylindrical sections with diameters d_1, d_2 and an average conical section with length Δ. The complexity of the problem is the possibility of collision of OD with the inner wall of pipe in transition. It is therefore necessary to determine the ratio between the radii of the platform base and the steps of movement. Obviously, to move inside the end sections must satisfy the following inequalities: $[2l_0/3^{0.5}] \leq [(d_1 - \delta)] \leq [2\, l_m/3^{0.5}]$; $[2l_0/3^{0.5}] \leq [(d_2 - \delta)] \leq [2\, l_m/3^{0.5}]$. To pass the transition from the middle section to the extreme with a smaller diameter d_2 without collision of the side bars with the inner wall of the pipe,

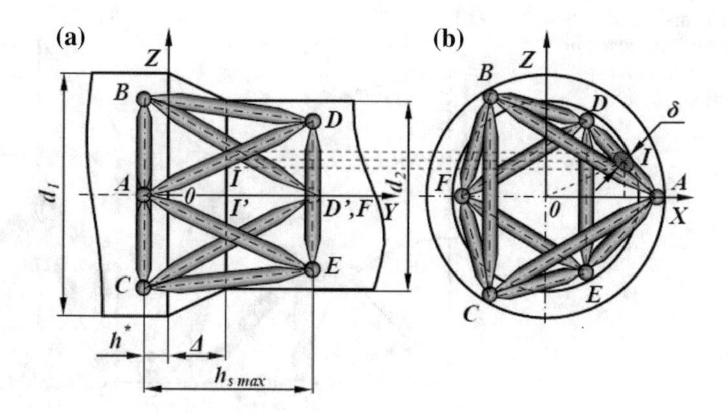

Fig. 7 The main parameters of OD when it is moved in a pipe of variable cross-section

it is necessary to observe restrictions on the maximum distance between the faces *ABC* and *DEF*: $h_{s\max} = [3^{0.5}(d_2 - \delta)(h^* + \Delta)]/4Z_I$.

The example of the algorithm for moving of OD in the pipe of variable cross-section for the cases when $h^* \le 0$ and $0 < h^* < \Delta$ is shown in Fig. 5b.

2.2.4 The Algorithm for the Movement of OD in T-Branch

Let the pipe have the T-branch consisting of two cylindrical pipes: the main diameter *D* and the tie-in diameter d (Fig. 5d). Then for the transition of the OD from the main tube in the tie-in the maximum length of the rods must satisfy the conditions: $l_{\max} > [D^2 + (d - \delta)^2/4]^{0.5}$; $(2l_0/3^{0.5}) \le (D - \delta) \le (2\ l_m/3^{0.5})$. The step of the movement of OD Inside pipes of T-branch is equal $h = H_{\max} - H_{\min}$ (Fig. 5a, b).

2.2.5 The Algorithm for the Movement of OD in the Elbow of Constant Cross-Section

The algorithm for the movement of the OD in the elbow of constant cross-section and the main its parameters are shown in Figs. 5f and 8. The main parameters of the OD and the elbow are following: the angle α is the angle between the faces *ABC* and *DEF*, equal step of the OD; $D'E'F'$ is projection of the face *DEF* on the face *ABC*, obtained by rotating the face *DEF* relative to the axis *OZ* by the angle α. When moving the OD in a bend, its triangular faces ABC and DEF must be equilateral and perpendicular to the plane *ZOY*, and the radius of the bend must pass through the medians of these faces.

As a result of the analysis, the following conditions for the passage of the OD inside of the elbow with the constant cross section without collision of the side rods with the inner wall of the pipe are determined.

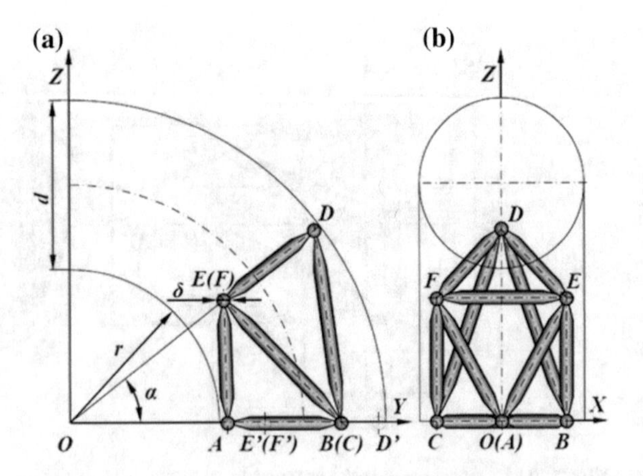

Fig. 8 The main parameters of OD when it is moved in the elbow of constant cross-section

- The following Inequality must be fulfilled $(2l_{min}/3^{0.5}) \leq (d - \delta) \leq (2l_{max}/3^{0.5})$.
- To move the OD without impacting the side rods AE and AF with the inner wall of the pipe, it is necessary that the angle *OAE* is greater than 90°. From this condition we obtain the following restriction for the angle α:

$$\alpha < \alpha_{\Pi} = arccos[(4r + 2\delta)/(4r + d + 2\delta)],$$

  where α_{Π} is the angle depending on the geometric parameters of the elbow and the vertices of OD.
- The movement of OD inside the elbow is limited by the lengths of the lateral rods *AE*(*AF*) and *BD*(*CD*) ($|AE| \geq l_{min}$, $|BD| \leq l_{max}$) and the following range of permissible values of the angle α between the platform and the base of the OD:

$$\alpha \geq \alpha_{\min} = arccos[(4d^2 + 32r^2 + 11\delta^2 + 8dr - 2d\delta + 32r\delta - 16l^2_{\min})/ 4(4r + d + 2\delta)(2r + \delta)];$$

$$\alpha \leq \alpha_{\max} = arccos[(28d^2 + 32r^2 + 11\delta^2 + 56dr - 34d\delta + 32r\delta - 16l^2_{\max})/ 2(4r + 4d - 2\delta)(4r + 3d - 2\delta)].$$

  The maximum possible step of moving OD inside the elbow is $\alpha = A_{max} - A_{min}$. In this case, the minimum number of steps is $S_{min} = (\Psi/\alpha) + 1$. Here Ψ is the angle between the end planes of the elbow.

After the passage of the elbow (not shown in the figures), the OD enters in the straight section of the pipe and continues to move with the maximum possible step $p = H_{max} - H_{min}$.

2.2.6 The Algorithm for the Movement of OD Along of Outside Surface of the Pipe

The algorithms for moving of OD along of outside surface of the tube (Fig. 5, g), including sections of transitions between pipes of different diameters and elbows, are similar to the above algorithms for moving the OD inside the pipes except for moving the OD on the T-branch when it is required to rearrange the OD from the main pipe on the tie-in and versa. The basic elements that define the geometry of the OD, in addition to its vertices, are the geometrically associated coordinates of the grips of the midpoints of the frontal *DEF* and the rear *ABC* faces.

2.2.7 The Algorithm for the Movement of OD in the Coaxial Pipe

Move the OD in the coaxial pipe can be carried out on the inner surface of the outer tube on the outer surface of the inner tube or in combination, when the front or the

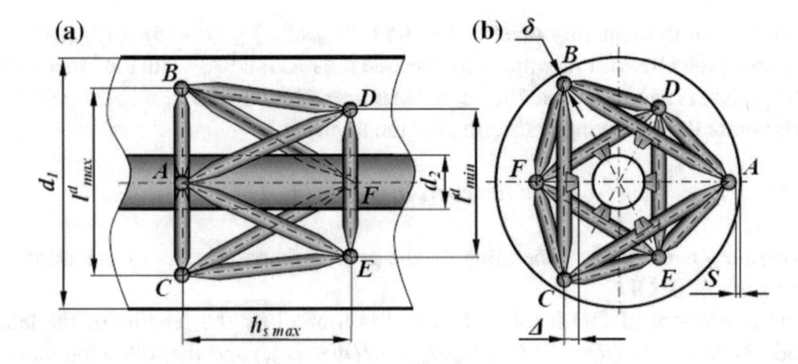

Fig. 9 The main parameters and the scheme of installation of OD inside the coaxial pipe at its movement on an external surface of an internal pipe

back face moves along the inner surface of the outer tube and, on the contrary, the opposite face moves along the outer surface of the inner tube. In this case, the basic elements that determine the geometry of the OD, in addition to its vertices, are also geometrically associated with them the coordinates of the contact surfaces of the grippers of the middle sections of the rods of the frontal DEF and the rear ABC faces (Fig. 9). Here, the OD has the ability to control the clearance between pipes also. In the case of misalignment of coaxial pipes can be carried out straightening operation with the help the OD to ensure the alignment of the pipes. Also, with the help of OD, installation operations can be carried out for coaxial placement of the connected sections of internal or external pipes during welding operations.

The algorithm of displacement of OD is similar to the algorithms of its displacements along the inner and outer surfaces of pipes. In this case, there are the following restrictions on the movement of the OD in the space between pipes.

When moving OD on the inner surface of the outer pipe, the dynamic length of the rods of the frontal *DEF* and rear *ABC* faces must meet the following conditions:

$$l^d_{\min} < (d_2 + 2\Delta)3^{0.5};\ l^d_{\max} = (d_1 - \delta)3^{0.5}/2.$$

When moving the OD on the outer surface of the inner pipe, the dynamic length of the rods of the frontal *DEF* and rear *ABC* faces must meet the following conditions:

$$l^d_{\min} < (d_2 + 2\Delta)3^{0.5};\ l^d_{\max} = (d_1 - \delta - 2S)3^{0.5}/2.$$

Here $l^d_{\min}$ and $l^d_{\max}$ are the minimum and maximum dynamic lengths of the rods of the frontal *DEF* and rear *ABC* faces, d_1 and d_2 are the inner diameter of the outer and outer diameter of the inner pipes, S is the technological clearance, Δ is the distance from the contact surface of the grippers of the middle sections of the rear and frontal faces, $h_{s\max}$ is the distance between the frontal *DEF* and rear *ABC* faces of the OD with the maximum extension of its side rods. It should be noted that the actual maximum length of the rods should be greater than their maximum dynamic

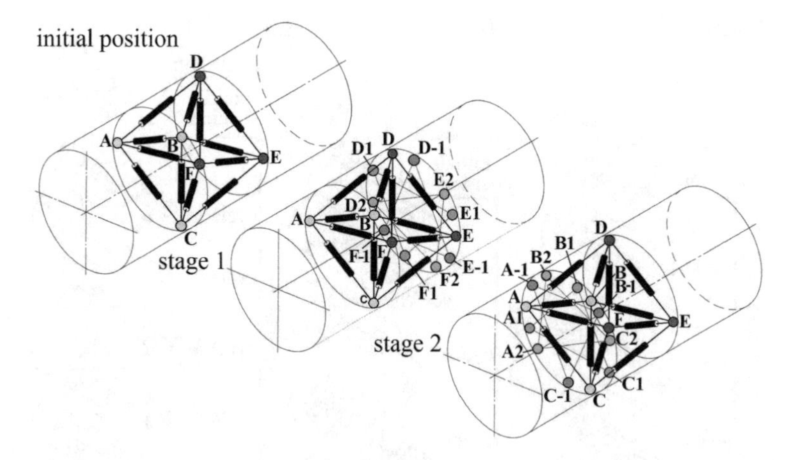

Fig. 10 Possibility of control over of physicomechanical properties, geometrical shape of contact surface and displacement trajectory

length, and the minimum real length of the rods is not more than their minimum dynamic length.

2.2.8 Other Capabilities of OD

The novel concept of OD may be used also in many other fields. The examples of applications of OD and its new functional capabilities are established below, e.g. [10, 11, 16–19]:

- *Example* 1. Possibility of control over of physical-mechanical properties, geometrical shape of contact surface and displacement trajectory (Fig. 10).
- *Example* 2. Travel of long objects in pipe (Fig. 11).
- *Example* 3. Travel of object in pipe with possibility of vibroprotection and positioning (Fig. 12).
- *Example* 4. Possibility of hole drilling in the end of pipe wall (Fig. 13).
- *Example* 5. Possibility of battering in pipe of wall end (Fig. 14).
- *Example* 6. OD can connect together, forming some novel mobile self-reconfigurable structures (swarm systems) for various applications (Fig. 15).

Example 1 Example of control of geometrical shape of contact surface by the OD is given in Fig. 10. In this mode, in the motion of the OM 1 (*ABCDEF*) (Fig. 4) of the adaptive mobile parallel spatial robot (Fig. 10), each longitudinal displacement of the rear face ΔABC and frontal face ΔDEF is preceded by alternating discrete rotations in both directions, with specified increment, relative to the direction of motion.

For each discrete position, mechanical contact is established between the radial limiters at points 7 (Fig. 4) of those faces and the internal contact surface or between the radial limiters at the midpoints of the rods in the frontal and rear faces and the

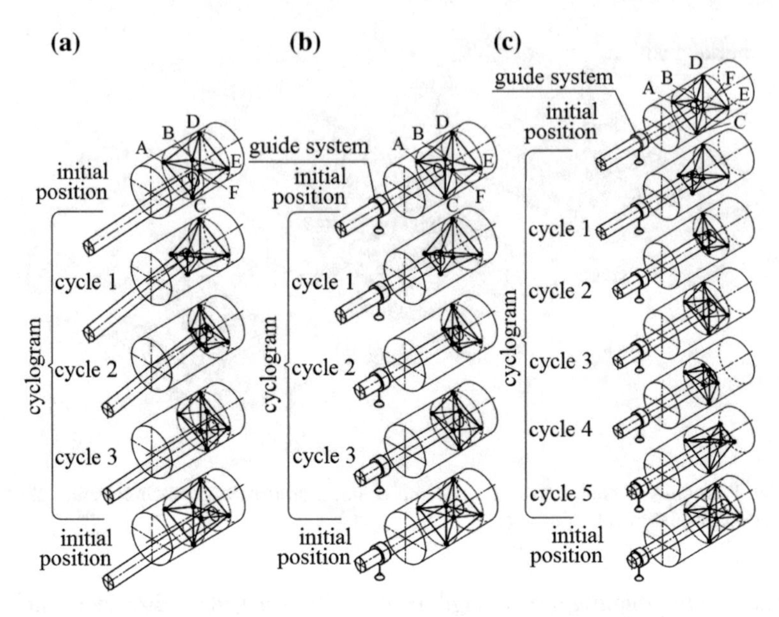

Fig. 11 Moving of long objects inside the pipeline: **a** without the guide system, **b** with the guide system but without moving OM, **c** with the guide system and moving OM

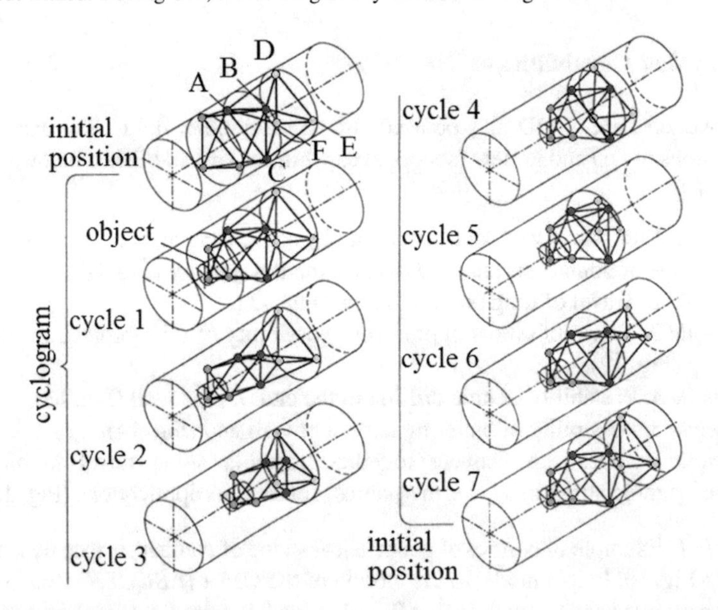

Fig. 12 Motion of object in pipe with possibility of vibroprotection and positioning by double OD

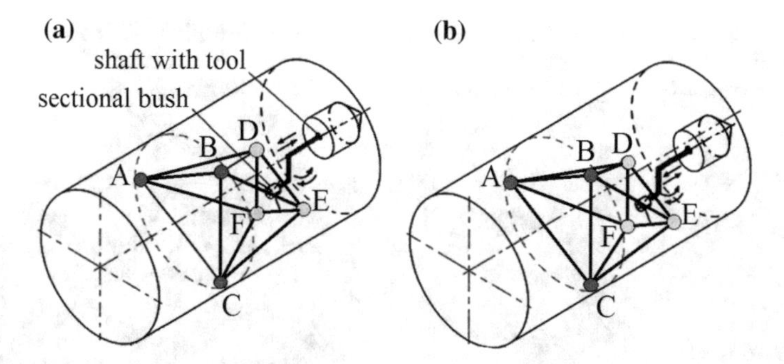

Fig. 13 Rotation of a machining tool by means of OM (*ABCDEF*): **a** rotation coaxial with the module's symmetry axis; **b** eccentric rotation

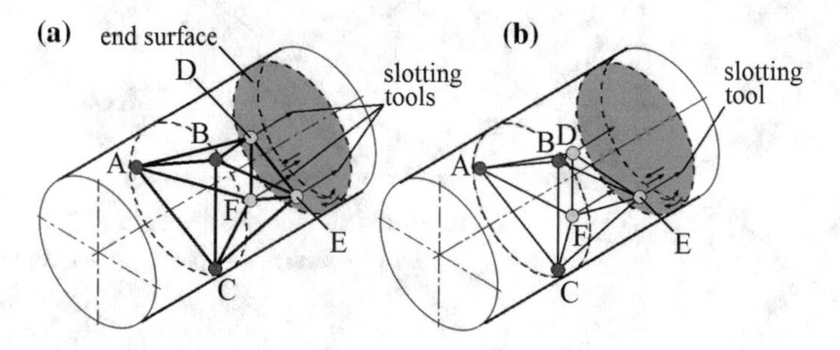

Fig. 14 Organization of simultaneous impact and vibrational effects on the end wall by slotting tools (**a**) and one time impact (**b**)

external constant surface. Their coordinates are determined in the basic coordinate system. These values permit judgments regarding the geometric form of the internal or external contact surfaces. The specified contact forces with the internal surface are determined from the readings of sensors 3; and the specified contact forces with the external surface are determined from the readings of sensors 4. The position of points 7 of the frontal and rear faces (ΔABC, ΔDEF) are determined from the readings of the relative displacement sensors 5 of linear drives 2 for the rods of the lateral faces and position sensors 9 at points 7.

In mode of control of elastic properties of contact surface (Fig. 10), the radial limiters at points 7 (Fig. 4) of the faces come into contact with an internal surface, with a force specified by the readings from sensors 3. Then, their position is determined in the basic coordinate system on the basis of the readings from sensors 5 and 9. Next, the force on the radial limiters is increased to another specified value, and their position is determined in the basic coordinate system. This procedure is then repeated with the initial force. The difference between the coordinates of the radial limiters at points 7 permits judgments regarding the elastoplastic properties of the contact surface. At specified contact force between the radial limiters and the internal surface, the temperature is measured by means of sensors 8, and the electrical resistance between

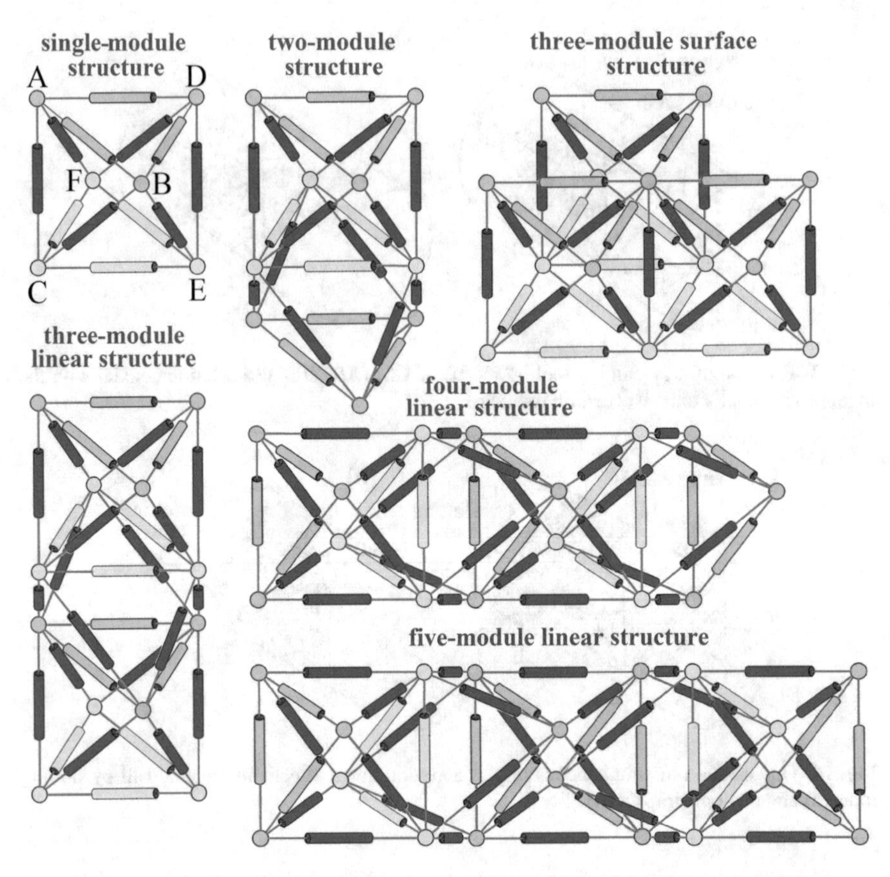

Fig. 15 Some examples of multi-module reconfigurable SEMS basis on the group of OD

them is recorded. These readings permit judgments regarding the physical properties of the contact surface. Analogously, the temperature and electrical resistance may be recorded for an external surface. Vibrational diagnostics may also be organized for an internal contact surface, with identification of the presence of mechanical defects (such as cracks in pipes). In that case, periodic acceleration of linear drives 2 for the rods of the rear or frontal faces permits impact and vibrational influence on the radial limiters at points 7 of these faces at the contact surface. Vibrational diagnostics of the object may then be based on the readings of acceleration sensors 10.

Example 2 In Fig. 11, we show examples of the motion of round or oval objects by OM 1 (*ABCDEF*) without (a) and with (b) a guide system. In this mode, OM 1 (Fig. 4) moves within a closed surface and fixes the frontal face (ΔDEF, say). The end of the extended object (such as pipe, rod, or cable) is placed in the rear face ΔABC (Fig. 11, initial position), and linear drives 2 are turned on in reverse. The length of the rods in the rear face is reduced until the object is captured by force specified by the readings of the force sensors 4 at the radial limiters at the midpoints of the rods in rear face ΔABC. Then, at a command from control system 11, linear drives 2 are switched off, and the coordinates of points 7 are determined, in the basic coordinate system. After the object is fixed (Fig. 11, cycle 1), coordinated decrease in length of the rods in the lateral faces (ΔABD, ΔBDE, ΔBCE, ΔCEF, ΔACF, ΔADF) shifts the rear face ΔABC together with the object into the closed surface by some fixed distance, which is recorded with respect to the basic coordinate system (Fig. 11, cycle 2). Then the length of the rods in the rear face is increased until the radial limiters at the midpoints, together with the object, are completely released (Fig. 11, cycle 3) and the object is unable to move in the opposite direction (Fig. 11a, cycle 3). Then the length of the rods in the lateral faces is increased to its initial value (Fig. 11, initial position). Thereafter, the motion of the object is repeated as many times as is necessary, and the total length traversed at the end of the process is determined. Note that the motion of the object may be conducted without (Fig. 11a) or with a guide system, as shown in Fig. 11b. Without a guide system, additional effort is required to prevent inverse motion of the object in cycle 3.

With a guide system (Fig. 11b), that is unnecessary. If required, motion in a combination of modes 1 and 4 is possible. In that case, the object will move with simultaneous independent motion of octahedral module 1 within the closed system. In contrast to mode 4 (Fig. 11b), the length of the rods in the rear face ΔABC increases not until the object is released but until fixing of points 7 of the rear face ΔABC at the closed internal surface with specified force, in accordance with the readings of sensors 3. Then linear drives 2 are switched off, and the coordinates of points 7 are calculated from the readings of sensors 5 and 9. Next, at a command from control system 11, linear drives 2 of rods in the frontal face ΔDEF are switched on in reverse, and their length is reduced until capture of the object by the radial limiters at the midpoints of these rods, with a force specified by the readings from sensors 4. Then linear drives 2 are switched off, and the coordinates of points 7 are calculated from the readings of sensors 5 and 9. At that point, the length of the rods in the lateral faces is increased, and the object is moved within the closed surface. Then the length of the rods in the frontal face ΔDEF is increased until the points 7 of the rear face ΔABC are fixed at the reduced length of the closed internal surface with specified force, in accordance with the readings of sensors 3. Then linear drives 2 are switched off, and the coordinates of points 7 are calculated from the readings of sensors 5 and 9. Thus, in this mode, not only the extended object but the octahedral module 1 will be moved. As a result, the distance traveled by the object relative to the closed internal surface will be increased, with a fixed number of cycles (operations).

If necessary, a combined movement can also be arranged, including a conjunction of moving the OM 1 and a long object (Fig. 11c). As the result, the distance of movement of the long object relative to the inner surface of the pipe increases with the same number of cycles.

Example 3 Motion of objects in the pipe by means of paired octahedral modules, with simultaneous positioning and vibrational protection of the objects (Fig. 12). In this mode, the octahedral modules are paired to form of a common face ΔABC. One OM (*ABCDEF*) performs motion over a closed internal surface of the pipe, analogously to movement of OD in pipe of constant cross-section (Fig. 5a). The other OM captures the object, analogously to the capture of the long objects in pipe (Fig. 11). The second OM also positions the object and ensures vibrational protection during motion over the closed internal surface. The capture force and spatial position of the object, as well as the impact and vibrational influence of OM (*ABCDEF*) on the OM with the object are monitored by means of force sensors 4 (Fig. 4), spatial position sensors 9 (Fig. 4), and three-axial acceleration sensors 10 (Fig. 4) mounted on radial supports of the rear face (the common face of the moving OM and the clamping OM) of the clamping module, as well as the relative displacement sensors 5 (Fig. 4) and relative velocity sensors 6 (Fig. 4) mounted at the rods of the lateral faces.

Vibrational protection of the object relies on the linear drives 2 (Fig. 4) in the rods of the lateral faces of the clamping module. It involves coordinated change in the length of those rods at a command from control system 11 (Fig. 4), on the basis of the readings of sensors 10 (Fig. 4) mounted at points 7 (Fig. 4) of the frontal and rear faces of the clamping module and also sensors 5 (Fig. 4) and 6 (Fig. 4) at each of the rods of the lateral faces of the clamping module.

Example 4 Rotation and supply of a machining tool (a drill or bit, for instance) by means of OM (*ABCDEF*) (Fig. 13). In this case, the tailpiece of the tool takes the form of a crankshaft with a rotating bush at the end that is incapable of axial motion. (The rotating bush is not shown in Fig. 13). The rotating bush is clamped by the radial limiters at points of the rods in the frontal face. Each limiter may be rigidly connected with one section of a three section bush intended for capture of the rotating tailpiece bush. The required force is determined on the basis of the sensors 4 in the frontal face. The radial limiters at the rear face are fixed at the internal contact surface, as in previous modes. Then, coordinated change in the length of the rods at the lateral faces brings the cutting section of the tool to the machining point, fixes the axis of tool rotation, and ensures the required cutting force, determined from the readings of sensors 3 at the rods of the lateral faces.

Next, coordinated change in the length of the rods at the lateral faces moves the axis of rotation of the clamped bush over a circle perpendicular to the axis of rotation; the circle radius is equal to the crankshaft radius. In tool rotation, coordinated increase in length of the rods at the lateral faces ensures its longitudinal supply with specified force; the generation of impact and vibration effects is possible in combination with tool rotation. In that case, the spatial position, cutting force, and impact and vibration effects are monitored by means of sensors 9 and 10 (Fig. 4) at the radial limiters of the frontal face and sensors 3, 5, and 6 (Fig. 4) at the rods in the lateral faces. The axis of tool rotation may be coaxial with the symmetry axis of OM 1 (Fig. 13a) or eccentric (Fig. 13b).

Example 5 Organization of impact and vibration effects by a slotting tool on the end surface (Fig. 14) of a tubular profile by means of OM 1 (Fig. 4).

In this mode, the slotting tools are established at each point of the frontal face. (The clamping of the slotting tool's tailpiece is not shown in Fig. 14). The radial limiters of the rear face are fixed at the internal contact surface as in previous modes. Then, coordinated change in the length of the rods at the lateral faces brings the working sections of the slotting tools into contact with the end surface at the machining site, and machining begins at specified frequency, amplitude, and force. In one time or double action, where necessary, the sequence of action may be modified. In each case, the spatial position, cutting force, and magnitude of the tools' impact and vibration effects are monitored by means of sensors 9 and 10 (Fig. 4) at the radial limiters of the frontal face and by sensors 3, 5, and 6 (Fig. 4) at the rods of the lateral faces. Real time operation is possible thanks to the use of a control system 11 based on neural computer 12 and corresponding software 13 (Fig. 4).

Example 6 OM 1 (Fig. 4) is the basic element not only in the single-module SEMS, but also in constructing multi-module spatial robot systems (Fig. 15). Accordingly, all the faces of OM 1 (Fig. 4) may be connected by an operator to similar modules. Junction faces are included in the additional octahedral modules, depending on the direction of growth. Each of the faces to which another module is connected forms a triangle consisting of rods with linear drives 2 that are hinged at the ends 7. Each pair of parallel faces of OM 1 may be used as the rear face (facing away from the motion) or frontal face (facing into the motion): for example, ΔABC and ΔDEF. The axis passing through the centers of these faces is in the direction of motion of OM 1.

Also the group of autonomous OD can connect together without any operator, forming some novel mobile self-reconfigurable structures (swarm systems) for various collective applications in conditions of incomplete certainty. To do this, all vertices of each OM 1 (Fig. 4) must be equipped with docking devices. The different types of docking devices for modular mobile robots, including the OD, are given in [20].

3 Conclusions

Means of expanding the functional capabilities of parallel spatial robots are considered. The promising design is an adaptive mobile parallel spatial robot of the Octahedral dodekapod. The benefits of the Octahedral dodekapod with respect to the other types robots are outlined, along with possible applications.

The Octahedral dodekapod may be used as the base element not only in the single-module SEMS, but also in the multi-module ones Octahedral dodekapod can connect together, forming some novel mobile self-reconfigurable structures (swarm systems) for various individual and collective applications in conditions of incomplete certainty.

The proposed adaptive mobile parallel spatial robot may be used at the macroscopic level for use on land, underground, underwater, in medicine, and in the aerospace industry. It may also be used at the microscopic level, for example, for in-veins movements and manipulations. This research provides the basis for the development of up to date parallel spatial robots and for the expansion of their functional capabilities.

Information regarding the practicality of adaptive mobile parallel spatial robots on the basis of currently existing technology may be found in [10, 11, 18].

References

1. Bekhit, A., Dehghani, A., Richardson, R.: Kinematic analysis and locomotion strategy of a pipe inspection robot concept for operation in active pipelines. Int. J. Mech. Eng. Mechatrons. **1**(1), 15–27 (2012)
2. Gradetsky, V.G., Knyazkov, M.M., Fomin, L.F., et al.: Mekhanika miniatyurnykh robotov (The miniature robot mechanics). Nauka, Moscow (2010)
3. Roh, S.G., Choi, H.R.: Development of differential-drive in-pipe robot for moving inside urban gas pipelines. IEEE Trans. Rob. **21**(1), 1–17 (2005)
4. Almonacid, M., Saltarén, R.J., Aracil, R., et al.: Motion planning of a climbing parallel robot. IEEE Trans. Robot. Autom. **19**(3), 485–489 (2003)
5. Sakamoto, S., Hara, F., Hosokai, H., et al.: Parallel-Link Robot for Pipe. In: Proceedings of the 31st Annual Conference of the IEEE Industrial Electronics Society IECON 2005, Raleigh, North Carolina, USA, 6–10 Nov 2005
6. Saltaren, R., Aracil, R., Reinoso, O.: Analysis of climbing parallel robot for construction applications. Comput.-Aided Civ. Infrastruct. Eng. **19**, 436–445 (2004)
7. Urdaneta, M., Garcia, C., Poletti, G., et al.: Development of a novel autonomous robot for navigation and inspect in oil wells. Control Eng. Appl. Inform. **14**(3), 9–14 (2012)
8. Enner, F., Rollinson, D., Choset, H.: Motion estimation of snake robots in straight pipes. In: Proceedings of the 2013 IEEE International Conference on Robotics and Automation (ICRA 2013), Karlsruhe, Germany, 6–10 May 2013
9. Saltaren, R., Aracil, R., Reinoso, O., et al.: Climbing with parallel robots. In: Habib, M.K. (ed.) Bioinspiration and Robotics: Walking and Climbing Robots, pp. 209–226. I-Tech, Vienna, Austria, EU (2007)
10. Sayapin, S.N., Sineov, A.V.: Adaptive mobile 3D manipulator robot and method of organizing displacements and control over physical-mechanical properties, geometrical shape of contact surface and displacement trajectory hereby. Russian Federation Patent 2,424,893,11 Jan 2009
11. Sayapin, S.N.: Parallel spatial robots of dodecapod type. J. Mach. Manuf. Reliab. **41**(6), 457–466 (2012)
12. Egorov, I.N., Kadkhim, D.A.: Primeneniye mobilnykh robotov pri bnutritrubnoy diagnostike truboprovodov s peremennym secheniem (Application of mobile robots for in-pine inspection of pipelines with variable cross-section). Neftegazovoe delo. **3**, 73–85 (2011) http://www.ogbus.ru
13. Merlet, J.-P.: Parallel Robots. Springer, Dordrecht (2006)
14. Gough, V.E.: Contribution to discussion of papers on research in automobile stability, control and tyre performance. Proc. Auto Div. Inst. Mech. Eng. **171**, 392–394 (1956–1957)
15. Stewart, D.: A Platform with six degrees of freedom. Proc. Inst. Mech. Eng. **180**(1), 371–386 (1965). 15
16. Sayapin, S., Siniov, A.: Siniov A (2008) The adaptive spatial mobile robot—manipulator and way of diagnostics of physical and mechanical properties and the geometrical form of a surface of contact and trajectory of movement with his help. In: Denier, J., Finn, M., Mattner, T. (eds.) Abstracts of the XXII International Congress of Theoretical and Applied Mechanics (ICTAM 2008), pp. 25–29. Australia, Adelaide (2008)

17. Sayapin, S., Karpenko, A., Hiep, D.X.: Dodekapod as universal intelligent structure for adaptive parallel spatial self-moving modular robots. In: Waldron, K.J., Tokhi, M.O., Virk, G.S. (eds.) Nature-Inspired Mobile Robotics, pp. 163–170. World Scientific, Singapore (2013)
18. Sayapin, S.N.: Mobile parallel robot-manipulator "octahedral dodekapod": History, present and future. J. Eng. Autom. Probl. **3**, 36–60 (2018)
19. Sayapin, S.N.: Novye mekhanizmy kosmicheskoy robototekhniki (New mechanisms of space robotics). In: Glazunov, V.A. (ed.) Novye mekhanizmy v sovremennoy robototekhnike (New mechanisms in modern robotics), pp. 207–231. Tekhnosphera, Moscow (2018)
20. Sayapin, S.N.: Principles of docking of modular mobile robots based on SEMS in their group interaction. In: Gorodetskiy, A.E., Tarasova, I.L. (eds.) Smart Electromechanical Systems: Group Interaction. Studies in Systems, Decision and Control, vol. 174, pp. 125–135. Springer, Cham, Switzerland (2019)

System Analysis and Management in Group Robotics Based on Advanced Cat Swarm Algorithm. Lower Level Hierarchy

Anatoliy P. Karpenko and Ilia A. Leshchev

Abstract *Problem formulation*: Nowadays, in many applications, there is a problem of developing mathematical models to control groups of complex dynamic objects that interact with each other in the process of achieving a common goal. As such, we consider the problem of extrema localization of an unknown scalar physical field. The tasks of detecting zones of radioactive, chemical, biological or other kinds of terrain contamination, temperature and salinity of the seas, and other similar problems can be set in this form. The article is a continuation of the authors' work, which considers a system of decentralized control of a group of robots designed to solve this class of problems and based on the use of modified Cat Swarm Algorithm. Decentralized control systems, as objects of mathematical modeling, are an important part of modern complex dynamic systems consisting of autonomous objects that work together on the basis of the organization of complex behavior of these objects in an unpredictable environment. Such systems significantly increase the control efficiency of dynamic systems at the strategic, tactical and executive levels. The relevance of the study is attributable to the imperfection of the existing models of distributed control systems. Aim of the study: The article is dedicated to the development of new mathematical models and approaches that can be used in solving problems of control of distributed dynamic systems in their group application in real environments. The main task is to manage the coordinated behavior of the group members, that separately solve a common task. It is assumed that the control is carried out in an uncertain and changing environment, which requires rapid adaptation to these changes. The main purpose of the study is to synthesize an effective situational control system for a group of robots designed to localize the extrema of an unknown scalar physical field. *Results*: On the basis of the situational theory, a mathematical model of the control system of a group of robots is developed. Using several test functions, a computational experiment was conducted to study the effectiveness of the proposed control system. The results of the computational experiment, which showed the prospects of the developed mathematical support and software, are presented.

A. P. Karpenko (✉) · I. A. Leshchev (✉)
Bauman Moscow State Technical University, Moscow, Russia
e-mail: apkarpenko@mail.ru

I. A. Leshchev
e-mail: leshchevia@gmial.com

© Springer Nature Switzerland AG 2020

A. E. Gorodetskiy and I. L. Tarasova (eds.), *Smart Electromechanical Systems*, Studies in Systems, Decision and Control 261, https://doi.org/10.1007/978-3-030-32710-1_5

Keywords Situational control · Global optimization · A group of robots · Control algorithm · Swarm algorithm · Decentralized system · Modified cat swarm algorithm · A robot

1 Introduction

We consider a class of search problems that can be reduced to the problem of extrema localization of an unknown scalar physical field by using a group of robots. Detection of zones of radioactive, chemical, biological or other kinds of terrain contamination, zones of harmful algae, zones of turbulence, temperature and salinity of the seas, and other similar problems can be set in this form [1–4]. This paper is a continuation of the paper [5] which considers the decentralized control system for a group of robots designed to solve this class of problems.

The methodological basis of the study is a bionic approach, that is, an approach based on technical copying of effective solutions found by living species in the process of their natural selection. In relation to the class of problems under consideration, implementation of this approach is reduced to constructing a control system for a group of robots using such swarm algorithms as the particle swarm algorithm, the bees algorithm, the ant colony algorithm, and so on [2–4].

As a basic swarm algorithm, we use the proposed in [5] the deep modification of the cat swarm optimization algorithm (*CSO*) [6–8] called *ACSO* (*Advanced CSO*). The algorithm takes into account peculiarities of the used hardware implementation of a group of robots.

In contrast to [5], we assume that the control system is poorly defined. Generally speaking, the main features of such systems are [9]: (a) uniqueness; (b) lack of a formalized purpose of existence; (c) lack of optimality; (d) dynamics (evolution of the object in time); (e) incompleteness of description; (f) presence of free will. We will show that the control system of a group of robots has, at least, features (b)–(e).

Lack of a formalized purpose of existence. At the verbal level, the purpose of existence of the group control system under consideration can be defined as follows: to localize the extrema of the given unknown scalar physical field with maximum probability under specified resource constrains. The dynamics of the system and its incomplete description do not allow to formalize this goal within the framework of Classical Control Theory (see below).

Lack of optimality. For lack of a formalized purpose of the system existence, it is impossible to construct an *objective* control criterion, that is, the criterion of optimality turns out to be largely subjective, depending on the decision-maker's (DM) preferences.

Dynamics of the system is caused by the following circumstances: change in the characteristics of the scalar field under study; parametric and/or structural change of the control object, for example, due to breakdowns of robots, etc.

Incompleteness of description is due to the following: inability to identify all possible situations in which the system may operate; lack of an accurate functioning model for each of the robots, inability to quantify many features of their functioning, etc.

It should be noted that fact of dynamics of the control system under consideration implies the need for this system to be adaptive.

Poorly defined control systems are the subject of study of the situational theory [10]. The basic concepts of the theory are the concepts of the *current situation* Q and the *full situation* S. The current situation Q in the control object is a set of its current state determined by the state vector and the state of the environment. The full situation is a set of the current situation Q in the object and the knowledge of a set of acceptable situations D. Therefore, $S = \langle Q, D \rangle$.

In accordance with the methodology of situational control, with the help of an appropriate classifier, the set of all possible full situations is divided into r classes D_k, $k \in [1{:}r]$, each of which is set in accordance with one of the admissible controls U_k. Thus, the elementary act of control can be represented in the following form:

$$S_k\colon Q_j \underset{U_k}{\rightarrow} Q_l.$$

This formula means that if the situation Q_j arises in the control object and it is admissible to use the control U_k, then, being applied, this control turns the situation Q_j into the situation Q_l. Transformation rules of this type are called logic-transformation rules (LT rules).

We consider the situational control of a group of robots at two hierarchical levels. The executive subsystem of one robot is the control object at the lower (*local*) level, while the group of robots is the control object at the upper (*global*) level. Here we study the local level of control. The global level will be considered in future publications.

Just like in the paper [5], we exclude the "tactical" problem of overcoming obstacles, confining ourselves to the problem of "strategic" robot control.

2 Problem Statement and Basic Definitions

We use the following notations [5]: $\Pi = \{X | X^- \le X \le X^+\} \in R^2$ is the study region (parallelepiped) of the search space, where $X^- = (x_1^-, x_2^-)$, $X^+ = (x_1^+, x_2^+)$ are the boundaries of the region, and the inequalities are understood component by component; t is the number of the current iteration of the search; $X_i(t) = X_i = (x_{i,1}, x_{i,2})$ is the current position of the robot M_i; $X_i'(t) = X_i' = \left(x_{i,1}', x_{i,2}'\right)$ is the position of this robot at the next iteration; X^* is the desired vector of optimal coordinates X which delivers the maximum value φ^* of the fitness function $\varphi(X)$ formalizing the intensity of the scalar field under study.

We consider the deterministic maximization problem

$$\max_{X \in \Pi \subset R^2} \varphi(X) = \varphi\left(X^*\right) = \varphi^*.$$

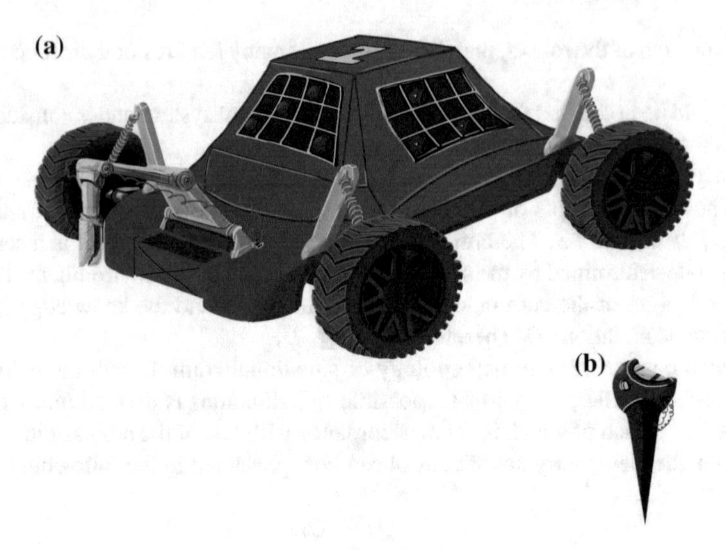

Fig. 1 Schematic image of M-robot (**a**); schematic image of C-robot (**b**)

We assume that in the initial state, the considered group of robots $M_i, i \in [1{:}n]$ is homogeneous. Each of the devices M_i is *a mother robot* (M-robot) that carries a fixed number ν of small ejected (non-returnable) *satellite robots* $C_{i,j}$ (C-robots); $j \in [1{:}\nu]$ (Fig. 1).

Each of the M-robots can measure values of the scalar field $\varphi(X) = \varphi(x_1, x_2)$ (test) at its stop points, the total number of which does not exceed $\hat{t}$. Each of the satellite robots can perform a *single* test at a distance from the robot M_i, not exceeding δ_X in each of the measurements x_1, x_2.

In robot $M_i, i \in [1{:}n]$, we allocate the executive subsystem (the control object) O_i and the local control subsystem (*Local Control System, LCS*) denoted as LCS_i. Thus, $M_i = \{O_i, LCS_i\}$; $i \in [1{:}n]$.

We use the following two strong assumptions: (1) the test results of M- and C-robots are accurate, that is, the measured values of the function (2) the robots can determine their coordinates x_1, x_2 with an error which is many times less than the value δ_X, so that these errors can be ignored.

We enter the following definitions.

C-vicinity of the robot M_i is the area $\Omega_i^C(t) = \Omega_i^C$ of the search space that satisfies the condition

$$\Omega_i^C = \{X = (x_1, x_2) | abs(x_j - x_{i,j}) \le \delta_X, \quad j = 1, 2\}, \quad i \in [1{:}n], \qquad (1)$$

in which, at a given moment t, the robot M_i can distribute its C-robots $C_{i,j}$. Here, $(x_{i,1}, x_{i,2}) = X_i$ is the vector of the current position of this robot.

M-vicinity $\Omega_i^M(t) = \Omega_i^M$ of the robot M_i is a set of points of the search space, the Euclidean distance $\|\cdot\|$ of which to the robot M_i does not exceed the value r_X:

$$\Omega_i^M = \{X | \|X_i - X_i\| \le r_X\}, \quad i \in [1:n].$$

Here, the value r_X (the radius of the vicinity) means the guaranteed range of communication channels between M-robots.

A set of neighbors $N_i(t) = N_i$ of the robot M_i is formed by M-robots currently located in the vicinity Ω_i^M:

$$N_i = \left\{M_j | X_j \in \Omega_i^M, j \in [1:n], j \neq i\right\}, \quad i \in [1:n].$$

The search region Π_i is one of the rectangular subareas that are nearest to the robot $M_i, i \in [1:n]$ and not yet explored by it and by any of its neighbors. We assume that the sides of the region Π_i are parallel to the coordinates $0x_1, 0x_2$.

The trace $Tr_i(t) = Tr_i$ of the robot M_i is a set of coordinates of the search space points in which the robot *previously* performed tests, as well as the results of these tests:

$$Tr_i = \{(X_i(\tau), \varphi_i(\tau)), \tau \in [0:t-1]\}, \quad i \in [1:n].$$

The extended trace $\overline{Tr_i}(t) = \overline{Tr_i}$ of the robot M_i is a combination of the trace of this robot and the traces of all its satellite robots:

$$\overline{Tr_i} = \left\{Tr_i \cup (X_{i,j}, \varphi_{i,j}), j \in [1:\nu']\right\}, \quad i \in [1:n].$$

Here, $\varphi_{i,j} = \varphi(X_{i,j})$; $\nu' = \nu'(t-1) \in [0:\nu]$ is a part of C-robots $C_{i,j}$ used (dispatched) by the robot M_i by the given moment *t*.

The criterion for ending the search is achievement by each of the M-robots of a given number of iterations $\hat{t}$. In this case, all the M-robots transmit information about all traces stored in their memory to the central control system.

3 Structure of the Situational Control System

The control object at the local level is the executive subsystem O_i of the robot $M_i; i \in [1:n]$; the control object at the global level is the group control object $O = \{M_i; i \in [1:n]\}$ [9]. The lower level is controlled by the local control system (*LCS*). The upper level is controlled by the *global control system* (*GCS*) which is the *coordinator* in terms of the theory of hierarchical multilevel systems [9]. Thus, the lower level of the hierarchy is defined by the set $\langle O_i, LCS_i \rangle$, and the upper level is defined by the set $\langle O, GCS \rangle$. The overall structure of the control system is shown in Fig. 2, where *DM* is the *Decision Maker*.

From the point of view of the DM, the global control system is a *decision support system* (*DSS*). In addition to the coordinator, this system includes a *basic simulation model* of the M-robot (*BSM*), as well as an adaptive *simulation model* (meta-model)

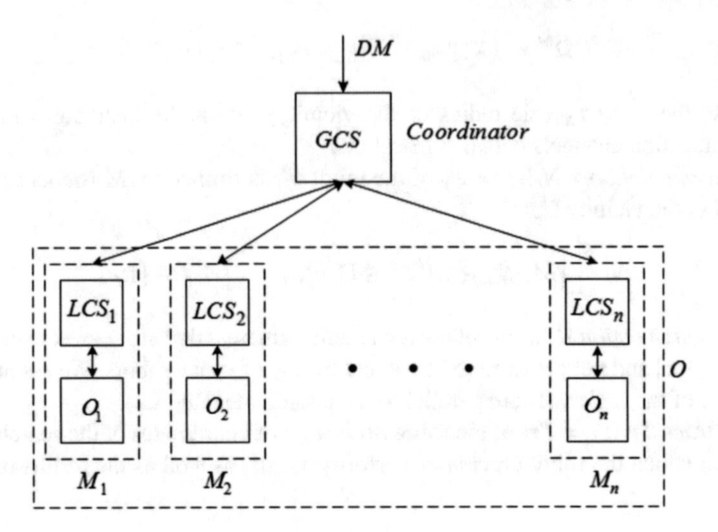

Fig. 2 Structure of the control system for a group of robots

$\Phi(X)$ of the fitness function $\varphi(X)$ [9]. Based on the *BSM* model, the *GCS* system generates n adaptive simulation models $SM_i(t) = SM_i, i \in [1{:}n]$ of M-robots, so the model SM_i reflects the current state of the robot M_i. The system *GCS* builds a simulation model $\Phi(X)$ on the basis of all current information on the results of tests performed by the M- and C-robots.

We confine ourselves to considering the local level of control.

The current local situation (L-situation) $Q_i = Q_i(t)$ in the robot M_i at the moment t is a set of information on the characteristics, current and previous states of the executive subsystem O_i of this robot:

$$Q_i = \langle P_i^O, X_i, \varphi_i, \overline{Tr}_i \rangle, \quad i \in [1{:}n].$$

Here, P_i^O is an integer vector of relevant characteristics of the subsystem O_i; the values X_i, φ_i determine the current state of the robot; the extended trace $\overline{Tr}_i$ is its previous states. The vector P_i^O contains information on the current subsystem structure O_i, that is, information on performance of the subsystem components and, in particular, the remaining number of the C-robots.

The full L-situation $S_i = S_i(t)$ in the robot M_i at the same moment t includes the following data: the current state $Q_i = Q_i(t)$ of this robot; information on the characteristics P_i^{LCS} of its control subsystem LCS_i; a classified set of current admissible situations $D_i = [D_{i,j}, j \in [1{:}r_i]]$; a set of admissible controls $\{U_{i,j}\} = U_i$ for each of the subsets $D_{i,j}$. Thus, the situation S_i is determined by the set

$$S_i = \langle Q_i, P_i^{LCS}, D_i, U_i \rangle, \quad i \in [1{:}n].$$

In these designations and agreements, the elementary act of control at the lower level of the hierarchy (control of the robot $M_i \in M$) at the moment t has the form of the following LT_L-rule:

$$\begin{cases} Q_i \in D_{i,j}, \ \ Q_i \underset{U_{i,j}}{\rightarrow} Q_i'; \\ \ \ else, \quad \ Q_i \underset{rand}{\rightarrow} Q_i'; \end{cases} \quad i \in [1{:}n], \ \ j \in [1{:}r_i].$$

Here, $Q_i' = Q_i(t+1)$; *rand* means that the choice of control occurs randomly from the number of permissible controls in the current situation.

The overall structure of the situational local control subsystem of the M-robot is shown in Fig. 3.

The task of the Analyzer is to assess the current situation Q_i. If this situation does not require a change in the robot control, the Analyzer does not send it for further processing. Otherwise, the description of this situation is sent to the Classifier. On the basis of the set of valid current situations D_i, the Classifier refers the situation Q_i to one or more classes $D_{i,j}$, each of which corresponds to the control $U_{i,j}$, and transmits this information to the Correlator. If the specified class is the only one, the Correlator transfers the control $U_{i,j}$ to the robot (its executive subsystem). Otherwise, the choice of control is carried out by the Extrapolator. If the Correlator and the Extrapolator cannot make a decision, the control selection can be made by a random selection block.

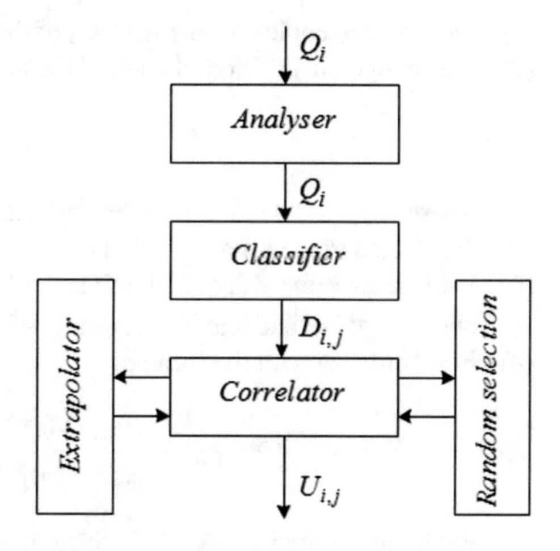

Fig. 3 Structure of the situational local control subsystem LCS_i of the robot $M_i, i \in [1{:}\ n]$

4 Situational Cat Swarm Optimization Algorithm ACSO-S

4.1 The Original Cat Swarm Optimization Algorithm

The essence of the *CSO* algorithm comprises three stages: initialization of the population, seeking, tracing [10].

Initialization of the population is carried out according to the following scheme.

(1) We initialize the values of all free parameters of the algorithm.
(2) In the search area Π, we create a population $M = \{M_i, i \in [1{:}n]\}$ of cats (M-robots) that are uniformly randomly distributed in this area.
(3) We set the initial velocities ΔX_i^0, $i \in [1{:}n]$ of the cats M_i that are uniformly randomly distributed in the region $\Pi_\Delta = \{\Delta X | \Delta X^- \le \Delta X \le \Delta X^+\} \subset R^2$ where components the of two-dimensional vectors $\Delta X^-, \Delta X^+$ are free parameters of the algorithm.
(4) Random cats in the population M numbering 0.8 n are declared in the state of *seeking*, and the others are in the *tracing* mode.
(5) For each of the cats $M_i \in M$ we calculate the value $\varphi_i^0 = \varphi(X_i^0)$ of the fitness function and find cats with the minimum φ^{worst} and maximum φ^{best} values of the function; $i \in [1{:}n]$.

The seeking mode implements a local search. The scheme of the seeking process for the cat $M_i, i \in [1{:}n]$ is as follows.

(1) We create ν' copies $X_{i,j}$, $j \in [1{:}\nu']$ of the current coordinates X_i of the cat M_i.
(2) We change the position of each of the points ν' according to the expression

$$X'_{i,j} = (1 + U_2(-1;1)\delta_X)X_{i,j}, \tag{2}$$

where $U_2(-1;\ 1)$ is a two-dimensional vector of random values uniformly distributed in the interval $[-1;1]$.
(3) We calculate the value of the fitness function $\varphi_{i,j} = \varphi(X'_{i,j})$ for each of the obtained points and find the maximum φ_i^{best} and minimum φ_i^{worst} of these values.
(4) in accordance with the formula

$$p_{i,j} = \frac{\varphi_{i,j} - \varphi_i^{worst}}{\varphi_i^{best} - \varphi_i^{worst}};\quad j \in [1{:}\nu'] \tag{3}$$

we calculate probability of choosing the point $X'_{i,j}$ as the next position of the cat M_i.
(5) Based on the probabilities (3), we choose the new position X'_i of the cat M_i using the roulette wheel selection method.

The tracing mode involves global search, and for the cat M_i, $i \in [1{:}n]$ it is determined by the following sequence of steps.

(1) We update velocity (increment of coordinates) of the cat by the formula

$$\Delta X_i' = \Delta X_i + cU_2(0;\,1)\big(X^{best} - X_i\big), \tag{4}$$

where c is a constant (free parameter); X^{best} is the position of the current globally best cat in the population.

(2) We check whether components of the velocity $\Delta X'$ are within the range of permissible values. If these limits are violated, the value of the corresponding velocity component is set equal to the maximum permissible value.

(3) The position of the cat is updated in accordance with the expression

$$X_i' = X_i + \Delta X_i'. \tag{5}$$

4.2 Modified ACSO Algorithm

The *ACSO* algorithm is proposed in our study [5]. The algorithm uses the following new concepts, compared to the *CSO* algorithm.

Search region Π_k is one of the nearest to the M-robot non-studied rectangular search subareas Π. In the case of rectilinear movement of the M-robot, the algorithm for determining the search region is presented in [5].

Knowledge μ *of the region* Π_k is defined as the ratio of the area of its studied part to the total area:

$$\mu(\Pi_k) = \mu_k = \frac{l_k \sigma}{\sigma_k}.$$

Here, l_k is the total number of tests performed in the region Π_k, which the M-robot "knows" about; $\sigma = \pi\,\delta_d$ is the area of the region that is considered to be studied if the test is somehow or other conducted in its center; σ_k is the total area of the region Π_k; δ_d is the free parameter of the algorithm.

Attractiveness χ *of the region* Π_k is calculated by means of the value

$$a_k = \frac{m_k}{m_\Sigma},$$

which represents the ratio of the number m_k of intersections of this region by perspective directions to the total number m_Σ of such directions [5]. We assume that attractiveness of the region Π_k is inverse to the value a_k:

$$\chi(\Pi_k) = \chi_k = \frac{1}{a_k}.$$

Perspective η *of the region* Π_k is an additive convolution of knowledge and attractiveness of the region:

$$\eta(\Pi_k) = \eta_k = \mu_k + \lambda\chi_k, \tag{6}$$

where λ is the weight factor.

Inadmissible approach of M-robots. If in the process of evolution of M-robots M_{i_1}, M_{i_2} in the search space the Euclidean distance between them is less than the value ρ_X, this situation is interpreted as an inadmissible approach of these robots. The rule of behavior of the robots M_{i_1}, M_{i_2} in this situation is considered in [5].

Landing of C-robots is carried out under the conditions of landing [5]. The operation algorithm of the M-robot in this case is presented in [5].

Inter-robot communication is determined by the rules for selecting the moments of time when neighboring M-robots exchange information with each other, as well as by the rules for determining the content of this information [5].

The following are the peculiarities of the *ACSO* algorithm which distinguish it from the *CSO* algorithm.

The seeking mode of the ACSO algorithm implies that not all $\boldsymbol{\nu}$ satellite robots participate in the local search on the iteration t but only a part of them which, in general, is different for each of the M-robots [5].

The tracing mode of the *ACSO* differs significantly from the same mode of the *CSO* algorithm [5].

(1) Any of the M-robots can switch to the tracing mode in the following two situations:

- in the seeking mode on the iteration t, it turned out that the values of the fitness function at all the $\nu'(t)$ points differ by no more than by the value δ_φ (the situation is interpreted as localization of one of the local maxima of the function $\varphi(X)$ with accuracy δ_X);
- the Euclidean distance between some two M-robots turned out to be less than the value ρ_X.

(2) The M-robot M_i in the tracing mode moves to the point with coordinates

$$X_i(t+1) = X_i^c + d_i Norm_2(0, \sigma), \tag{7}$$

where X_i^c is the center of the region Π_i; $Norm_2(0, \sigma)$ is a two-dimensional vector of normal random numbers with zero mean and standard deviation which is equal to σ. Movement takes place only if the region Π_i is large enough, that is if the inequality $d_i > 2\delta_X$ is true.

4.3 Situational ACSO-S Algorithm

The main distinctive feature of the *ACSO-S* algorithm is that the tracing mode of this algorithm is based on the situational analysis.

We distinguish the following classes of full situations S_i, $i \in [1{:}n]$ of the robot M_i:

- $D_{i,1} = D_{i,1}(\nu')$—the robot M_i is fully functional, the remaining number of C-robots equals to ν';
- $D_{i,2} = D_{i,2}(\nu', V_{\min}, V_{\max})$—functionality of the robot is limited by increasing its minimum and (or) reducing its maximum possible movement velocities $V_{\min}$, $V_{\max}$ respectively;
- $D_{i,3} = D_{i,3}(\nu', R_{\min}, R_{\max})$—functionality of the robot is limited by increasing the minimum and (or) reducing the maximum possible radii of the trajectory of the robot $R_{\min}$, $R_{\max}$ respectively;
- $D_{i,4} = D_{i,4}(\nu', V_{\min}, V_{\max}, R_{\min}, R_{\max})$—functionality of the robot is reduced due to limitations of its velocity and the radius of the trajectory;
- $D_{i,5} = D_{i,5}(\nu')$—functionality of the robot is reduced to a complete loss of mobility (it is possible to use only C-robots).

The control $U_{i,j}$, $j \in [1{:}5]$ is determined as the rule of choosing the region Π_i into which the M-robot moves if the full situation S_i belongs to the class $D_{i,j}$.

The control $U_{i,1}$ is defined in the study [5]. We will consider the rule of control formation $U_{i,3}$.

The scheme of the research process for the cat M_i, $i \in [1{:}n]$ received the control $U_{i,3}$ for execution is as follows.

(1) In accordance with the values $R_{\min}$ and $R_{\max}$, a set of route arcs is formed; the length of the arcs is equal to the maximum distance that the robot can overcome in one iteration (Fig. 4);

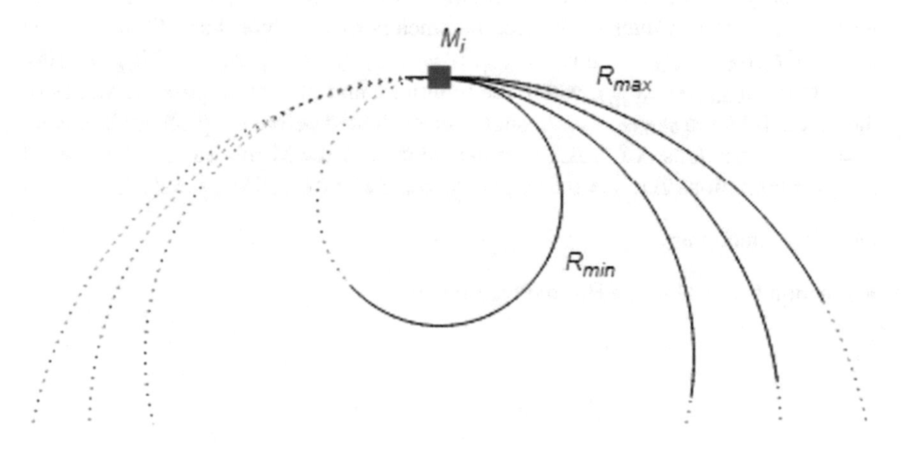

Fig. 4 Scheme of formation of the route arcs of the robot M_i, $i \in [1{:}\ n]$

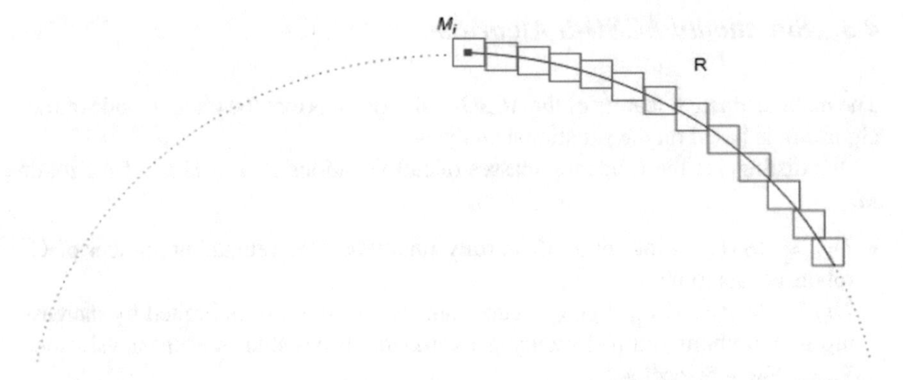

Fig. 5 Scheme of formation of the route arcs of the robot M_i, $i \in [1{:}\ n]$

(2) Each route arc is divided into square discretes with a diameter of $2\delta_X$ (Fig. 5);
(3) In accordance with the rule (5), the perspective of each discrete is evaluated;
(4) The discrete with the highest value of perspective is moved to the center.

5 Software Implementation and Computational Experiment

The software implementation of the *ACSO* algorithm is performed in the integrated development environment of Microsoft Visual Studio which allows to connect third-party extensions at various levels. The object-oriented programming language C# is used, which allows to quickly create a graphical user interface and graphs of varying complexity.

The study was carried out in the multi-start mode (the number of starts is 100) with the following values of the free parameters of the algorithm: $|S| = 10$; $c = 2$; $\tilde{m} = 8$; $\delta_X = \rho_X = \Delta x_i = \frac{d}{100}, i = 1, 2$; $r_X \approx \frac{d}{10}$; $n_C = 25$; $\hat{t} = 100$; $\delta_\varphi = 0,01(\varphi_{\max} - \varphi_{\min})$, where the minimum and maximum possible values of the fitness function $\varphi_{\max} - \varphi_{\min}$ should be evaluated before the problem is solved; $\sigma = \frac{d}{100}$; restrictions ΔX^-, ΔX^+ on the velocity of the M-robot s_i are determined from the conditions $\Delta x_{i,k}^- = 0.5\, x_k^-$, $\Delta x_{i,k}^+ = 0.5\, x_k^+$; $i \in [1{:}|S|]$, $k = 1, 2$.

Objective functions:

- function inversed to the Himmelblau function;

$$x_k^- = -6, \quad x_k^+ = 6, \quad k = 1, 2;$$

As an experiment, efficiency of the algorithm for the following situations was studied:

(1) The robots do not have alternative controls; the number of functioning robots is reduced.
(2) Functionality of several robots is limited according to the situation $D_{i,3}$. The control $U_{i,3}$ is transferred to execution.

Estimated probability of the extrema localization of the objective function is used as a criterion for evaluating efficiency of the algorithm.

Some results of computational experiments are presented in Figs. 6 and 7.

In Table 1 and the corresponding Fig. 6, the probability $\tilde{p}$ of the extrema localization of the objective function is used as a criterion for evaluating efficiency of the algorithm. The table and figure show almost 100% probability of the global

Fig. 6 Evaluation of the algorithm *ACSO-S* efficiency for the situation 1. ▬, ▬, ▬, ▬—probability of localization of four, three, two and one extremum of the objective function, respectively. ▬—Probability of localization of the extreme value of the objective function

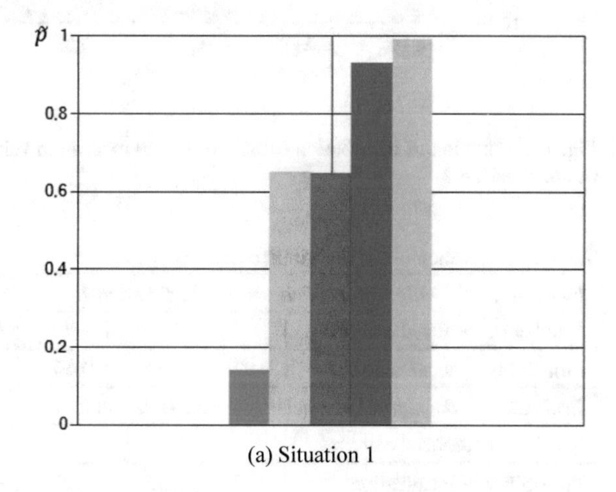

(a) Situation 1

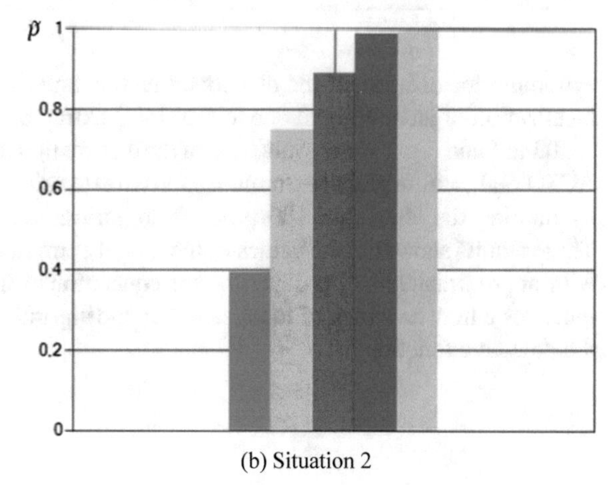

(b) Situation 2

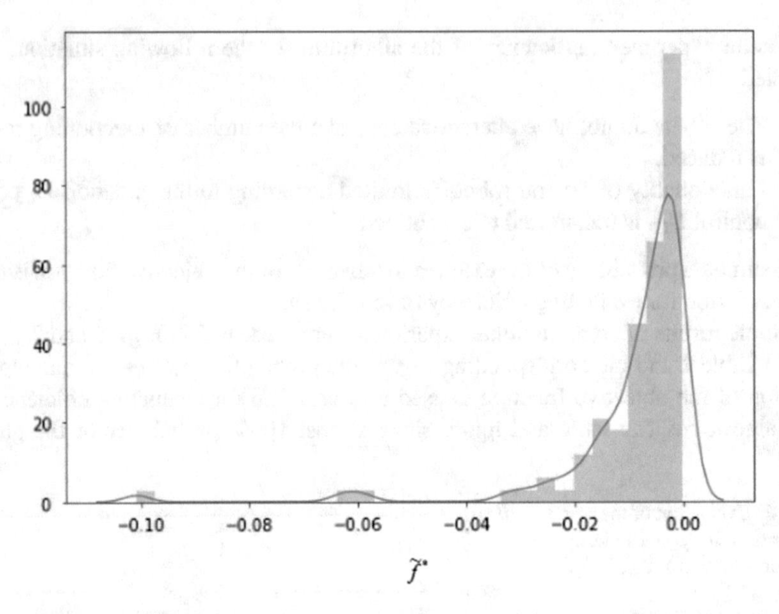

Fig. 7 Evaluation of distribution density of global extremum values localized by ACSO-S algorithm: situation 2

Table 1 Evaluation of the algorithm efficiency

Evaluation of the algorithm efficiency for the situation 1					
Number of localized extrema	1	2	3	4	Objective function
Probability of localization	0.92	0.64	0.66	0.14	0.99
Evaluation of the algorithm efficiency for the situation 2					
Number of localized extrema	1	2	3	4	Objective function
Probability of localization	0.98	0.85	0.73	0.39	0.99

extremum localization of the objective function and sufficiently high probability of localization of an additional one to four local extrema.

Table 1 and Fig. 7 show evaluation of the distribution density of the localized by the ACSO-S algorithm global extremum approximate values $\tilde{f}^*$ of the problem, obtained by running the algorithm 100 times from various random initial robot locations. These results show that the values of the global extremum obtained by the algorithm with approximately 0.8 probability are concentrated in the range (−0.4; 0). This indicates a high accuracy of localization by the algorithm of the global extremum of the objective function.

6 Conclusion

The presented results of the computational experiments show the prospects of algorithmic and software solutions. Since the control system of a group of robots is poorly defined, the approach based on the theory of situational control is appropriate for the synthesis of such a system. The peculiarity of the synthesized control system is its hierarchy: the executive subsystem of a robot is the control object at the lower (local) level, and a group of robots is the control object at the upper (global) level. The study considers only the local level of control. The relevant task is to synthesize the control system at the global level which is a decision support system. To solve this problem, it is necessary to develop a basic simulation model of an M-robot (p. 2), as well as adaptive simulation models of each of the M-robots that can reflect its current state. An important separate task is to design a subsystem that visualizes all information on the system in a form that is convenient for the DM to make decisions.

References

1. Pettersson, L.M., Durand, D., Johannessen, O.M., Pozdnyakov, D.: Monitoring of harmful algal blooms. Praxis Publishing, London (2012)
2. White, B.A., Tsourdos, A., Ashokoraj, I., Subchan, S., Zbikowski, R.: Contaminant cloud boundary monitoring using UAV sensor swarms. In: Proceedings of the AIAA Guidance, Navigation, and Control Conference, San Francisco, USA, pp. 1037–1043 (2005)
3. Hayes, A.T., Martinoli, A., Goodman, R.M.: Distributed odor source localization. IEEE Sens. J. **2**(3), 260–271 (2002)
4. Lilienthal, A., Duckett, T.: Creating gas concentration grid maps with a mobile robot. In: Proceedings of the IEEE/RSJ Conference on Intelligent Robots and Systems, Las Vegas, USA, pp. 118–123 (2003)
5. Karpenko, A.P., Leshchev, I.A.: Nature-inspired algorithms for global optimization in group robotics problems. In: Gorodetskiy, A., Tarasova, I. (eds.) Smart Electromechanical Systems. Studies in Systems, Decision and Control, vol. 174. Springer, Cham (2019)
6. Xing, B., Gao, W.-J.: Innovative Computational Intelligence: A Rough Guide to 134 Clever Algorithms. Springer International Publishing Switzerland, 451 p (2014)
7. Sharkey, A.J.C.: Swarm robotics and minimalism. Connect. Sci. **19**(3), 245–260 (2007)
8. Swarmanoid: Towards humanoid robotic swarms. http://www.swarmanoid.org/publications_byyear.php
9. Pospelov, D.A.: Situational Control. Theory and Practice. Nauka, Moscow, 284 pp (1986)
10. Chu, S.C., Tsai, P., Pan, J.S.: Cat swarm optimization. In: Yang, Q., Webb, G. (eds) PRICAI 2006: Trends in Artificial Intelligence. PRICAI 2006. Lecture Notes in Computer Science, vol. 4099. Springer, Berlin (2006)

Dynamic Switching of Multi-agent Formation in Unknown Obstacle Environment

Stanislav L. Zenkevich, Anaid V. Nazarova and Jianwen Huo

Abstract *Problem statement*: In the smart electromechanical system (SEMS), formation control of multi-agent system is a typical group coordination problem, which has broad application prospects in the fields of geographic survey, environmental investigation, security rescue, mine clearance and target defense. In the case of performing a certain task, multi-agent system need to transform its formation to avoid obstacles according to environmental constraints in real time. This paper introduces a method of dynamic switching formation structure of multi-agent system under unknown obstacle conditions. For example, multi-agent system which is currently in the form of a triangle formation need to change its formation into a straight line as it pass through a narrow, long passage. *Purpose of research*: This paper proposes a strategy for dynamic switching formation of multi-agent system in unknown environment. In this strategy, the Polya enumeration theorem is used to build the set of multi-agent system formation, and controlling the switch of formation with different obstacles conditions is based on the finite state machine theory. In the case of formation switching, the collision avoidance strategy between different agents is considered when the motion trajectory intersects. In addition, the target point assignment problem of one agent switching from the current formation structure to the target one is solved by the minimum distance method in this paper. *Results*: A formation switching control system of multi-agent system was developed, which includes a network of interacting finite-state machines and a set of formation topologies, and an underlying control strategy. In order to verify the correctness and operability of the proposed method, simulation and real experiments were carried out. The results show that the multi-agent system can switch from one topology to another according to the shape of the obstacle during the movement, and avoid collision with each other during the formation switching. *Practical significance*: According to the proposed

S. L. Zenkevich—Deceased.

A. V. Nazarova (✉) · J. Huo
Bauman Moscow State Technical University, Moscow, Russia
e-mail: avn@bmstu.ru

J. Huo
e-mail: huojianwen@hotmail.com

© Springer Nature Switzerland AG 2020

A. E. Gorodetskiy and I. L. Tarasova (eds.), *Smart Electromechanical Systems*, Studies in Systems, Decision and Control 261, https://doi.org/10.1007/978-3-030-32710-1_6

solution, the respond ability of group robots to changing environments in performing tasks is improved.

Keywords Multi-agent system · SEMS · Formation switching · Finite state machine · Polya enumeration theorem · Collision avoidance strategy

1 Introduction

Nowadays, agent coordinated control has become one of the hot issues of multi-agent systems. Multi-agent can accomplish some tasks that a single agent cannot solve with coordinated control, which is usually more efficient. Multi-agent formation control is a typical multi-agent coordination problem, which is widely used in geological exploration, military reconnaissance, rescue and transportation.

During the execution of the mission, the formation controller moves the team members to maintain a certain topological structure (for example: line, column, triangle, diamond, etc.) toward the target. However, in the environment of cluttered obstacles, it is difficult for multi-agent to maintain a certain formation to reach a predetermined target, and therefore the formation of multi-agent must be changed as the environment changes. So far, different methods have been proposed to solve the problem of multi-robot formation conversion. When multi-agent moves indoors, Levi et al. [1] used the Petri-net method to describe the behavior of multi-robot formation topological structure conversion, and realized the transformation of multi-robot topological structure, and it successfully avoided the obstacles. However, Balch and Arkin [2] proposed a behavior-based multi-robot formation maintenance method; the behavior system integrates four formations (line, column, diamond, wedge) and three object-oriented behaviors. A system using this technology is able to navigate to the target point and avoid obstacles while maintaining the formation, but this method has limitations of the type of formation, and in some circumstances, it is not optimal to reach the target point. Literature [3] performed robot formation through the analysis of the behavior of the biological system. The behavior of doves is brought in the multi-robot formation control. Doves model was established by using the directed graph and the artificial potential field theory, and the model was applied to the independent formation of multi-agent to ensure the multi-robot adaptively changes the formation structure during the execution of the task. This paper builds a dove model using directed graphs and artificial potential field theory, and applies this model to multi-agent autonomous formations. This ensures that the multi-robot adaptively changes the formation structure during the execution of the task.

The above-mentioned research on formation conversion is based on behavioral methods, and research work on many other methods has been carried out, such as the leader-follower method. Liu et al. [4] developed a method of formation retention and transformation. The method identifies the state and function of each agent and can determine the most appropriate topological structure to pass through the gap or avoid obstacles. However, their approach does not reflect the adaptability of the formation structure required for the cluttered area. Urcola and Montano [5] using the spring-damper model to control the formation conversion, so that the formation topological structure can flexibly adapt to the shape and constraints of the surrounding environment. Chen et al. [6] used the receiving-horizon leader-follower (RH-LF) control framework to solve the multi-robot formation control problem, and studied the trajectory tracking of a group of robots along with the formation. Gómez et al. [7], Dai et al. [8] provides a formation control solution with a defined formation structure and a dynamically adaptable environment. The solution enables multi-agent to safely avoid obstacles, pass narrow passages, and includes a collision-free strategy for the multi-robot formation conversion.

Based on graph theory, Olfati-Saber and Murray [9], Hong et al. [10], Hendrickx et al. [11], DeVries et al. [12] use graph theory to construct the network dynamics of multi-agent topology, and derives a reasonable distributed control algorithm to guide the agent to the expected configuration. The network dynamics of fixed and switched topologies are given in [9] and it explain the relationship with Laplacian graphs. The undirected graph and the weighted adjacency matrix are used in [10] to describe the relationship between the multi-agent and the set of formations of all possible topologies, and the topological structure is converted by a conversion signal. Hendrickx et al. [11] studied the construction and conversion of two-dimensional continuous graphs, which can be applied to multi-agent formation coordination and control. DeVries et al. [12] invokes the attributes of similarity conversion, mapping the kernel of the Laplacian operator of unweighted, connected, undirected graph to the kernel of the Laplacian operator of the weighted directed graph with equal topological edges, and the converted Laplacian operator is used to derive the distributed controller in continuous time, which uses the underlying inter-agent communication topology to direct the agent to the expected configuration.

Here, we mainly focus on the formation transformation control of multi-agent in the case of different obstacles. For example, Ani Hsieh et al. [13] proposed a multi-agent formation switching controller from a simple closed geometric formation to an required two-dimensional geometric pattern; the controller can be used to deploy multi-agent to surround buildings or the fenced area and automatically switch the formation structure while avoiding collisions or maintaining specified relative distance constraints. A multi-agent formation switching controller with continuous state is designed in [14]. At the same time, the optimal path for searching the agent to the target point according to the switching parameters is described. In [15], the multi-agent formation switching control strategy is designed by the path projected by the three-dimensional matching graph, and the global minimum travel distance and the shortest jump distance are analyzed simultaneously. The algorithm can solve the problem that multi-agents do not intersect in the formation switching and can effectively direct multi-agent from the source location to the target location, which enables a seamless transition from the old formation to the new formation. Honig et al. [16] studied the formation control of multi-agent in known environments, and analyzed the planning of multi-agent moving from the initial formation to the collision-free path of a given target formation.

In this paper, our analysis relies on Polya theorem of discrete mathematics, finite state machine theory and control theory. Based on the finite state machine theory to solve a set of intelligent logic control tasks in [17], we use the Polya theorem to enumerate all the formation structures of multi-agent. This makes it possible to control the switching and movement of the multi-agent formation in accordance with the shape of the specific obstacle in an unknown environment. Therefore, the method of formation switching in this paper is different from the method of establishing the parameter matrix of the typical formation structure of the linear, column, triangle, and diamond shapes proposed in [2, 18–21]. Moreover, unlike the literature [13–16], the paper uses the minimum distance method to describe the assignment of target points when a single agent switches to the target formation. The paper also provides a collision avoidance solution when multi-agent motion trajectories intersect.

The rest of this paper is organized as follows. Section 2 describes the establishment of the formation and the assignment of the target points of the target formation structure. Section 3 discusses strategies for multi-agent to avoid collisions when formation is switched. In Sect. 4, a behavioral mechanism is described for multi-agent motion and formation switching. After that, Sect. 5 presents simulations and experiments to verify the validity of our results. Finally, Sect. 6 presents a discussion of the conclusions and future work.

2 Basic Issues

In this section, we will discuss how to use the Polya enumeration theorem to build formation topological structure set and to describe the assignment of target points for a single agent when switching to the target formation.

2.1 Formation Topological Structure Set

In this paper, we consider using graph theory $G = (V, E)$ to represent the base map corresponding to a group of N-agents formation, where $V = \{V_1, V_2, \ldots, V_N\}$ is the set of vertices in graph G, $E \subset V \times V$ is a set of edges in the graph G. Therefore, each vertex in V represents a corresponding agent, such as V_i to agent i. Each edge $(i, j) \in E(i, j \in V$ and $i \neq j)$ in E describes the relationship between the agent R_i and R_j. Next, a polynomial is constructed by using the Polya enumeration theorem to calculate the number of non-isomorphic graphs to construct formation switching set. Considering that the permutation group on the vertices of the N-th order is exactly the symmetric group S_N, the group induces a permutation group with the edge set E of $k_n (n = \frac{N(N-1)}{2})$ in a natural way, which is expressed as $S_N^{(2)}$. It is known that each permutation α in $S_N^{(2)}$ is uniquely written as the product of disjoint periods, so for each integer k from 1 to n, we denote that $J_k(\alpha)$ is cycle number of the

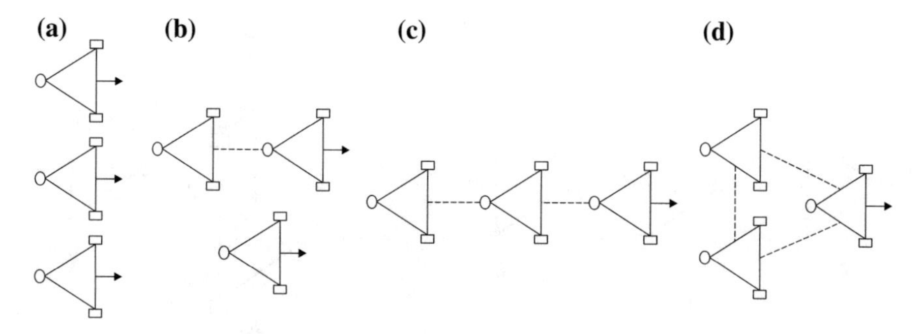

Fig. 1 The basic formation topological structure set of three agents

length k in the decomposition of a non-cross cycle α. Finally, the loop index of the permutation group $S_N^{(2)}$ is a polynomial in the variable $Z(S_N)$, which is determined by the following equation [22]

$$Z(S_N) = Z(S_N^{(2)}) = \frac{1}{N!}\sum_{\alpha \in A}\prod_{k=1}^{n} S_k^{J_k(z)} \tag{1}$$

The loop index of the symmetric group S_N is described above, and now we turn to the enumeration of graphs. Undoubtedly, the most common enumeration of graphs is $c(x) = 1 + x$, then the graph G enumeration function with N agents can substitute $c(x^k)$ into $Z(S_N)$, as follows:

$$g_N(x) = Z\left(S_N^{(2)}, 1 + x\right) = \frac{1}{N!}\sum_{\alpha \in A}\prod_{k=1}^{n}\left(1 + x^k\right)^{j_k(\alpha)} = \sum_{k=1}^{n} a_k x^k \tag{2}$$

The coefficients of the polynomial a_k describe the number of structures with k edges. Using expression (2), we can get topological structure set of multi agent. For example, when N = 3, $g_N(x) = \frac{1}{6}\left[(1 + x)^3 + 3(1 + x)(1 + x)^2 + 2(1 + x^3)\right] = 1 + x + x^2 + x^3$, which means that there is a case in the formation set with no edges (Fig. 1a), one case with one edge (Fig. 1b), one case with two edges (Fig. 1c) and one case with three edges (Fig. 1d). Therefore, three agents can get 4 types of topological structure.

2.2 *Target Point Allocation of Target Formation*

Assuming that N agents initially move in the OXY plane according to a certain formation F, the position of the formation F is $F = \{r^1, r^2, \ldots, r^N\}$, which is given by the Cartesian coordinate set, in which $r^i = (x^i, y^i)^T$ is the Cartesian coordinates of the agent R_i in the OXY plane. Next, we need to determine that a single agent i moves

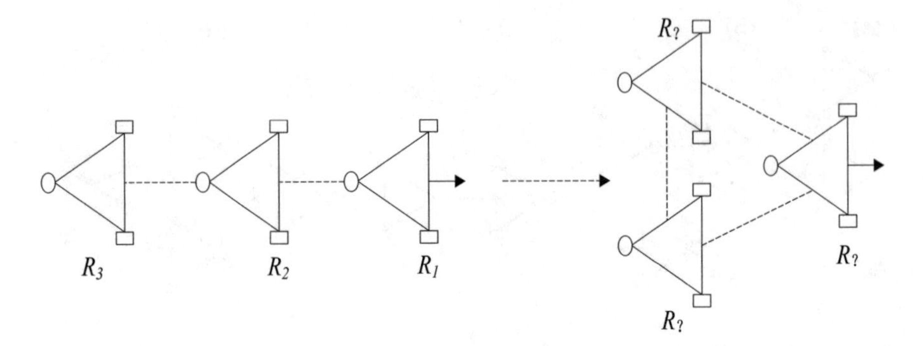

Fig. 2 Target point allocation diagram

from a position r_o^i of the current formation F_o to a position r_g^i in the target formation F_g (as shown in Fig. 2). We define the variable set $X = \left[x_{og}\right]$, $x_{og} \in \{0, 1\}(o, g \in V)$, and $x_{og} = 1$ means that there is a correlation between the position of the current formation structure F_o and the target formation structure F_g. Otherwise $x_{og} = 0$, in solving this problem, there are two parameters that will be changed: one is the specific role x_{og} assigned to each agent in the topological structure, and the other is the position of single-agent at the target formation r_g^i; but the two parameters are related to each other, and if x_{og} is determined, the target position can be determined. In order to find these two parameters, we use the agent to move from F_o to F_g with the smallest moving distance. The specific form is as follows:

$$\sum_{o,g=1}^{V} \sum_{i=1}^{N} x_{og} \left\| r_o^i - r_g^i \right\|^2 \rightarrow min \tag{3}$$

The problem of solving Eq. (3) has the following options, such as an enumeration method and an auction algorithm for allocation. If the number of agents in the formation is small, we can use the enumeration and sorting algorithm to get the parameters x_{og} or r_g^i.

3 Collision Avoidance Strategy

In this section, we discuss the mechanisms by which multi-agents avoid collision when switching formation. It is assumed that the multi-agent moves in a straight line from the current formation position F_o to the target formation position F_g during the formation switching. As you can see from Sect. 2, the initial and target positions of all agents are known, and the motion path from the initial position to the target position can be planned. Then, we use the principle of Skew lines, that is, in the three-dimensional space of position (x, y)-time (t) (Fig. 3), the time of passing multi-agent

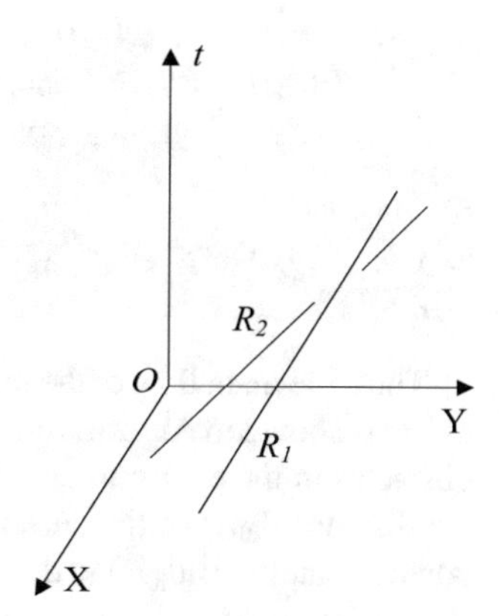

Fig. 3 Multi-agent position (x, y)–time (t)

through a certain position is different, thereby avoiding collision with each other in the intersections of a certain path. This means that when the agent moves along the intersecting trajectory, in a certain time interval $[t_0, t_1], [t_2, t_3], \ldots$, the agent stops or slows down to let other agents pass the intersecting path.

Suppose the kinematic equation of the agent R_i is

$$\begin{cases} \dot{x}^i = v_x^i \\ \dot{y}^i = v_y^i \end{cases} \tag{4}$$

Then, the relationship between the position of the agent and the time can be obtained by $r^i = r_o^i + v^i t, t \in T_i$, in which T_i is the time required for the agent R_i to move along the trajectory to the target formation; $v^i = \frac{r_g^i - r_o^i}{|r_g^i - r_o^i|} v^i$, in which v^i is a scalar. Furthermore, the distance $d_{ij}(t)$ between multi-agents is obtained as

$$d_{ij}^2(t) = (r^i - r^j)^T \cdot (r^i - r^j) \tag{5}$$

Algorithm 1 (*Multi-agent mutual collision avoidance algorithm*)

Input: The current position F of the multi-agent, the target formation position F_g, the speed v, the safety distance d between the agents;

Output: speed v^i;

Begin

(1) Calculate the distance $d_{ig}(t)$ from the current position F to the target formation position F_g using the expression $d_{ig}^2(t) = (r^i - r_g^i)^T \cdot (r^i - r_g^i)$;
(2) if $d_{ig}^2(t) = 0$,
Calculating the distance $d_{ij}(t)$ between the multi-agents using the expression (5);
if $d_{ij}(t) \le d$,

if $d_{ig}(t) \leq d_{jg}(t)$, multi-agent R_i waits, $v^i = 0$;
else multi-agent R_j waits, $v^i = 0$;

else $v^i = v$;
Else moves to the target position F_g and the algorithm ends.

Thus, the main flow of the proposed multi-agent mutual collision avoidance mechanism is shown in Algorithm 1. In addition, it is worth noting that in the case of static obstacles in the environment, Algorithm 1 needs to change some parameters. First, the distance $d_{i0}(t)$ of the current position F of the agent R_i to the obstacle is calculated; secondly, if $d_{i0}(t) < d$, multi-agent is controlled to bypass the static obstacle R_i.

4 Multi-agent Dynamic Formation Switching Strategy

In this section, we discuss the use of finite state machines to dynamically switch the formation of multi-agent in different obstacle environments. First, the state of the multi-agent in the obstacle environment is divided into three states: initial, motion, and formation switching. In the initial state, we assume that each agent is in stationary state at this time; motion state, in which multi-agent moves according to the current topological structure; formation switching state in which the multi-agent is controlled to switch from the current formation structure to another formation structure; of course, the formation set switched in the formation switching state is obtained by Formula (2). Then, using the algorithm of [17] to perform logical control. For example, when the state machine receives the command "wait" from the top layer of the system, the command "ready" is issued to each agent. At this moment, multi-agent is static. When the state machine receives the signal "movement", it sends a command "go" to the multi-agent. At this time, the multi-agent needs to follow a certain control law to ensure that it moves according to the current formation structure. When the command "transformation" appears, the state machine outputs the command "transform" to multi-agent. At this time, multi-agent selects the formation structure to be switched in set according to the shape of the obstacle in the environment, and then completes formation switching. When the command "stop" arrives, the state machine outputs the command "pause", and multi-agent stops moving. The whole process requires commands from the top of the system or from the operator with a certain time interval. Because the nature of the obstacles encountered by multi-agent during the movement is different, it is possible to keep the current formation constantly moving, or it may continuously switch formation. Of course, when using a finite state machine for formation switching, the top layer of the system is required to use the sensor to create an environmental map or to sense obstacle shapes. In addition, this paper does not consider the control law of the lower layer of the system. For example, according to a certain control law, multi-agent can maintain the current formation (linear, column, triangle, diamond, etc.) movement.

5 Experiments and Results Analysis

Simulation Experiment In order to verify the correctness and operability of the proposed method, we carried out simulation experiments in the ROS environment. During the simulation experiments, we imported the known environment map and four agents into the ros_stage, and used four types of formation structures according to the obstacles in the map. The specific simulation experiment plan is shown in Fig. 4. It includes: 1—The top layer of the system, the intelligent layer or the human-machine operation interface. At this layer, it sends the command/oper_msg, including the start and stop commands, and the command U that controls the state change. 2—Logical layer, on this layer there is a set of pre-prepared formation set. The task of this layer is to change the state of the finite state machine after receiving the command from the top layer, and send the command Z and the target point coordinates of the formation structure to the lower layer; 3—The tactical layer, the output of which are the agent motion control signals generated according to the relevant control law and it will feed back state execution to the logic layer. 4—Execution layer, which is the main environment and robot model.

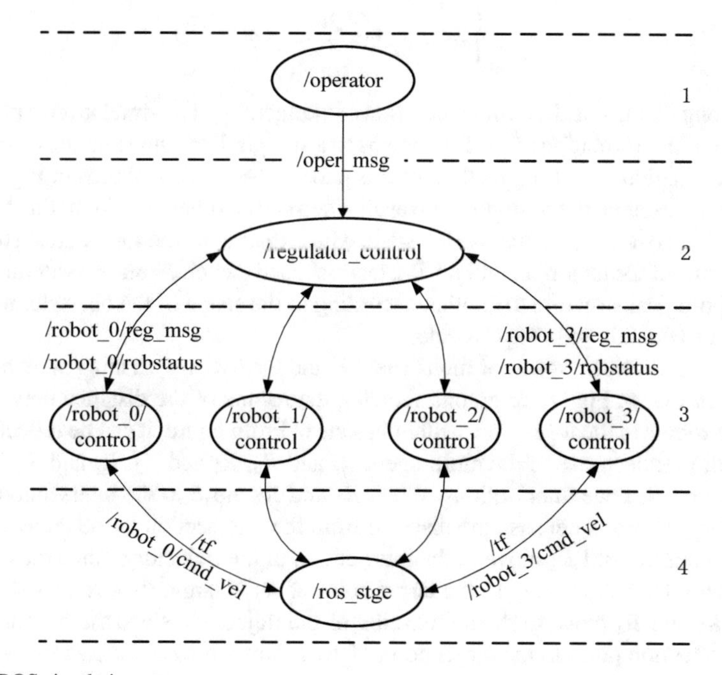

Fig. 4 ROS simulation structure

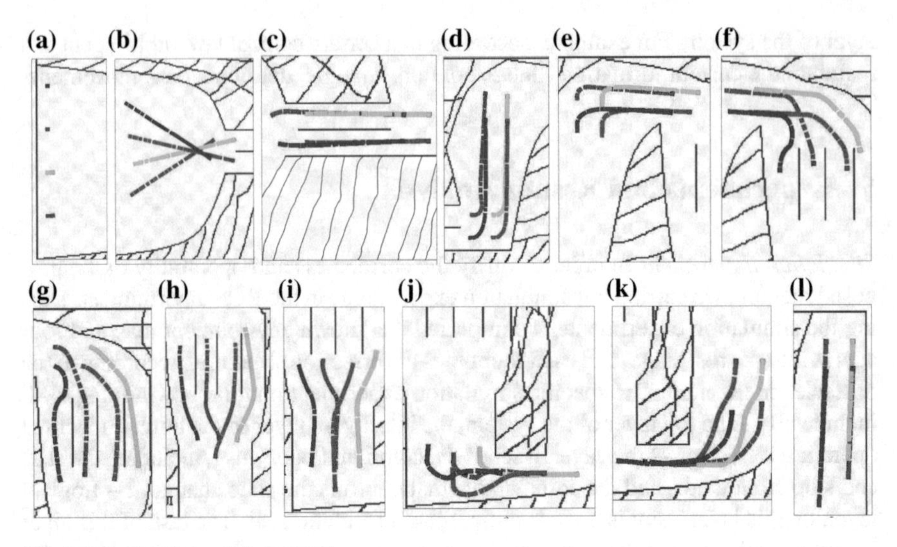

Fig. 5 Results of simulation experiments

In the third layer (tactical layer) of the simulation experiments, we use the relative distance and relative direction of the multi-agent as the control law, that is

$$\begin{cases} v = d_{ig} \\ \omega = \arctan \frac{y^i - y_g^i}{x^i - x_g^i} - \theta^i \end{cases} \tag{6}$$

Among them, θ^i is the current direction of the agent R_i. The simulation experiment sets the initial formation of the four agents to a straight line, and simulates according to the formation switching method of this paper. The result is shown in Fig. 5. The electronic resource of the simulation result video can be found in [23]. In Fig. 5a is the initial state; c–e, g, i, l is the motion state, which maintains the topological structure of the current formation; b, f, h, j, k is a team-shaped switching state, switching from one topological structure to another according to the shape of the obstacle, which is in order to quickly pass the obstacle.

The speed change result of multi-agent in the formation switching state b is presented in Fig. 6; Fig. 7 are graphs showing the results of the distance between the ground robots in the formation switching state b. From figure, it can be reflected that the motion trajectories of the multi-agent R_2 and R_3, R_1 and R_2, R_3 and R_4, R_1 and R_3, R_1 and R_4 have intersections. When R_2 and R_3 move to the intersection of the trajectory, R_3 waits because the distance from the intersection to the target point is equal; when R_1 and R_2 move to the intersection of the trajectory, since the distance from the intersection point to the target point of R_2 is larger than R_1, so R_1 Waits; when R_3 and R_4 move to the intersection of the trajectory, since the distance from the intersection point to the target point of R_3 is larger than R_4, R_4 waits; when R_1 and R_3 move to the intersection of the trajectory, due to the intersection point to the

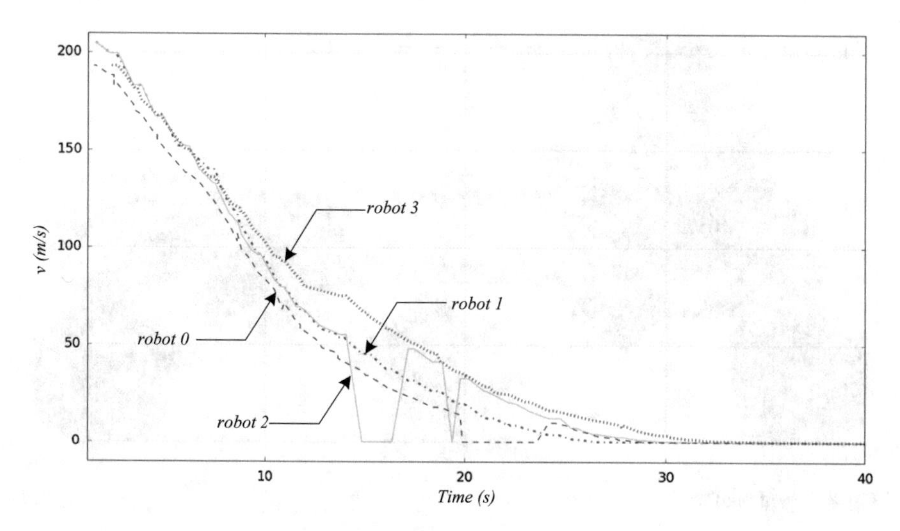

Fig. 6 Speed results of multi-agent motion along the trajectory

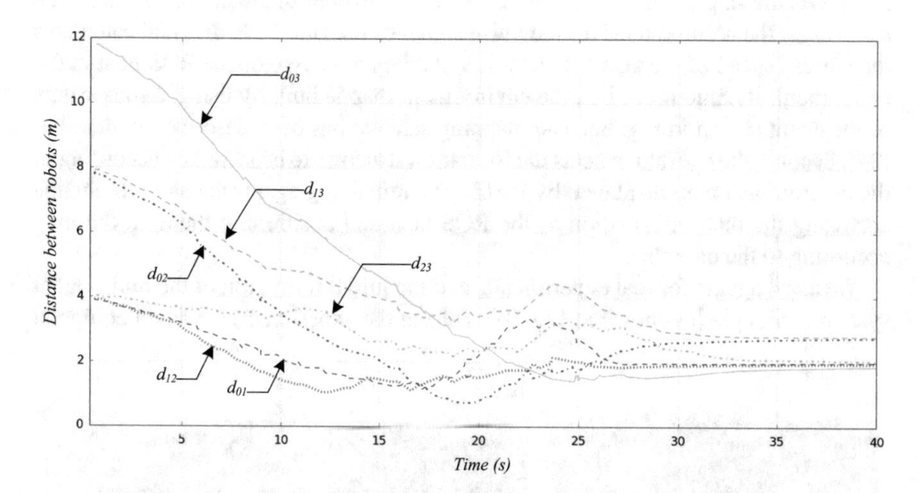

Fig. 7 Distance results between multi-agents

target point of R_3 is larger than R_1, and thus R_1 waits; when R_1 and R_4 move to the intersection of the trajectory, R_4 waits because the distance from the intersection to the target point is equal. The correctness and effectiveness of the avoiding mutual collision algorithm of multi-agent system can be reflected from Figs. 6 and 7.

Real Experiments This paper uses the agent platform shown in Fig. 8 for a real experiment. The agent shown in Fig. 8a is equipped with a lidar and a Hall code disc, and the function of the agent is to collect environmental data by using a laser sensor. The sensor of the agent shown in Fig. 8b is only a Hall code disc. All of the top layer systems of the agent are the Raspberry Pi operating system with Ubuntu. The lower

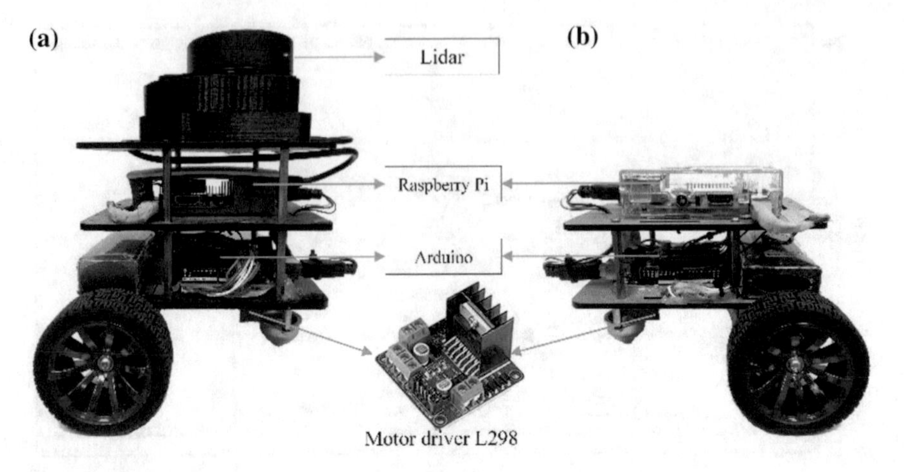

Fig. 8 Agent platform

layer uses the single-chip arduino to output PWM to control the motor speed. At the same time, the arduino receives the pulse signal of the Hall code disc and calculates the current speed of the agent. In addition, the laptop is used as the ROS host in the experiment. Its function: First, the environment map is built by using the laser data of the agent shown in Fig. 8a. The mapping scheme has been described in detail in [24]. Second, the operator selects the formation structure to be switched according to the environment map displayed by RVIZ. The remaining agents act as ROS clients, accessing the map information of the ROS host and positioning them on the map according to the odometer.

We use 4 agents for real experiments, and the initial formation of the multi-agent system is triangle (as shown in Fig. 9a). During the experiment, the agent equipped

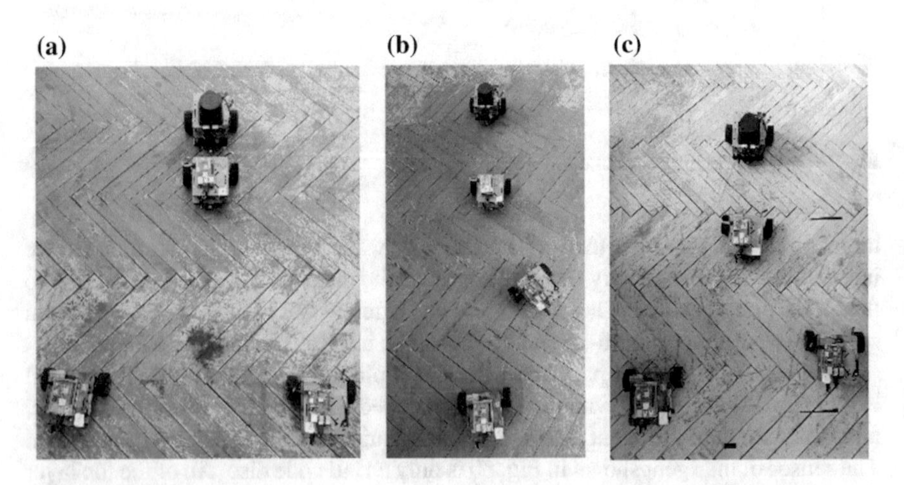

Fig. 9 Real experiment results

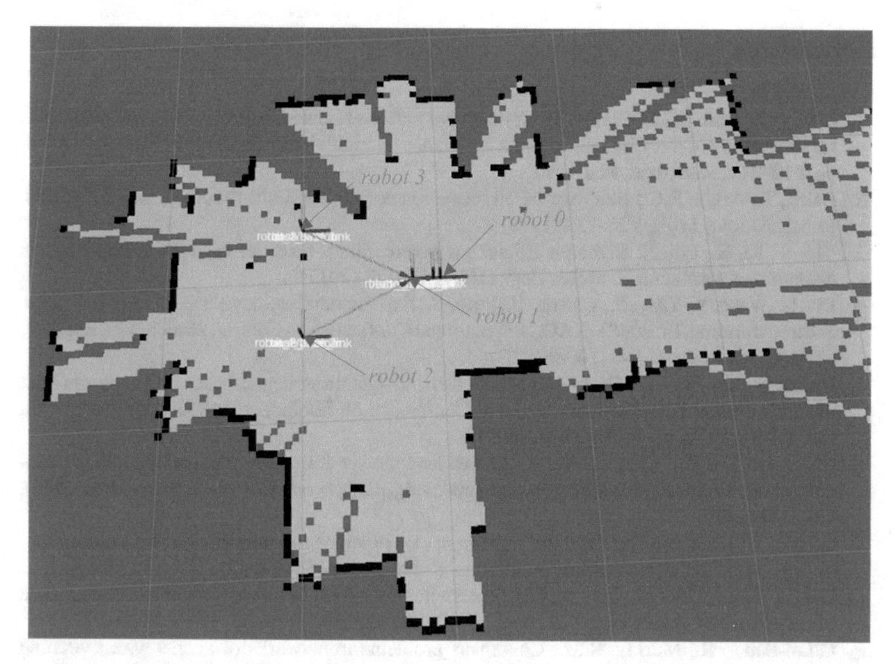

Fig. 10 The result of the mapping

the laser sensor was always in the lead position, and the result of the mapping is shown in Fig. 10. At the logical layer of the program, two formation structures of switching (line type and triangle type) and the aimed positions of target formation are set. Using the formation switching method provided in this paper, after 2 s, the formation of the multi-agent system becomes linear (as shown in Fig. 9b); after 3 s, it becomes triangular (as shown in Fig. 9c).

6 Conclusion

This paper proposes a strategy for dynamic formation switching of multi-agent in an unknown environment. This strategy uses the Polya enumeration theorem to build all the formations set of multi-agents, so that in the case of encountering different obstacles, the finite state machine theory is used to control the switching of the formation. In the case of formation switching, this paper considers the collision avoidance strategy of multi-agent when the motion trajectory intersects. In addition, this paper also solves the problem of target point allocation of the agent switching from the current formation structure to the target formation structure by using the minimum distance method. This paper does not consider the control of the formation switching and maintains a variety of formations (such as lines, columns, triangles, diamonds, etc.), which is the research direction that needs to be carried out later.

References

1. Levi, P., Muscholl, M., Bräunl, T.: Cooperative mobile robots Stuttgart: architecture and tasks. In: Proceeding of the 4th International Conference on Intelligent Autonomous Systems, IAS-4, pp. 310–317, Karlsruhe, Mar 1995
2. Balch, T., Arkin, R.C.: Behavior-based formation control for multi-robot teams. IEEE Trans. Robot. Autom. **14**(6), 926–939 (1998)
3. Yin, S., Li, H., Liu, J.: Research article new pigeon flocks behavior used for multiple robots autonomous formation. J. Softw. Eng. **11**(2), 210–216 (2017)
4. Liu, L., Wang, Y., Yang, S., Watson, G., Ford, B.: Experimental studies of multi-robot formation and transforming. In: 2008 UKACC International Conference on Control, Manchester, England, United Kingdom, pp. 1–6 (2008)
5. Urcola, P., Montano, L.: Cooperative robot team navigation strategies based on an environment model. In: 2009 IEEE/RSJ International Conference on Intelligent Robots and Systems, St. Louis, MO, USA, pp. 4577–4583 (2009)
6. Chen, J., Sun, D., Yang, J., Chen, H.: Leader-follower formation control of multiple non-holonomic mobile robots incorporating a receding-horizon scheme. Int. J. Robot. Res. **29**(6), 727–747 (2010)
7. Gómez, J.V., Lumbier, A., Garrido, S., Moreno, L.: Planning robot formations with fast marching square including uncertainty conditions. Robot. Auton. Syst. **61**(2), 137–152 (2013)
8. Dai, Y., Kim, Y., Wee, S., Lee, D., Lee, S.: A switching formation strategy for obstacle avoidance of a multi-robot system based on robot priority model. ISA Trans. **56**, 123–134 (2015)
9. Olfati-Saber, R., Murray, R.M.: Consensus problems in networks of agents with switching topology and time-delays. IEEE Trans. Autom. Control **49**(9), 1520–1533 (2004)
10. Hong, Y., Hu, J., Gao, L.: Tracking control for multi-agent consensus with an active leader and variable topology. Automatica **42**(7), 1177–1182 (2006)
11. Hendrickx, J.M., Fidan, B., Yu, C., Anderson, B.D.O., Blondel, V.D.: Formation reorganization by primitive operations on directed graphs. IEEE Trans. Autom. Control **53**(4), 968–979 (2008)
12. DeVries, L., Sims, A., Kutzer, M.D.M.: Kernel design and distributed, self-triggered control for coordination of autonomous multi-agent configurations. Robotica **36**, 1077–1097 (2018)
13. Ani Hsieh, M., Kumar, V., Chaimowicz, L.: Decentralized controllers for shape generation with robotic swarms. Robotica **26**(45), 691–701 (2008)
14. McClintock, J., Fierro, R.: A hybrid system approach to formation reconfiguration in cluttered environments. In: 16th Mediterranean Conference on Control and Automation, Ajaccio, France, pp. 83–88 (2008)
15. Liu, L., Shell, D.A.: Multi-robot formation morphing through a graph matching problem. distributed autonomous robotic systems. Springer Tracts Adv. Robot. **104**, 291–306 (2014)
16. Honig, W., Satish Kumar, T.K., Ma, H., Koenig, S., Ayanian, N.: Formation change for robot groups in occluded environments. In: 2016 IEEE/RSJ International Conference on Intelligent Robots and Systems (IROS), Daejeon, South Korea, pp. 4836–4842 (2016)
17. Zenkevich, S.L., Nazarova, A.V., Zhu H.: Logical control a group of mobile robots. In: Smart Electromechanical Systems. Studies in Systems, Decision and Control, vol. 174, pp. 31–43 (2019)
18. Hsu, H.C.-H., Liu, A.: Multi-agent based formation control using a simple representation. In: IEEE International Conference on Networking, Sensing and Control, Taipei, Taiwan, china, pp. 276–281 (2004)
19. Yang, L., Cao, Z., Tan, M.: Dynamic formation control for multiple robots in uncertain environments. Robot **32**(2), 283–288 (2010). (in Chinese)
20. Ren, L., Wang, W., Du, Z., Tang, D.: Dynamic and optimized formation switching for multiple mobile robots in obstacle environments. Robot **35**(5), 535–543 (2013). (in Chinese)
21. Wu, H., Luo, Y., Zhou, Q., Yin, Y.: Obstacle avoidance strategy of wheeled robot formations based on time efficiency. Inf. Control **46**(2), 211–217 (2017). (in Chinese)
22. Harary, F., Palmer, E.M.: Graphical Enumeration. Academic Press (1973)

23. Video with Simulation Results. Available at: https://drive.google.com/file/d/1YQiVN4WReV3i1Etyb5QTqCMQqMxfXobW/view. Accessed 16 Nov 2018
24. Zenkevich, S.L., Zhu, H., Huo, J.: Experimental study of motion of the mobile robots moving in the convoy type formation. Mekhatronika, Avtomatizacia, Upravlenie **19**(5), 331–335 (in Russian) (2018)

Methods and Algorithms of Situational Control

Situational Control of the Group Interaction of Mobile Robots

Andrey E. Gorodetskiy, Irina L. Tarasova and Vugar G. Kurbanov

Abstract *Problem statement*: solving the problems of situational control of mobile robots in the group play an important role for effective collaboration when working together. The article deals with the problem of joint movement of a group of mobile robots. *Purpose*: setting the problem of situational control in the group interaction of mobile robots and analysis of the principles of situational control of a group of robots. *Results*: a generalized mathematical description of the problem of situational control in the group interaction of mobile robots is obtained, various structural approaches to the organization of situational control of a group of robots are analyzed. A generalized mathematical description of the planning of situational control of a group of robots is given. *Practical significance*: the possibility of using the proposed mathematical formulation of the problem of situational control on the example of the passage of a group of robotic cars through the intersection is shown.

Keywords Situational control · Groups of intelligent robots · Environment of choice · Deterministic · Stochastic and not fully defined constraints · Synthesis of the algorithm for finding the optimal solution

1 Introduction

The problem of control in the group interaction of objects is a global problem, actual for many spheres of life. Wherever there is a group of living or technical objects that

A. E. Gorodetskiy · I. L. Tarasova (✉) · V. G. Kurbanov
Institute of Problems of Mechanical Engineering, Russian Academy of Sciences, St. Petersburg, Russia
e-mail: g17265@yandex.ru

A. E. Gorodetskiy
e-mail: g27764@yandex.ru

V. G. Kurbanov
e-mail: vugar_borchali@yahoo.com

V. G. Kurbanov
Saint-Petersburg State University of Aerospace Instrumentation, Saint Petersburg, Russia

© Springer Nature Switzerland AG 2020

A. E. Gorodetskiy and I. L. Tarasova (eds.), *Smart Electromechanical Systems*, Studies in Systems, Decision and Control 261, https://doi.org/10.1007/978-3-030-32710-1_7

need to work together to do some work or solve some task, this problem arises. In the technical field, it is most relevant in robotics. At the same time, optimizing the interaction of a group of robots in the performance of a joint task is now becoming increasingly difficult. It requires taking into account the complexity and intelligence of technical robot control systems, providing a constant expansion of the scope of such robotic systems (RS). In complex RS based on Smart Electromechanical systems (SEMS) [1], it is necessary to analyze not only the behavior of an individual robot, but also the behavior of a group of robots interacting with it. The complex RS under consideration has appropriate behavior due to the presence of the Central nervous system in SEMS [2]. A group of interacting robots can be transport robots performing coordinated movements, robots collectors performing joint operations, etc.

When creating such groups, specialists are primarily faced with the need to assess the ability of the group to make the right decisions under uncertainty. It is necessary to take into account the static and dynamic features of each robot and account features of decision-making in the Central nervous system of individual robots. For this purpose, computer simulation with identification and evaluation of decisions in complex dynamic systems under conditions of uncertainty [3] of the environment is usually used. This environment will eventually be filled with robots and all sorts of "smart" systems. Naturally, it will then require these robots and artificial intelligence systems to have "instincts" that allow them to avoid collisions with obstacles and each other while driving. However, if these instincts are too strong, robots will be too slow, which will adversely affect the effectiveness of their actions. To eliminate this situation, it is necessary to develop algorithms that will constantly strive to find the optimal balance between speed and security. This will allow the robots to always operate with high efficiency.

In addition, the task of controlling a group of robots has additional complexity due to the need to ensure coordination between robots. In complex RS, each robot must satisfy its kinematic and dynamic equations, as well as existing phase constraints, including dynamic constraints that ensure no collisions between robots.

The need to compare different trajectories of interacting robots (development scenarios) with each other, as well as restrictions on the practical feasibility of algorithms for processing data about objects and the environment lead to the fact that the area of determining the characteristics of the state in one way or another is discretized. For example, it is translated into a logical form by fuzzification of data [4]. Then they are analyzed using algebraic logical expressions [5, 6] and optimized by mathematical programming methods in ordinal scales or generalized mathematical programming [7].

The creation and development of situational control systems requires a lot of resources to collect information about the objects and control environment, their dynamics, control methods, as well as to systematize this information within the semiotic model. Therefore, it is considered that the method of situational control is advisable to apply only in cases where other methods of formalization lead to the problem of too large (for practical implementation) dimension.

It should be noted that the control technology developed by scientists currently does not guarantee the absolute safety of the movement of robots, which is confirmed by several incidents that occurred during the tests. Existing control technology for cars also does not guarantee this. They are only trying to minimize the likelihood of a collision. For example, Magnus Eggerstedt thinks that: "The existing control systems of such cars are very conservative, they will not allow the car to move in the presence of even the slightest danger. All this will lead to the fact that on the roads on which the robot cars will move, there will always be congestion and traffic jams, which can not be eliminated independently, even the most highly intelligent automatic systems" [8].

2 Setting the Task of Controlling the Movement of a Group of Robots

In solving the problems of control of the group interaction of robots, much attention is paid to the issues of their self-organization and maintenance of homeostasis within the group. For example, solving the problems of formation of formations [9], working out coordinated movements [10], joint search and transportation of objects [11, 12], etc. However, the mechanisms considered in these tasks do not guarantee the completeness of the tools that are necessary to meet all the tasks that the group may be faced with. There is a need for another level of control, which would be an interface between the group and the operator, setting the target tasks [13, 14], and which can be attributed to the optimization tasks of situational control [15, 16]. In this case, the operator can be not only a person, but also a computer program that makes decisions.

Consider the task of translating a group of robots $A = \{a_1, a_2, \ldots, a_n\}$ located at time t_0 at the points $S = \{s_1, s_2, \ldots, s_n\}$ the surrounding limited space $L^3 \subset E^3$, (E^3—three-dimensional Euclidean space) to target points $F = \{f_1, f_2, \ldots, f_n\}$ of this space by the time t_f in the minimum time:

$$T = t_f - t_0 \rightarrow \min, \quad (1)$$

with a minimum probability of collision of robots:

$$P_A \rightarrow \min. \quad (2)$$

Usually, when using various mathematical optimization methods, the condition (2) is replaced by an inequality of the form:

$$\sum_{i,j} m_{ij}(t_k) \leq M, \quad (3)$$

where: M is the maximum number of collisions, i, j are robot numbers from 1 to n ($i \neq j$), k—the number of the moment in time from the time interval T, the value $m_{ij}(t_k)$ is determined from a Boolean expression:

"if at time t_k a trajectory r_i of the robot a_i intersects with the trajectory r_j of the robot a_j that is $r_i \cap r_j \neq 0$, so $m_{ij}(t_k) = 1$, else $m_{ij}(t_k) = 0$".

The search for a solution to the problem is carried out in an environment of choice, which changes over time t, that is, it is dynamic $O(t)$. It can be split Into $O(t_k)$ layers with some constant or variable step h_k:

$$O(t) = \{O(t_0), \ldots, O(t_k), \ldots, O(t_f)\}. \quad (4)$$

Each layer $O(t_k)$ ($k = 0, 1, \ldots [t_f / h_k]$) contains the surrounding space L^3, divided into cells $e_q(t_k)$ with constant or variable steps h_x, h_y, h_z along the axes X, Y, Z. Here q is the cell number, $q = 1, 2, \ldots, Q$.

$$O(t_k) = \{e_1(t_k), e_2(t_k), \ldots, e_Q(t_k)\}. \quad (5)$$

Each cell $e_q(t_k)$ is characterized by the presence or absence of a_i robots, and obstacles $B_i(t_k)$, in the form of: prohibiting the movement of signs $v_i(t_k)$, traffic lights $w_i(t_k)$, road markings $\gamma(t_k)$, etc.

$$B_i(t_k) = \{v_i(t_k), w_i(t_k), \gamma(t_k), \ldots\}. \quad (6)$$

In addition, each cell is characterized by a matrix of interaction of the robot with the environment $G(t_k) = \{G_1(t_k), G_2(t_k), \ldots, G_\nu(t_k)\}$, describing the influence of the cell environment on the dynamic state of the robot, it is a collection of cells—the traffic rules of type: if—then.

The complexity of the problem, which requires the use of methods of situational control, is that the parameters and functions characterizing the cells $e_q(t_k)$, can be deterministic, stochastic, or not fully defined.

An example of the task can be the task of driving a group of robot cars through the intersection.

In this case, the surrounding space L^3 of dimension L_x, L_y, L_z is allocated in the vicinity of the intersection along the X, Y, Z axes. It is divided at the beginning of the control t_0 into cells $e_q(t_0)$ with constant steps h_x, h_y, h_z along the axes X, Y, Z. At points $S = \{s_1, s_2, \ldots, s_n\}$ of the surrounding space L^3 and the corresponding cells contain robot cars $A = \{a_1, a_2, \ldots, a_n\}$. Each of them is characterized by speed, acceleration and target points $F = \{f_1, f_2, \ldots, f_n\}$ this space. They need to arrive at the time t_f in the minimum time T. Moreover, the number of possible collisions of robot-cars must satisfy the inequality (3).

Cell sizes (steps h_x, h_y, h_z) are selected larger than the dimensions of the largest robot-car. Each cell $e_q(t_k)$ is characterized by the presence or absence of robot-cars a_i, and obstacles $B_i(t_k)$. In addition, each cell is characterized by the interaction of the robot-car with the environment in the form of matrices $G(t_k) = \{G_1(t_k), G_2(t_k), \ldots, G_\nu(t_k)\}$, describing the influence of the cell environment (road surface, humidity, temperature, etc.) on the dynamic state of the robot. In the linear setting is the transfer function of the vehicle by the disturbance. The set of cells is characterized by the rules of movement $R_m(t_k)$ through the intersection of the type: if – then. These rules are determined by the type of crossroad.

For example:
When passing through the intersection
If: "*There are no traffic lights*
No additional characters
Each street has 1 lane
It is necessary to turn to the right
In front of the intersection there is a robot car
It moves at a speed greater than controlled"
So "*Controlled robot car continues to move to the intersection (moves to the next cell) without braking*"

The selection environment $O(t)$ containing cells and robot cars changes over time t, that is, it is dynamic. It can be broken into $O(t_k)$ layers with some constant or variable h_k step depending on the dynamic properties of robot cars and perturbing properties of the cell environment. Then the optimization problem (1) with constraints (3) and other logical, logical-probabilistic and logical-linguistic constraints can be solved sequentially for each layer of the environment of choice $O(t)$. However, this approach does not guarantee the minimization of the travel time through the intersection of the whole group of cars $A = \{a_1, a_2, \ldots, a_n\}$, since the selection environment at each subsequent step depends on the decisions made in the previous steps and can change over time.

3 Principles of Situational Control

In the considered formulation of the problem, the following structural approaches to the organization of situational control of a group of robots are possible: centralized control of a_i ($i = 1, 2, \ldots, n$) robots with operator K (Fig. 1), decentralized control of robots a_i without isolation of the leader robot (Fig. 2), decentralized control of a_i robots with allocation of the leader's work $a_L \in A$ (Fig. 3), combined control of a_i

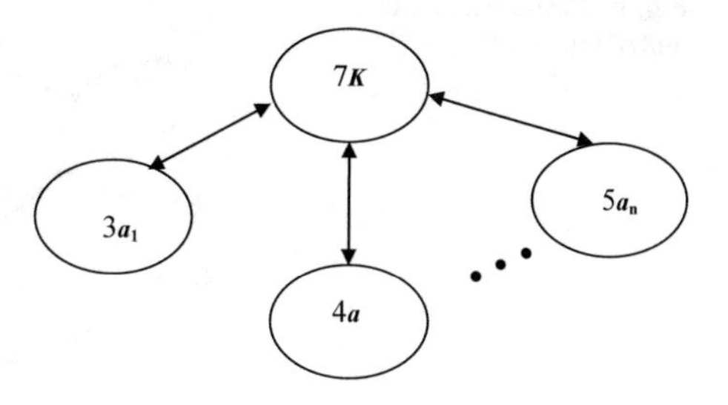

Fig. 1 Centralized control

Fig. 2 Decentralized control

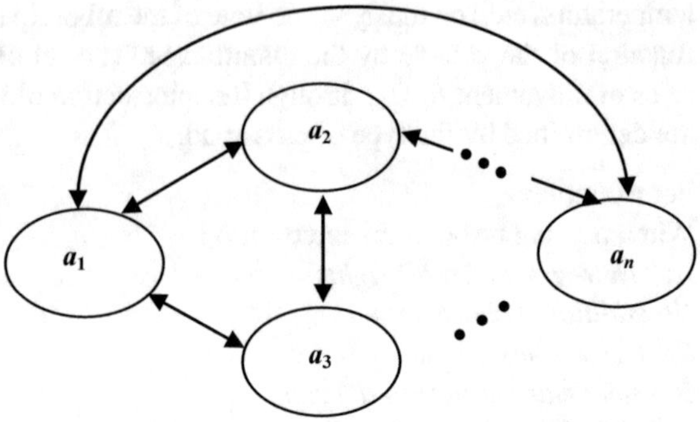

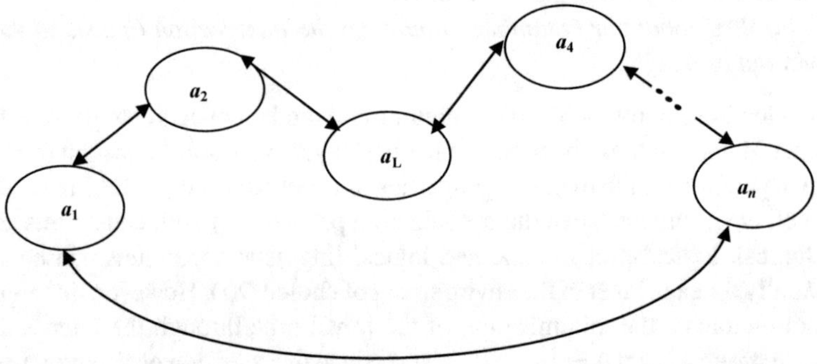

Fig. 3 Decentralized control with a leader

robots with the operator K without highlighting the robot of the leader a_L (Fig. 4) and with the allocation of the robot of the leader a_L (Fig. 5).

Centralized control with the operator requires the following functions [13]:

- interface between the team robots and the operator;
- creation of a database and a database of rules on the environment and robots;
- decomposition of the task into subtasks;
- building a joint action plan of the group of robots;
- distribution of areas of responsibility between robots.

Fig. 4 Combined control with coordinator

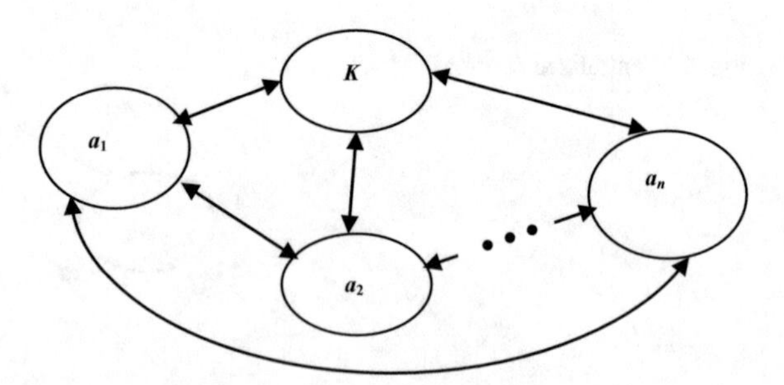

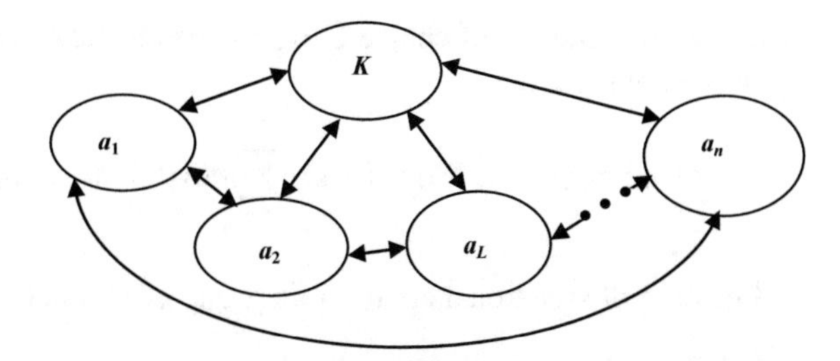

Fig. 5 Combined control with coordinator and leader

Most of these functions are typical of any other control structure. In order to ensure that the structure in question has the property of completeness (that is, if the goal is achievable, the sequence of actions to the goal will be found), access of all robots of the group to the database and operator knowledge is required. This imposes certain restrictions on the access time to the shared database of each member of the group and it requires calculation and dynamic redistribution of priorities on the sequence of control of the members of the group of robots.

With decentralized control without a leader, each robot is engaged in a logical conclusion [17]. This will make it difficult or impossible to access the databases and knowledge of other team members. This will lead to a violation of the coordination of movements of the group members. In addition, it will be possible for conflicts to occur with independent decision-making by each member of the group.

The equivalence of roles in the group in decentralized control is eliminated when choosing a leader [18]. In this case, the leader coordinates the movements of the group members and eliminates conflicts, and the rest of the robots play a supporting role of remote databases and knowledge and query relays. At the same time, the effectiveness of movement control of a group member will depend heavily on the degree of its remoteness from the leader, and the entire group on the number of group members.

Under combined control with the operator and without isolation of the robot leader, group members can share information about their own condition and planning movements. This allows them to analyze commands or movement plans coming from the operator and reject unacceptable ones that lead to conflicts with neighbors. However, the equality of group members requires operator intervention when setting priorities to resolve conflicts. This can lead to slower decision-making in the group.

Under combined control with the operator and with the allocation of robot of the leader group members can also exchange information about their own condition and planning movements. But the presence of a leader allows you to eliminate contradictions and set priorities within the group with the help of a leader. This increases the security of group management, but there is a dependence of the effectiveness of conflict control on the degree of remoteness of the conflicting robots from the leader.

With any structural approach to the organization of situational control of a group of robots, it is necessary robots together to collect information about the environmental parameters, the current state of individual robots of the group, the planned actions of members of the group of robots, etc. After collecting the information in the general case, the environment of choice $O(t_k)$ can be characterized at some point in time t_k by the following tuple:

$$O(t_k) = \langle A(t_k), S(t_k), F(t_k), \sum_{i,j} m_{ij}(t_k), M, B_i(t_k), G(t_k), Rm(t_k) \rangle. \quad (7)$$

Planning of situational control of a group of robots is:

- division of group tasks into subtasks:

$$O(t_0) \underset{U(t_1)}{\Rightarrow} O(t_1) \cdots O(t_0) \underset{U(t_f)}{\Rightarrow} O(t_f), \quad (8)$$

where: $U(t_k) = \{u_{a_1}(t_k), u_{a_2}(t_k), \ldots, u_{a_n}(t_k)\}$, $u_{a_i}(t_k)$—control action applied to a_i robot at the moment of time t_k, $k = 0, 1, \ldots, f$.

- distribution between robots of group of solutions of subtasks so that the solution of a group problem was carried out in the minimum time taking into account the available restrictions, including on information interaction.

In general, the solution of the group control problem will be the synthesis of the search algorithm, so ordered set $\varpi \subset \Omega$ of the many alternative combinations of the controls $U(t_k)$ the best combinations of the control laws each member of the group of robots $u_{a_i}(t_k) \in U(t_k)$ based on the estimated quality Q given the system preferences E and entourage choice $O(t_k)$.

$$\varpi \subset 2^U \times Q^U, \quad (9)$$

where: 2^U—denotes the set of all subsets U, and Q^U—set of all quality ratings (tuples of length from 2 to $|U|$), ×—sign of Cartesian product.

4 Conclusion

Wherever there is a group of complex intelligent technical objects that must work together to perform some work or solve some problem, there is a problem of finding the optimal algorithm of situational control.

The natural restriction on the synthesis time of the algorithm for finding the optimal solution in real time imposes restrictions on the number of members of the controlled group and the distances between them associated with the dynamics of the environment of choice and the dynamics of controllability of the robots themselves.

Classification of situations makes it much easier and faster to plan the situational control of a group of robots. Classification of situations is to attribute the current situation to one or more classes corresponding to some control.

If the resulting solution to the classification problem is unique and the selected class of situations requires some certain impact on the objects, then the objects are served associated with this control class. At the same time, one of the prerequisites for the effective application of the situational approach should be observed: the number of possible control decisions is significantly less than the number of possible situations [19]. Otherwise, you need to solve the problem of analysis of the current situation in the environment of choice $O(t_k)$. It becomes a problem of estimating the previous control with the purpose of making decisions about the change plan situational control. Adjusting the plan of situational control at each subsequent step does not necessarily lead to the construction of the optimal plan, since in the aggregate the selected step-by-step path to the goal will not be guaranteed to be optimal (the control process may be non-Markov). Therefore, it is necessary before making a decision on the plan of situational control to conduct a simulation of control of a group of robots, for example, on the basis of fuzzy mathematical modeling of poorly formalized processes and systems [4]. This allows for step-by-step construction of the path to go back and discard the ineffective parts of the track (see Fig. 6, where bold denotes the best way). Naturally, such a search for the optimal plan of situational control based on modeling requires additional computing power and time from the control system. Significant progress in this direction can provide parallelization of calculations, i.e. simultaneous passage of all possible ways of the plan of situational control with the subsequent decision on optimality.

Acknowledgements This work was financially supported by Russian Foundation for Basic Research Grants 16-29-04424, 18-01-00076 and 19-08-00079.

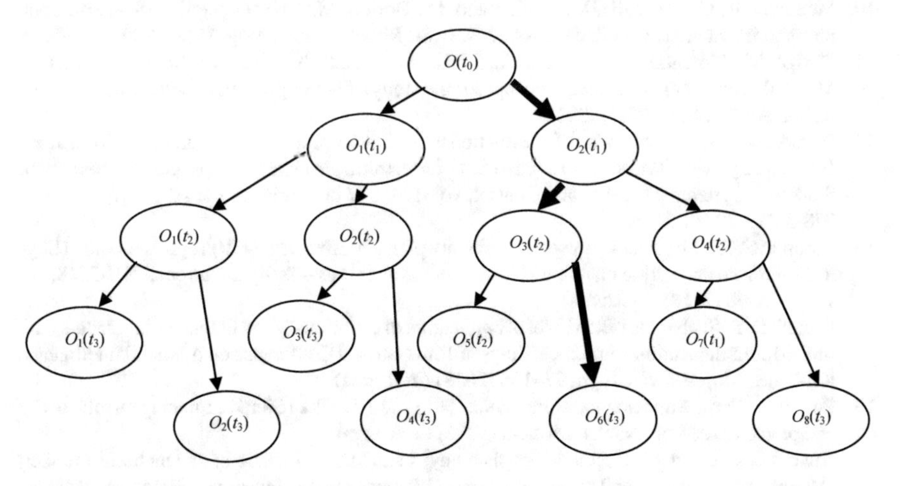

Fig. 6 Example of finding the optimal solution

References

1. Gorodetskiy, A.E.: Smart Electromechanical Systems. In: Studies in Systems, Decision and Control, vol. 49, 277 pp. Springer International Publishing, Switzerland (2016). https://doi.org/10.1007/978-3-319-27547-5
2. Gorodetskiy, A.E., Kurbanov, V.G.: Smart Electromechanical Systems: The Central Nervous Systems. In: Studies in Systems, Decision and Control, vol. 95, 270 pp. Springer International Publishing, Switzerland (2017). https://doi.org/10.1007/978-3-319-53327-8
3. Gorodetskiy, A.E., Kurbanov, V.G., Tarasova, I.L.: Decision-making in central nervous system of a robot. Inf. Control Syst. (1), 21–30 (2018). https://doi.org/10.15217/issnl684-8853.2018.1.21 (in Russia)
4. Gorodetsky, A.E., Tarasova, I.L.: Nechetkoe matematicheskoe modelirovanie ploho formalizuemyh processov i sistem (Fuzzy mathematical modeling of poorly formalized processes and systems). SPb. Publishing House of Polytechnical Institute, 336 pp. (2010) (in Russia)
5. Gorodetsky A.: Osnovy teorii intellektual'nyh sistem upravlenija (Fundamentals of the theory of intelligent control systems). LAP LAMBERT Academic Publishing GmbH@Co. KG Publ., 313 pp. (2011)
6. Kulik, B.A., Fridman, A.Y.: Logical analysis of data and knowledge with uncertainties in SEMS. In: Gorodetskiy, A.E. (ed.) Smart Electromechanical Systems. Studies in Systems, Decision and Control, vol. 49, pp. 45–59 (277 pp.). Springer International Publishing, Switzerland (2016). https://doi.org/10.1007/978-3-319-27547-5
7. Iudin, D.B.: Vychislitel'nye metody teorii priniatiia reshenii (Computational methods of decision theory), 320 pp. Nauka Publ., Moscow (1989) (in Russia)
8. Lee, S.G., Diaz-Mercado, Y., Egerstedt, M.: Multirobot control using time-varying density functions. IEEE Trans. Robot. **31**(2), 489–493 (2015). https://doi.org/10.1109/TRO.2015.2397771
9. Rubenstein, M., Ahler, C., Nagpal, R.: Kilobot. A low cost scalable robot system for collective behaviors. In: Proceedings of IEEE International Conference on Robotics and Automation (2012)
10. Mondada, F., Gambardella, L.M., Floreano, D., Dorigo, M.: The cooperation of swarm-bots: physical interactions in collective robotics. IEEE Robot. Autom. Mag. **12**(2) (2005)
11. Dorigo, M., Floreano, D., Gambardella, L.M., Mondada, F., Nolffi, S., Baaboura, T., Birattari, M., et al.: Swarmanoid: a novel concept for the study of heterogeneous robotic swarms. IEEE Robot. Autom. Mag. **20**(4) (2013)
12. Gorodetskiy, A.E., Tarasova, I.L.: Situational control a group of robots based on SEMS. In: Gorodetskiy, A.E., Tarasova, I.L. (eds.) Smart Electromechanical Systems: Group Interaction. Studies in Systems, Decision and Control, vol. 174, 337 pp. Springer. https://doi.org/10.1007/978-3-319-99759-9
13. Vorob'ev, V.V.: Logicheskij vyvod i ehlementy planirovaniya dejstvij v gruppah robotov (Logical inference and action planning elements in robot groups). Trudy konferencii KII-2018, no. 1, pp. 88–96 (2018) (in Russia)
14. Ivanov, D.Y., Shabanov, I.B.: Model of application of coalitions of intelligent mobile robots with limited communications. In: Proceedings of 16th National Conference on Artificial Intelligence KII-2018, Moscow, vol. 1, pp. 97–105 (2018) (in Russia)
15. Pospelov, D.A.: Situacionnoe upravlenie: Teoriya i praktika (Situation management: theory and practice), 286 pp. Nauka, Moskva (1986) (in Russia)
16. Kunc, G., O'Donnel S.: Upravlenie: sistemnyj i situacionnyj analiz upravlencheskih funkcij (Management: system and situation analysis of management functions). Progress, Moskva, 588 pp. (2002) (in Russia)
17. Karpov, V.E.: Control in static swarms. Problem statement. In: VII-th International Scientific-Practical Conference "Integrated Models and Soft Computing in Artificial Intelligence" (2013) (in Russia)

18. Vorob'ev, V.V.: Logical inference and action planning elements in robot groups. In: Proceedings of 16th National Conference on Artificial Intelligence KII-2018, Moscow, vol. 1, pp. 88–96 (2018) (in Russia)
19. Fridman, A.Y.: SEMS-based control in locally organized hierarchical structures of robots collectives. In: Gorodetskiy, A.E., Kurbanov, V.G. (eds.) Smart Electromechanical Systems: The Central Nervous System, Studies in Systems, Decision and Control, vol. 95, pp. 31–47. Springer International Publishing, Switzerland (2017)

The Formation of the Dish System of the Space Radio Telescope Antenna

Andrey E. Gorodetskiy, Vugar G. Kurbanov and Irina L. Tarasova

Abstract *Problem statement*: The formation of the dish system of the radio telescope antenna in the near-earth orbit can significantly improve its main characteristic, namely, the value of the aperture surface utilization factor (SUF). In this case, practically excluded its dependence on deformations of weight and wind effects. *Purpose*: Solving the problem of increasing the aperture efficiency (AE) of the space radio telescope by controlling the position of the reflecting shields of the dish system and the radiation receiver. *Results*: The method of formation in orbit with the satellites of the earth the dish system of the space telescope. The method of its adjustment and focusing with the help of controlled elements based on SEMS modules is developed. *Practical significance*: The automatic control system (ACS) is described, which allows forming, adjusting and focusing the dish system of the antenna of the space radio telescope. The ACS provides an increase AE of a space radio telescope with a significant variation in the operating frequency ranges of the received radiation and in the absence of the influence of the Earth's astronomical climate.

Keywords Space radio telescope · Sub dish · The utilization factor of the surface of reflecting shields · Control elements · SEMS · Alignment · Focus adjustment · Automatic control system

---

A. E. Gorodetskiy (✉) · V. G. Kurbanov · I. L. Tarasova
Institute of Problems of Mechanical Engineering, Russian Academy of Sciences, St. Petersburg, Russia
e-mail: g27764@yandex.ru

V. G. Kurbanov
e-mail: vugar_borchali@yahoo.com

I. L. Tarasova
e-mail: g17265@yandex.ru

V. G. Kurbanov
Saint-Petersburg State University of Aerospace Instrumentation, Saint Petersburg, Russia

© Springer Nature Switzerland AG 2020

A. E. Gorodetskiy and I. L. Tarasova (eds.), *Smart Electromechanical Systems*, Studies in Systems, Decision and Control 261, https://doi.org/10.1007/978-3-030-32710-1_8

1 Introduction

Radio telescopes usually have two-dish antennas. They are the electronic analog of the well-known in astronomical optics telescopes Cassegrain and Gregory. The two-dish antenna consists of a main-dish, a sub-dish and an irradiator (Fig. 1). A large dish is a paraboloid of rotation or a cut from it. A small dish can be part of hyperboloid of revolution of two nappes (in the Cassegrain system) or part of an ellipsoid (in the Gregory system) [1].

For normal operation of a two-dish Cassegrain antenna (Fig. 1a) it is necessary in one of the focuses of the small hyperbolic dish (F_1) to place the phase center of the irradiator, and the second focus (F_2) to combine with the focus of the parabolic dish. In this case, the main-dish will be irradiated as if some fictitious irradiator were in its focus. It should be borne in mind that the radiation pattern (DN) of the real and fictitious irradiators will be different.

Gregory's two-dish system differs from the Cassegrain system only in that the auxiliary dish is a cut from the ellipsoid of rotation (Fig. 1b). The latter has the property that if one of its foci F_1 put a concentrated source, the reflected rays from the inner surface will gather in its second focus F_2. Therefore, if the phase center of the irradiator is placed in one of the focuses of the ellipsoid F_1, and the second focus F_2 is combined with the focus of the parabolic dish, the system will work similarly to the previous one. Cassegrain antennas are more common because they have a shorter axial length and provide less phase distortion.

The main characteristic of the antenna system is the value of the aperture surface utilization factor (SUF). Its value strongly depends on the accuracy of the dish surface shapes and on the accuracy of the alignment of the dish focuses, which depend on the weight deformations and wind effects, as well as on the change in the length of the

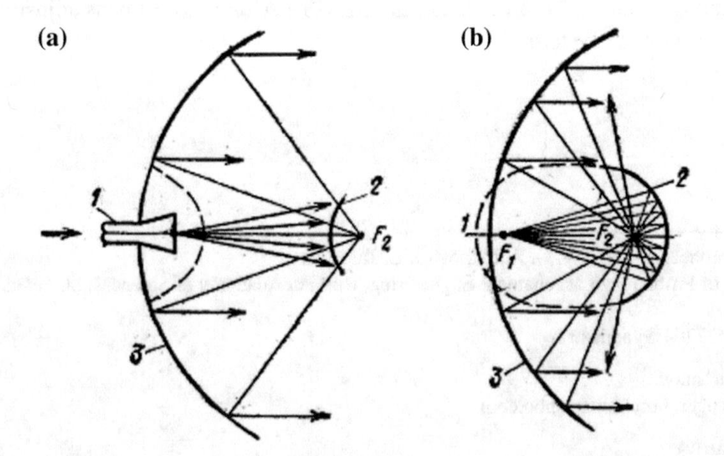

Fig. 1 Two-dish antennas: **a** Cassegrain system, **b** Gregory system; 1—irradiator, 2—sub-dish, 3—main-dish

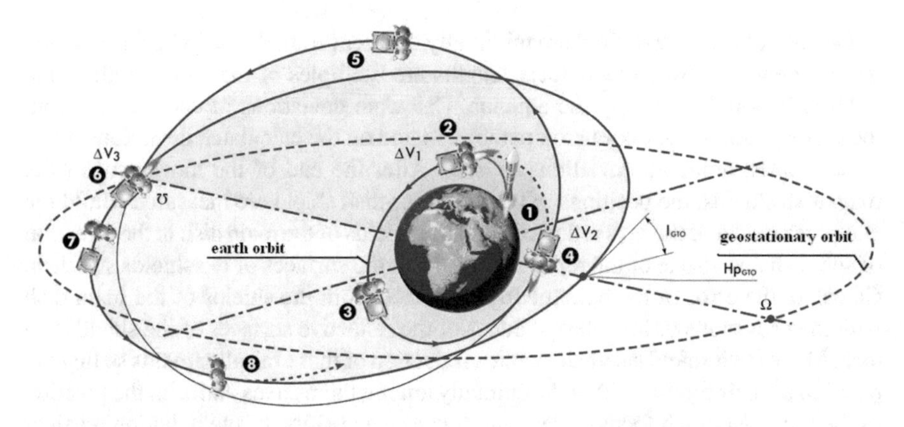

Fig. 2 Earth satellite orbits. 1, 2—launch phase of the satellite launch vehicle, 2, 3—phase of the near-earth orbit, 4, 5, 6—phase of the geostationary orbit and 6, 7, 8—phase of the descent or retention in orbit

received antenna radiation. The influence of these factors on antenna SUF is especially noticeable for large radio telescopes and small (millimeter) wavelengths [2, 3]. For example, when using the Effelsberg (Germany) radio telescope with a diameter of 100 m for receiving millimeter—wave signals, its SUF dropped to 10–20%, which made this radio telescope in the millimeter range completely ineffective [4].

Reducing the influence of these factors on the reduction of SUF of large radio telescopes is primarily associated with various ways of adapting the dish surfaces of antennas [5–7]. The essence of the proposed methods is to control the position of the shields that make up the dish surfaces, according to the results of measurements and calculations. Practically get rid of the influence of weight and wind disturbances can be due to the formation of the radio telescope antenna in space, for example in the geostationary orbit of the Earth (Fig. 2). Naturally, in this case, the formation of a dish system in orbit using artificial satellites is much more complicated and there are a number of problems associated with a large temperature drop, solar radiation, periodic adjustment of the position and orientation of satellites that require appropriate energy consumption, etc.

2 Antenna Formation in Near-Earth Orbit

To increase SUF with a significant spread of the operating frequency ranges of the received signals and large weight and wind deformations of the antenna design elements, the shields of the reflecting surfaces of the main dish and the counter-dish are mounted on the control elements [7, 8]. Measure the position of the boards forming the reflecting surface of the main dish of the antenna. According to the measured values of the positions of the main dish shields for each shield, an approximating paraboloid is built in the computer in such a way that the focal length and the position of the base of each paraboloid are minimally different from the neighboring one and the differences between their focal lengths are multiples of the wavelength of the radio radiation received by the antenna. Calculate deviations of each Board from the corresponding approximating paraboloid and on the calculated deviations move these boards, reducing deviations to zero. After the end of the movement of the main dish shields, the positions of each counter-dish shield are measured. Build the computer model of the rays reflected from the shields of the main dish in the direction of sub dish and stroke of the reflected rays from the surfaces of the shields sub dish. Calculate the error of the extreme rays reflected from the shields of the main dish with the provisions of the relevant edges of the reflective surfaces of the shields sub dish. Move each shield sub dish towards reduction of these misalignments so that the position of their focuses differed minimally among themselves and with the position of the secondary dish focus system and (or) to the position of the radiation receiver provided that the length of the beams (optical paths) from the primary focus to the reflective surfaces of the shields sub dish and the differences between them, as well as the length of the beams (optical paths) from the reflecting surfaces of the shields sub dish to secondary focus and the differences between them are a multiple of the wavelength of the radiation to accept.

The disadvantage of this method of antenna formation is the strong influence of the astronomical climate of the antenna installation location on the antenna SUF at the time of astronomical observations. Eliminate the influence of the astronomical climate of the Earth, mainly in the millimeter wavelength range can be due to the installation of reflecting surfaces of the antenna on the earth's satellites, output to earth orbit. In this case, the radiation receiver installed on the control element, control elements with shields of reflecting surface of main dish and controlled the elements with shields of sub dish set everyone on the satellite (Fig. 3) [9].

Wherein the controlled elements are used for the serial connection of modules SEMS [10, 11]. The use of SMS modules with parallel structures of the multi-link electric drive makes it possible to obtain the maximum accuracy of guidance of shields and receiver with minimal travel time due to the introduction of parallelism in the processes of measurement, calculation and movement. The use of high-precision ultrasonic motor in SEMS as electric drives provides the ability to operate in extreme conditions, including in outer space [12, 13]. The use of a serial SEMS connection makes it possible to position the reflecting shields at small displacements of the center of mass, which facilitates the stabilization of satellites. Small dimensions and controllable reconfigurability of the SEMS design provide high flexibility and maneuverability of the aerospace robotic systems designed on their basis. In addition, the supply of SEMS developed by a team of information and control systems similar to the Central Nervous System (CNS) of man [14], will allow robotic systems based on SEMS to make reflective decisions in conditions of uncertainty and thus increase their control and adaptability.

The formation of the dish antenna system in orbit is performed by the automatic control system (Fig. 4), containing a central control computer 23 located on the Earth, the output of which is connected to the system 24 of the satellite output into

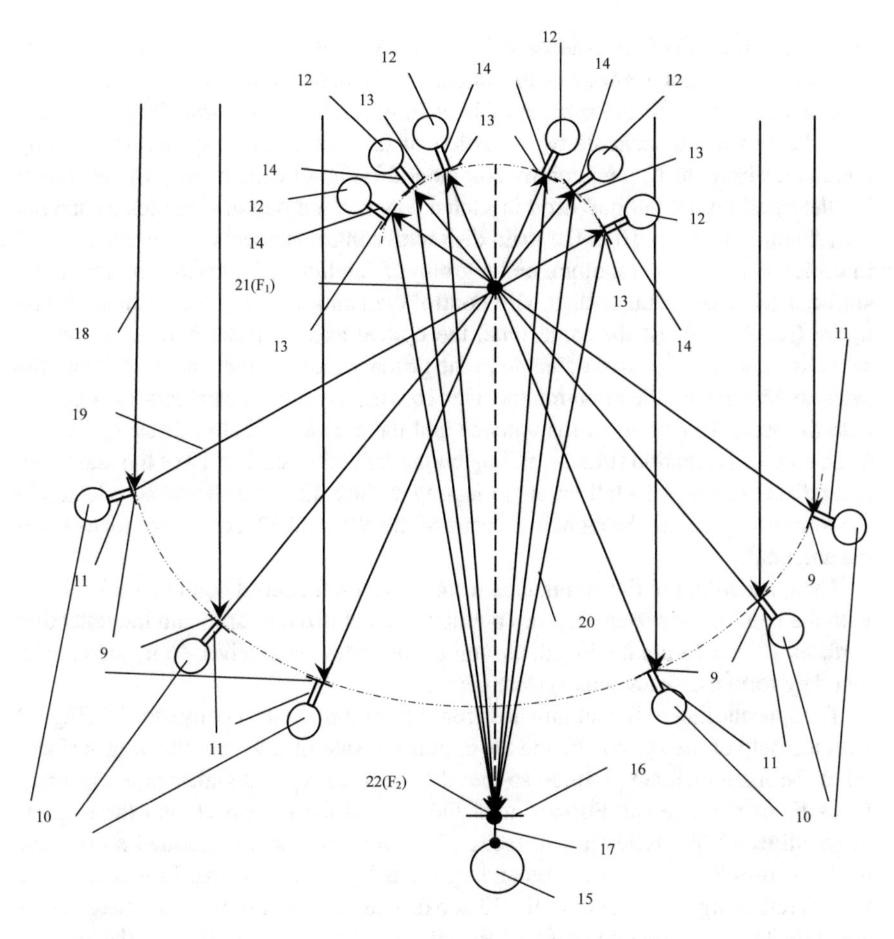

Fig. 3 Scheme of satellites location in orbit during the formation of the dish system

Fig. 4 Block diagram of the control system forming the antenna group of satellites

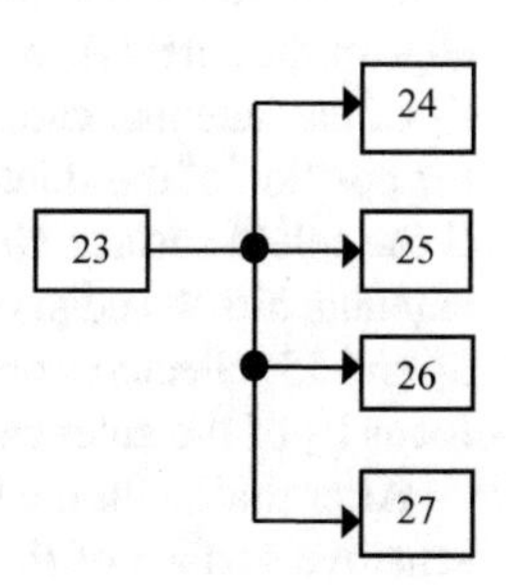

orbit, the system 25 of the satellite stabilization in orbit, the system 26 of the satellite orientation and the system 27 of the antenna alignment, as follows.

In accordance with the scheme of launching satellites into orbit (Fig. 2) and in accordance with the scheme of formation of the dish antenna system (Fig. 3) by commands from the Central control computer 23 (Fig. 4) with the help of the system 24, the satellites are put into orbit in such a way that a group of satellites 9 carrying the reflecting surfaces 10 of the main dish with control elements 11 formed a figure in orbit close to the paraboloid, and a group of satellites 12 carrying the reflecting surfaces 13 of the counter-dish with control elements 14 formed an ellipsoid-type figure (see Fig. 3). At the same time, the optical axes of these figures are on the same line and their focuses coincide in the primary focus of the antenna F1, and the satellite 15 carrying the radiation receiver 16 with the control elements 17 is placed near the secondary focus of the antenna and the ellipsoid F2. In this case, the rays of 18 received radiation (Fig. 3), falling on the reflecting surface 10 of the main dish and reflected as rays 19 fall on the reflecting surface 13 sub dish and is reflected in the form of rays 20 are brought into focus of the ellipsoid 22 (secondary focus F2 of the antenna).

Then, according to the commands from the Central control computer 23 (Fig. 4) with the help of the system 25, the satellites are stabilized in orbit and the reflecting surfaces 10 and 13 are deployed, as well as the radiation receiver 16 in such a way that they form a dish antenna system (Fig. 3).

Then, according to the commands from the central control computer 23 (Fig. 4) with the help of the system 26 and the engines of satellites 9, the reflecting surfaces 10 of the main dish are oriented so that the reflected rays 19 gather in the primary focus F_1 of the antenna. Further, with the help of the system 26 and the engines of satellites 12, the reflecting surfaces 13 of the sub dish are oriented so that the reflected rays 20 gather in the secondary focus F_2 of the antenna. In this case, the shields reflecting the surfaces of the 13 sub dish are oriented so that all reflected rays are collected in the secondary focus F_2 with equal phases. In this case, the antenna SUF will be the maximum. In particular, as shown in Fig. 5, the beam 32 from the edge of the shield A the surface 10 of the main dish passes through the primary focus F_1 of the antenna, enters the region D of the shield surface 13 sub dish agreed with the position of the shield surface 10 of the main dish, and the beam 33 from the edge B the shield surface 10 of the main dish passes through the primary focus F_1 of the antenna, hits the edge C of the shield surface 13 sub dish. At the same time, the rays 34 and 35 reflected from the shield 30 of the surface 13 are collected in the secondary focus F_2 of the antenna with equal phases.

After that, with the help of the system 26 and the engines of the satellite 15, the sensitive surface of the receiver 16 is oriented so that the focus F2 of the antenna falls on the sensitive surface of the receiver 16.

However, during the orientation of the dish surfaces of the antenna with the help of satellite engines, it is not possible to achieve a complete agreement between the position of the surface shields 10 of the main dish and the positions of the surface shields 13 of the sub dish (see Fig. 5). In particular, the rays 32 and 33 from the edge A and b of the shield of the surface 10 do not fall on the edges D and C, respectively,

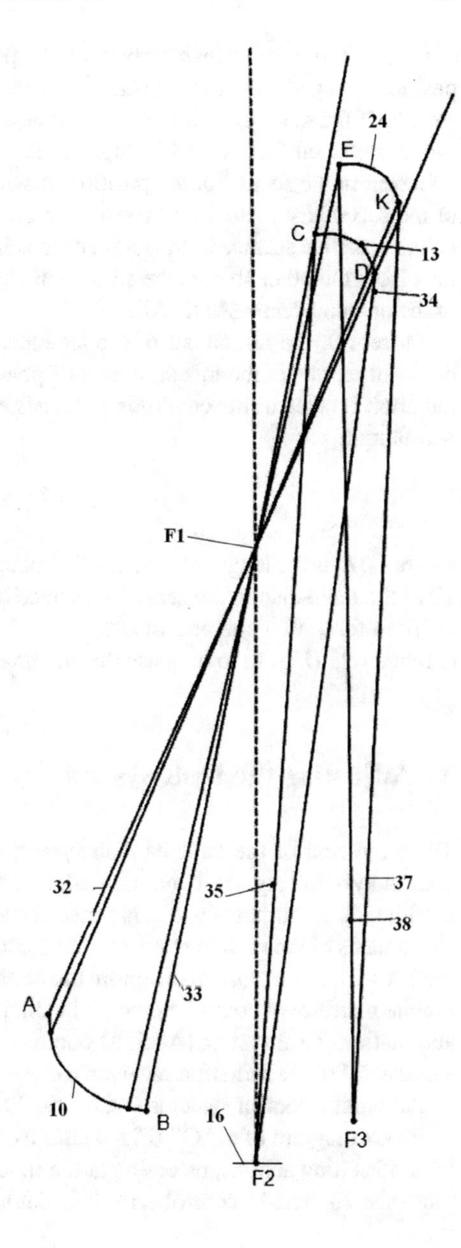

Fig. 5 The course of the beams in the dish antenna system

of the shield of the sub dish surface 13, the position of which is inconsistent with the position of the shield of the surface 10 of the main dish. Therefore, not all rays from shields of the main dish fall on the shields of sub dish and, in addition, as shown in Fig. 5, reflected from E and K edges of the shield surface 24, the rays 37 and 38 is collected in the focus F3, the position of which does not coincide with the position of the secondary focus F2 of the antenna. As a result, the rays can either never get on the sensitive surface of the radio, or to achieve it in an inconsistent phase with the rays from the other shields the surface of sub dish. Therefore, when the orientation of the antenna decreases its AE.

Moreover, the alignment of the antenna with only the engines of the satellites does not eliminate the misalignment of phases and changing the wavelength of the radiation is taken as the condition of consistency between the phases depends on the wavelength λ.

$$L_i - L_j = n\lambda \tag{1}$$

where L_i, L_j is the length of the radiation path from the adjacent shields of the surface 10 of the main dish to the sensitive surface of the receiver 16, n—is an integer.

Therefore, after the end of the orientation commands from the central control computer 23 (Fig. 4) to increase the antenna AE using the system 27, it is adjusted.

3 Adjusting the Dish System

The alignment of the antenna dish system in orbit is performed by the automatic control system (Fig. 6), containing a guidance control computer 28 located on the satellite 15, the output of which is connected to the automatic control system (ACS) 29 coordinates of the mobile platform of the control element 11 shields of the reflecting surfaces 10 of the main dish, automatic control system (ACS) 30 coordinates of the mobile platform of control elements 14 shields of reflecting surfaces 13 sub dish and automatic control system (ACS) 31 coordinates of the mobile platform of the control element 17 of the radiation receiver 16.

Automatic control systems (ACS) 29, 30 and 31 are identical.

Block diagram of a ACS (Fig. 7) has a control computer 39, containing a calculator 40 setting actions, receiving at the input of the task from the control computer guidance 28, and 41 control error calculator, one input of which is associated with

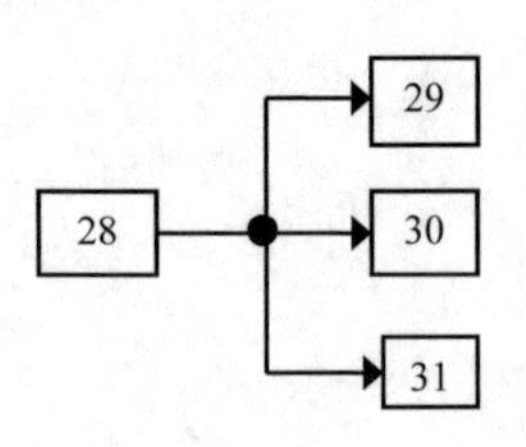

Fig. 6 Block diagram of the alignment system

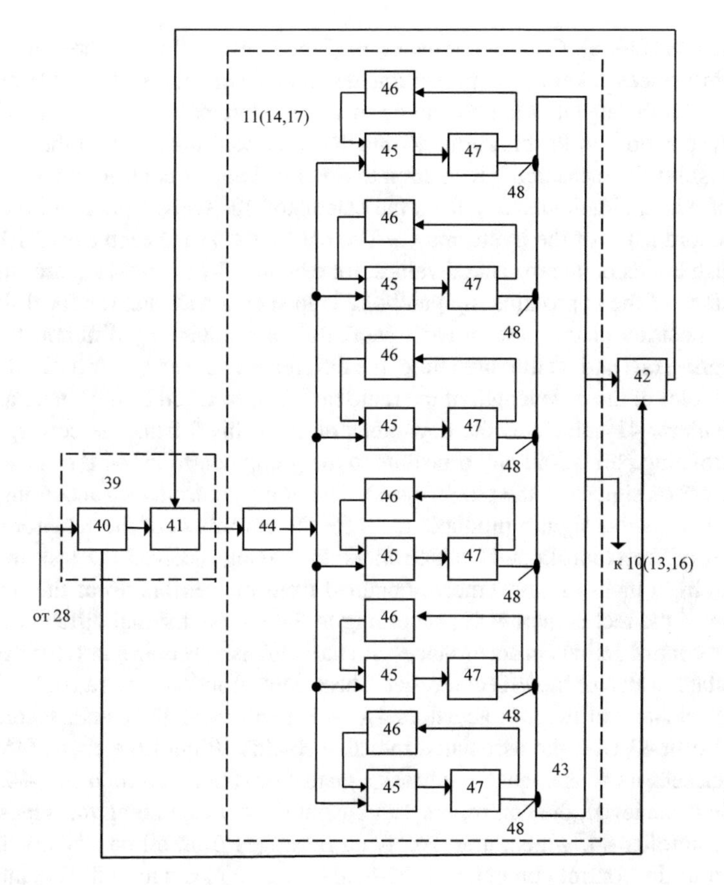

Fig. 7 Block diagram of automatic coordinate control system of moving control elements platforms

the output of the calculator 40, the other—with the output of the measurement system 42 coordinates of the mobile platform 43, and the output—with the input of the group controller 44 control element 11 or 14 or 17. The output of the group controller 44 is connected to the inputs of six controllers 45, the second inputs of which are connected to the outputs of the sensors 46 of the movement of the actuator legs 47, some outputs of which are connected to the inputs of the sensors 46 of the movement of the actuator legs 47, and others through the hinges 48 with the movable platform 43 connected to the input of the measurement system 42 of the platform coordinates. Depending on the purpose of the control element 11 or 14 or 17, either the reflecting surface 10 of the main dish or the reflecting surface 13 of the counter-dish or the radiation receiver 16 are attached to it.

The dish system is adjusted as follows.

At the beginning of the alignment, the automatic control system (ACS) 29 coordinates of the mobile platform of the control elements 11 shields of the reflecting surfaces 10 of the main dish is started by commands from the control computer of guidance 28 (Fig. 6). Control computer 39 (Fig. 7) ACS 29 using the computer 40 sets of influences, takes in the measurement system 42 of coordinates of the movable platform 43, the signal at the beginning of the measurements. The system 42 measures the positions of the platforms 43 and the associated shields 10 of the main dish and transmits the measured information to the calculator 41 control errors, the other input of which simultaneously from the calculator 40 receives the required values of the coordinates of the platforms 43. The calculator 41 for each shield 10 of the main dish builds on the measured values, for example, by the least squares method, the surface of the approximating paraboloids in such a way that the focal distance and the position of the base of each paraboloid are minimally different from the neighboring one and, at the same time, the differences between their focal distances are multiples of the wavelength of the radio radiation received by the antenna. Then, the calculator 41 calculates the deviations of each shield from the corresponding approximating paraboloid and transmits to the group controller 44 the corresponding correction signals of the platforms 43. The group controller 44 according to the received corrective signals produces tasks for the movement of the actuator legs 47 for each of the controllers 45. Controllers 45, having received the task to move, subtract from them the movements obtained from the sensors 46 of the feedback position of the feet-actuators 47, according to the received signal differences, they produce control actions in accordance with the established control law, for example, manipulated value of the PID controller. These control actions are transmitted to the electric actuators of the foot-actuators 47, which will move them and, accordingly, the platform 43 with the associated movable shields 10 until the signals from the feedback sensors 46 are equal to the task signals from the group controller 44. When equality is achieved, the controllers 45 transmit the corresponding messages to the group controller 44, which, after receiving messages from all controllers 45, will transmit to the control computer of the guidance 28 a message to the beginning of the automatic control system (ACS) 30 coordinates of the mobile control platforms 14 shields of the reflecting surfaces 13 of the sub dish (see Fig. 6).

Automatic control systems (ACS) 30 coordinates of the movable platform control elements 14 shields reflecting surfaces 13 sub dish work as ACS 29, but in this case, the system 42 measures the position of the platforms 43 and their associated shields 13 sub dish.

After installation of the shields 13 sub dish, i.e., after receiving messages from all the controllers 45, the ACS 30 will transmit to the control computer 28 of the guidance message at the beginning of the operation of the automatic control system 31 coordinates of the moving platform control element 17 of the detector 16 (Fig. 6). The control computer 39 (Fig. 7) ACS 31 via the computer 40 sets of influences, takes in the measurement system 42 of coordinates of the movable platform 43, the signal at the beginning of the measurements. The system 42 measures the positions of the platforms 43 and the associated radiation receiver 16 and transmits the measured information to the calculator 41 control errors, the other input of which simultaneously from the calculator 40 receives the required values of the coordinates of the platforms 43. The calculator 41 determines the required position of the center of the sensitive surface of the receiver 16 and the direction perpendicular to the focal plane of the antenna. Then, the calculator 41 calculates the deviations of the measured coordinates of the platform 43 from the required by condition of approval of the coordinates of the center of the sensitive surface of the receiver and its direction on the focal plane. The calculated deviations are transmitted by the calculator 41 to the group controller 44 in the form of platform 43 correction signals. Then, the group controller 44 on the received corrective signals produces tasks for the movement of the actuator legs 47 for each of the controllers 45, which, after receiving the task for the movement, subtract from them the movements obtained from the sensors 46 of the feedback of the position of the actuator legs 47, according to the received signal differences are produced in accordance with the established control law, for example in proportion to the integral-differential (PID), the control actions. These actions are transmitted to the electric actuators of the actuator legs 47, which will move them and, accordingly, the platform 43 with the associated receiver 16 until the signals from the feedback sensors 46 are equal to the task signals from the group controller 44. When equality is achieved, the controllers 45 transmit the appropriate messages to the group controller 44, which, after receiving messages from all controllers 45, will transmit to the control computer of the guidance 28 a message about the end of the antenna alignment and its readiness for astronomical observations.

When the frequency or wavelength of the radio emission received by the antenna changes, the ratio of the optical path lengths (1) achieved earlier in the alignment ceases to be performed. Therefore, the control computer of the guidance 28 transmits to the ACS 29, 30, 31 a new wavelength value, they make new calculations and produce signals for correcting the position of the shields 10, 13 and the receiver 16 entering the regulators for focusing the dish system.

Thus, the antenna alignment with the help of controlled elements provides an increase in the SUF of the space radio telescope with a significant spread of the operating frequency ranges of the received radiation and in the absence of the influence of the astronomical climate of the Earth.

4 Conclusion

To form a dish system of a large-diameter space radio telescope (up to 100 m), thousands of satellites with reflecting surfaces and one with a radiation receiver will need to be put into near-earth orbit. However, it is possible to gradually increase the diameter of the radio telescope and, accordingly, a gradual increase in the number of satellites of the group forming the dish antenna system.

The formation of the dish system of the radio telescope antenna in the near-earth orbit can significantly improve its main characteristic, namely, the value of the aperture surface utilization factor (SUF). In this case, practically excluded its dependence on deformations of weight and wind effects. However, the alignment of the antenna in the construction of the dish system is much more complicated, since it must be made in outer space. Accordingly, the focusing of the antenna becomes more complicated when the length of the received radiation changes.

When solving the problem of automatic antenna alignment and focusing in open space, it is advisable to install the shields of the reflecting surfaces of the dish antenna system on the controlled elements made in the form of SEMS modules.

The use of SEMS modules with parallel structures of the multi-link electric drive makes it possible to obtain high accuracy of control of the position of the shields and receiver with minimal travel time due to the introduction of parallelism in the processes of measurement, calculation and movement. The use of high-precision piezoelectric motors in SEMS provides the ability to operate in extreme open space conditions. The use of serial connection of SEMS in the controlled elements allows the positioning of shields with minimal displacement of the center of mass.

Acknowledgements This work was financially supported by Russian Foundation for Basic Research Grants 16-29-04424, 18-01-00076 and 19-08-00079.

References

1. Esepkina, N.A., Korol'kov, D.V., Parijskij, Y.N.: Radioteleskopy i radiometry (Radio telescopes and radiometers), 416 pp. Nauka Publication, Moscow (1973) (in Russian)
2. Gorodetskiy, A.E., Tarasova, I.L.: Detection and identification of dangerous space objects using adaptive matrix radio receivers. Informatsionno-upravliaiushchie sistemy **5**, 18–23 (2014) (in Russian)
3. Gorodetsky, A.E., Kurbanov, V.G., Tarasova, I.L., Agapov, V.A.: Problems of efficiency matrix receivers for radio images in astronomy. Radiotekhnika **1**, 88–96 (2015) (in Russian)
4. Artemenko, Y.N., Gorodetsky, A.E., Dubarenco, V.V., Kuchmin, A.Y., Tarasova, I.L.: Problems of creation of high-precision large millimeter-range radio telescopes. In: Proceedings of the Russian National Radio Astronomy Conference (WRC-2014) "Radio Telescopes, Instruments and Techniques of Radio Astronomy," PRAO ASC LPIA.E., Pushchino, 22–26 Sept 2014 (in Russian)
5. Razdorkin, D.Y., Romanenko, M.V.: The optimization algorithm of a two-dish antenna with dish of parabolic shields. Zhurnal radioelektroniki **4** (2000) (in Russian)
6. Gorodetsky, A.E., et al.: Sistema avtomaticheskogo navedeniya radioteleskopa (Automatic guidance system of the radio telescope). Patent USSR, no. 2319171 (2008)
7. Dubarenco, V.V., et al.: Sposob adaptacii otrazhayushchih poverhnostej antenny (Method of adaptation of the reflecting surfaces of the antenna). Patent USSR, no. 2518398 (2012)
8. Gorodetsky, A.E., Dubarenco, V.V., Kuchmin, A.Y., Agapov, V.A.: Control systems RT-70 of the sub dish (SUFFA) using parallel computing and measuring mechanical structures. In: Proceedings of the Russian National Radio Astronomy Conference (WRC-2014) "Radio Telescopes, Instruments and Techniques of Radio Astronomy," PRAO ASC LPI., Pushchino, 22–26 Sept 2014 (in Russia)
9. Gorodetskiy, A.E. (RU), Kurbanov, V.G. (RU), Tarasova, I.L. (RU): Method of formation of reflecting mirror surfaces of space radio telescope antenna. Patent RU, no. 2694813 (2019)
10. Gorodetskiy, A.E.: Smart electromechanical systems modules. In: Gorodetskiy, A.E. (ed.) Smart Electromechanical Systems, 277 pp. Springer International Publishing (2016). https://doi.org/10.1007/978-3-319-27547-5
11. Gorodetskiy, A.E.: Smart electromechanical systems architectures. In: Gorodetskiy, A.E. (ed.) Smart Electromechanical Systems, 277 pp. Springer International Publishing (2016). https://doi.org/10.1007/978-3-319-27547-5

12. Artemenko, Y.N., Gorodetsky, A.E., Doroshenko, M.S., Konovalov, A.S., Kuchmin, A.Y., Tarasova, I.L.: Problems of the choice of electric drives of space radio-telescope system dish system. Mehatronica, Avtomatizacia, Upravlenie **1**, 26–31 (2012) (in Russian)
13. Artemenko, Y.N. (RU), Gorodetsky, A.E. (RU), Dubarenko, V.V. (RU), Kuchmin, A.J. (RU), Agapov, V.A. (RU): Analysis of the dynamics of actuators automatic control systems of space radio telescope subdish. Informatsionno-upravliaiushchie sistemy **6**, 2–5 (2011) (in Russian)
14. Gorodetskiy, A.E., Kurbanov, V.G.: Smart electromechanical systems: the central nervous systems, 270 pp. Springer International Publishing (2017). ISBN 978-3-319-53326-1. https://doi.org/10.1007/978-3-319-53327-8

Hierarchical Neuro-Game Model of the FANET Based Remote Monitoring System Resources Balancing

Vladimir A. Serov, Evgeny M. Voronov and Dmitry A. Kozlov

Abstract The article discusses the basic principles of the methodology for the resource control optimizing of the remote monitoring system based on the FANET (Flying Ad Hoc Network) in real time. The hierarchical game model of optimization of control of system resources in the conditions of uncertainty on the basis of the coordinated stably-effective compromises is developed. The problem of synthesis of game algorithms of load balancing in communication channels based on neural networks of radial basis functions is solved. The developed situational model and neurofeedback algorithms provide structural adaptation of the system to the changing operating conditions and a given level of time delays in communication channels under the conditions of uncertainty of the input data queues.

Keywords FANET (Flying Ad Hoc Network) · Remote monitoring system · Unmanned aerial vehicle · Neural network control · Neural network of radial basis functions · Hierarchical game under uncertainty with the right of the first move · Coordinated stable-effective compromise · Robast quality assurance

1 Introduction

In solving the problems of situation analysis and control, remote monitoring systems (RMS), including mobile networks FANET (Flying Ad Hoc Network), consisting of unmanned aerial vehicles (UAV), satellite communication channels, control centers, are widely used [1–8].

The object of monitoring is usually a distributed group of objects that share RMS resources for data transfer. When evaluating the efficiency of RMS infrastructure,

V. A. Serov (✉) · D. A. Kozlov
MIREA—Russian Technological University, Moscow, Russia
e-mail: ser_off@inbox.ru

D. A. Kozlov
e-mail: kozlov.da97@gmail.com

E. M. Voronov (✉)
Bauman Moscow State Technical University, Moscow, Russia
e-mail: emvoronov@mail.ru

© Springer Nature Switzerland AG 2020

A. E. Gorodetskiy and I. L. Tarasova (eds.), *Smart Electromechanical Systems*, Studies in Systems, Decision and Control 261, https://doi.org/10.1007/978-3-030-32710-1_9

it is essential to take into account the following factors: high mobility of nodes and dynamically changing topology of FANET, distributed nature of the tasks to be solved, inconsistent (conflict) nature of interaction of RSM subsystems, functioning under uncertainty of input information flows. As it's known, these factors can be most fully taken into account in the framework of the game-theoretic direction of system analysis. In particular, currently game approaches to the development of methods of synthesis of topologies and control of mobile networks are actively used [9–18]. With increasing information and structural complexity of RMS, the requirements of optimal coordination of subsystems interaction in order to ensure a given level of efficiency of the system as a whole in a given range of operating conditions become crucial. The solution of this problem is possible within the framework of a hierarchical game approach based on the principle of coordinated stably effective compromises (COSTEC) [19–22]. Computational technology of COSTEC search in the hierarchical conflict multicriteria optimization under uncertainty tasks was considered in [23–26]. However, existing methods and algorithms of COSTEC search have high computational complexity and do not allow solving problems in real time, which directly affects the quality of remote monitoring.

A promising tool for solving problems of conflict control in real time is neuroevolutionary technology [27–29] and a technology based on the achievements of quantum Informatics [30]. In the present article the methodology of resources control optimization in real time on the basis of hierarchical game models and technology of neural networks of radial basis functions (RBF-networks) is developed.

The representation of the situation model in the form of a coalition structure allows to simulate all possible changes in the structure of RSM and topology of FANET in the process of functioning and changes in the conditions of conflict interaction of subsystems.

2 Hierarchical Game Model of Resource Balancing in Remote Monitoring System Based on FANET

One of the possible variants for building RSM infrastructure based on FANET is shown in Fig. 1. UAV teams M_1, M_2 monitor the current situation in the areas of the ground operation G_1, G_2. To transfer information to the control center (CC) each of the groups $M_1,\ M_2$ uses individual communication channel $FANET_1$ and $FANET_2$ respectively as the primary and shared satellite channel as a backup. Depending on the amount of information transmitted, the data streams from M_1, M_2 are distributed over the communication channels available to them.

Figure 2 shows a hierarchical structural scheme of data flow control in RSM channels, including levels of coordination, control, process.

In accordance with the above structural scheme, the information transfer process consists of two subprocesses: each of the groups M_1, M_2 forms together with

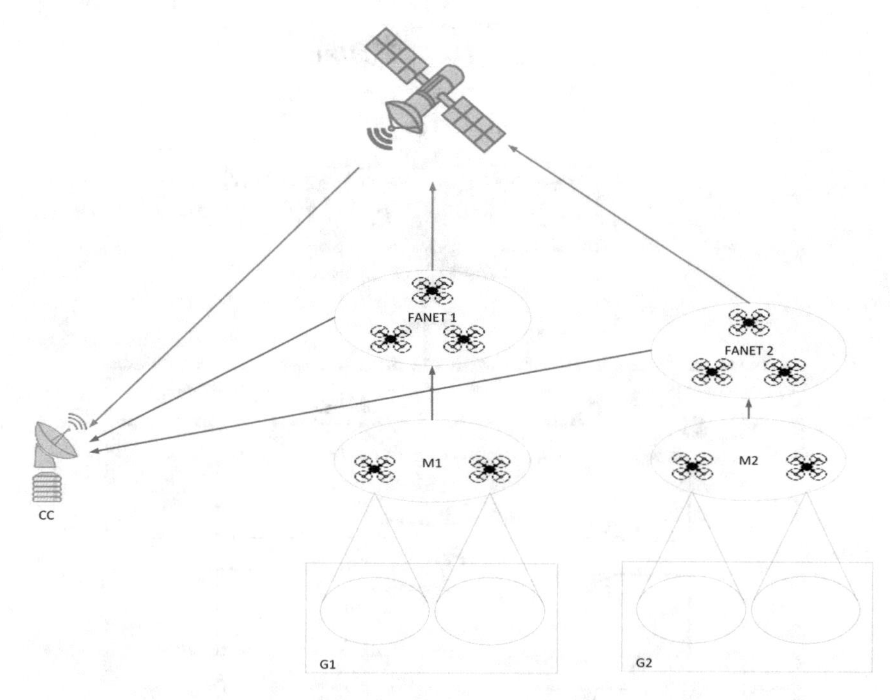

Fig. 1 RSM infrastructure based on FANET

the data CC a subscriber pair (S_1, D_1)—subprocess 1 and (S_2, D_2)—subprocess 2, respectively. Receivers D_1, D_2 collect and process data from sources S_1, S_2.

Subscriber pairs use channels F_1 (FANET 1) and F_2 (FANET 2) exclusively, respectively, and share a satellite channel F_c. The RSM infrastructure is expected to operate under uncertainty. An uncertain factor is the vector of input data flows in subscriber pairs (S_1, D_1) and (S_2, D_2)$\mathbf{z} = [q_1, q_2]^T \in \mathbf{Z}$, which is known only that it can take any value from the set $\mathbf{Z} = \mathbf{Q}_1 \times \mathbf{Q}_2$ of the form

$$\mathbf{Z} = \left\{ \mathbf{q} = [q_1, q_2]^T \,\middle|\, q_{1\min} \leq q_1 \leq q_{1\max};\, q_{2\min} \leq q_2 \leq q_{2\max} \right\}. \quad (1)$$

At the top level there is a coordinating Center, which has the right of the first move and forms a vector of priorities $\mathbf{v} = [\psi_1,\ \psi_2]^T$, according to which the resource of the common channel F_c is distributed among the subsystems. The set of valid priority values is given as

$$\mathbf{V} = \left\{ \boldsymbol{\psi} = [\psi_1, \psi_2]^T \,\middle|\, \psi_1, \psi_2 \geq 0,\, \psi_1 + \psi_2 = 1 \right\}. \quad (2)$$

The part of the shared channel resource P_{mci} allocated to service the source stream S_i, $i = \overline{1, 2}$, is calculated as

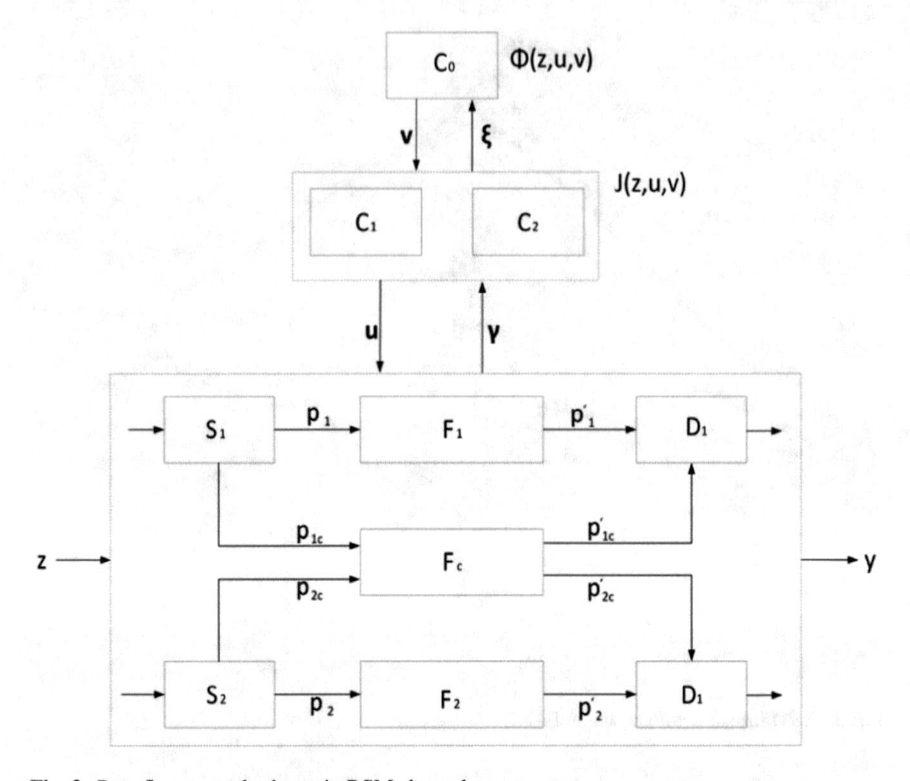

Fig. 2 Data flow control scheme in RSM channels

$$P_{mci} = \frac{\psi_i}{\psi_1 + \psi_2} P_{mc}, i = \overline{1, 2}, \tag{3}$$

where P_{mc} is the bandwidth of the shared channel F_c.

At the control level, there are subsystems C_1, C_2, that, with fixed priorities, form the vector $\mathbf{u} = [u_1, u_2]^T \in \mathbf{U}$ of distribution of input traffic through channels:

$$\begin{gathered} p_1 = q_1 u_1;\ p_2 = q_2 u_2;\ p_{1c} = q_1(1 - u_1);\ p_{2c} = q_2(1 - u_2), \\ \mathbf{U} = \left\{ \mathbf{u} = [u_1, u_2]^T \,\middle|\, 0 \leq u_i \leq 1, i = 1, 2 \right\}. \end{gathered} \tag{4}$$

The data flows distribution in the communication channels is calculated according to the table algorithm presented in Table 1.

The data transmission efficiency at the control level is evaluated by a vector indicator, the components of which characterize the time delays in the channels that determine the efficiency of the mission as a whole.

The performance indicator of the subscriber pair (S_i, D_i) is defined as

$$J_i(\mathbf{v}, \mathbf{u}, \mathbf{z}) = \max\{\tau_{ii}(\mathbf{v}, \mathbf{u}, \mathbf{z}), \tau_{ic}(\mathbf{v}, \mathbf{u}, \mathbf{z})\}, i = 1, 2. \tag{5}$$

Table 1 Distribution of data flows and time delays in channels

Channel	Stream on channel input	The condition of saturation of the channel (subchannel)	Flow in channel (subchannel)	Flow growth rate	Given flow growth rate	Time delay in channel (sub-channel)
1	$p_{11} = u_1 q_1$	$p_{23} = (1 - u_2) q_2$	$p'_1 = p_{11}$	$\Delta_1 = 0$	$\Delta_{1\Pi} = 0$	$\tau_{11} = \tau_0 \exp\left(\alpha \frac{p_{11}}{P_{m1}}\right)$
		$p_{11} > P_{m1}$	$p'_1 = P_{m1}$	$\Delta_1 = p_{11} - p'_1$	$\Delta_{1n} = \Delta_1$	$\tau_{11} = \tau_0 \exp(\alpha) + \tau_m \left(\exp\left(\beta \frac{\Delta_{1n}}{P_{m1}}\right) - 1\right)$
2	$p_{22} = u_2 q_2$	$p_{22} \leq P_{m2}$	$p'_2 = p_{22}$	$\Delta_2 = 0$	$\Delta_{2n} = 0$	$\tau_{22} = \tau_0 \exp\left(\alpha \frac{p_{22}}{P_{m2}}\right)$
		$p_{22} > P_{m2}$	$p'_2 = P_{m2}$	$\Delta_2 = p_{22} - p'_2$	$\Delta_{2n} = \Delta_2$	$\tau_{22} = \tau_0 \exp(\alpha) + \tau_m \left(\exp\left(\beta \frac{\Delta_{2n}}{P_{m2}}\right) - 1\right)$

(continued)

Table 1 (continued)

Channel	Stream on channel input	The condition of saturation of the channel (subchannel)		Flow in channel (subchannel)	Flow growth rate	Given flow growth rate	Time delay in channel (sub-channel)
3	$p_3 = p_{13} + p_{23}$, $p_{13} = (1 - u_1)q_1$ $p_{23} = (1 - u_2)q_2$	$p_3 \leqq P_{m3}$		$p'_{13} = p_{13}$	$\Delta_{13} = 0$	$\Delta_{13n} = 0$	$\tau_{13} = \tau_{23} = \tau_0 \exp\left(\alpha \frac{p_3}{P_{m3}}\right)$
				$p'_{23} = p_{23}$	$\Delta_{23} = 0$	$\Delta_{23n} = 0$	
		$p_3 > P_{m3}$	$p_{13} \leqq P_{m13}$	$p'_{13} = p_{13}$	$\Delta_{13} = 0$	$\Delta_{13n} = 0$	$\tau_{13} = \tau_0 \exp(\alpha)$
			$p_{23} > P_{m3} - p_{13}$	$p'_{23} = P_{m3} - p_{13}$	$\Delta_{23} = p_{23} - p'_{23}$	$\Delta_{23n} = \Delta_{23}\left(1 + \frac{p'_{13}}{p'_{23}}\right)$	$\tau_{23} = \tau_0 \exp(\alpha) + \tau_m \left(\exp\left(\beta \frac{\Delta_{23n}}{P_{m3}}\right) - 1\right)$
			$p_{13} > P_{m3} - p_{23}$	$p'_{13} = P_{m3} - p_{23}$	$\Delta_{13} = p_{13} - p'_{13}$	$\Delta_{13n} = \Delta_{13}\left(1 + \frac{p'_{23}}{p'_{13}}\right)$	$\tau_{13} = \tau_0 \exp(\alpha) + \tau_m \left(\exp\left(\beta \frac{\Delta_{13n}}{P_{m3}}\right) - 1\right)$
			$p_{23} \leqq P_{m23}$	$p'_{23} = p_{23}$	$\Delta_{23} = 0$	$\Delta_{23n} = 0$	$\tau_{23} = \tau_0 \exp(\alpha)$
			$p_{13} > P_{m13}$	$p'_{13} = P_{m13}$	$\Delta_{13} = p_{13} - p'_{13}$	$\Delta_{13n} = \Delta_{13}\left(1 + \frac{p'_{23}}{p'_{13}}\right)$	$\tau_{13} = \tau_0 \exp(\alpha) + \tau_m \left(\exp\left(\beta \frac{\Delta_{13n}}{P_{m3}}\right) - 1\right)$
			$p_{23} > P_{m23}$	$p'_{23} = P_{m23}$	$\Delta_{23} = p_{23} - p'_{23}$	$\Delta_{23n} = \Delta_{23}\left(1 + \frac{p'_{13}}{p'_{23}}\right)$	$\tau_{23} = \tau_0 \exp(\alpha) + \tau_m \left(\exp\left(\beta \frac{\Delta_{23n}}{P_{m3}}\right) - 1\right)$

Performance indicator of the coordinating Centre:

$$J_0(\mathbf{v}, \mathbf{u}, \mathbf{z}) = \max\{J_1(\mathbf{v}, \mathbf{u}, \mathbf{z}),\ J_2(\mathbf{v}, \mathbf{u}, \mathbf{z})\}. \tag{6}$$

The task of the channel balancing control optimizing in FANET is formalized in the form of a hierarchical game under uncertainty with the right of the first move

$$\boldsymbol{\Gamma} = \left\langle C_0, \mathbf{N}, \mathbf{V}, \mathbf{U}, \mathbf{Z},\ J_0(\mathbf{v}, \mathbf{u}, \mathbf{z}), \boldsymbol{\Gamma}'(\mathbf{v})\right\rangle, \tag{7}$$

$$\boldsymbol{\Gamma}'(\mathbf{v}) = \langle \mathbf{N}, \mathbf{U}, \mathbf{Z}, \mathbf{J}(\mathbf{v}, \mathbf{u}, \mathbf{z})\rangle. \tag{8}$$

In (7) C_0—the coordinating Center; $\mathbf{N} = \{1, 2\}$—a set of interacting subsystems—subscriber pairs; $\mathbf{V}$—a set of permissible values of the priority vector $\mathbf{v}$ of the form (2), formed at the coordination level; $\mathbf{U}$—a set of permissible values of subsystems control parameters vector of the form (4); $\mathbf{Z}$—a set of possible values of the uncertain factor of the form (1); $J_0(\mathbf{v}, \mathbf{u}, \mathbf{z})$—the Center efficiency indicator of the form (6); $\boldsymbol{\Gamma}'(\mathbf{v})$—game model of the form (8), formalizing the rules of subsystems conflict interaction—subscriber pairs under uncertainty at the control level with a fixed strategy of the Center $\mathbf{v}$.

3 Neural Network of Radial Basis Functions Architecture

The control law $\mathbf{B}^{opt}(q_1, q_2) = \{\mathbf{v}(q_1, q_2), \mathbf{u}(q_1, q_2)\}$, formed on the basis of the principle of coordinated stably effective compromises (COSTEC) [4], corresponds to the optimal balancing of FANET channels depending on the input load in the communication channels. The method of the control law $\mathbf{B}^{opt}(q_1, q_2)$ synthesis is based on the use of the library of genetic algorithms of multi-criteria conflict optimization (GAMCO) [23–26], has a very high computational complexity, which does not allow a real-time balancing.

To solve this problem in real time, a neural network technology based on the use of RBF-networks has been developed, which is a further development of the methodology described in [5, 6]. The RBF network architecture is shown in Fig. 3.

To determine the matrix of synaptic weights $\mathbf{W} = \left[w_{st}, s = \overline{1, k}, t = \overline{1, 4}\right]$, the problem of RBF-network training is solved. The training set is formed in the form

$$\mathbf{L} = \left\{\left(\left(q_1^j, q_2^j\right), \left(u_1^j, u_2^j, \psi_1^j, \psi_2^j\right)\right),\ j = \overline{1, |\mathbf{L}|}\right\}, \tag{9}$$

where each training pair corresponds to the optimal mode of balancing RSM resources at fixed values of the input load: $\psi_1^j = \psi_1^{opt}\left(q_1^j, q_2^j\right)$, $\psi_2^j = \psi_2^{opt}\left(q_1^j, q_2^j\right)$, $u_1^j = u_1^{opt}\left(q_1^j, q_2^j\right)$, $u_2^j = u_2^{opt}\left(q_1^j, q_2^j\right)$. The GAMCO library is used to form the training set.

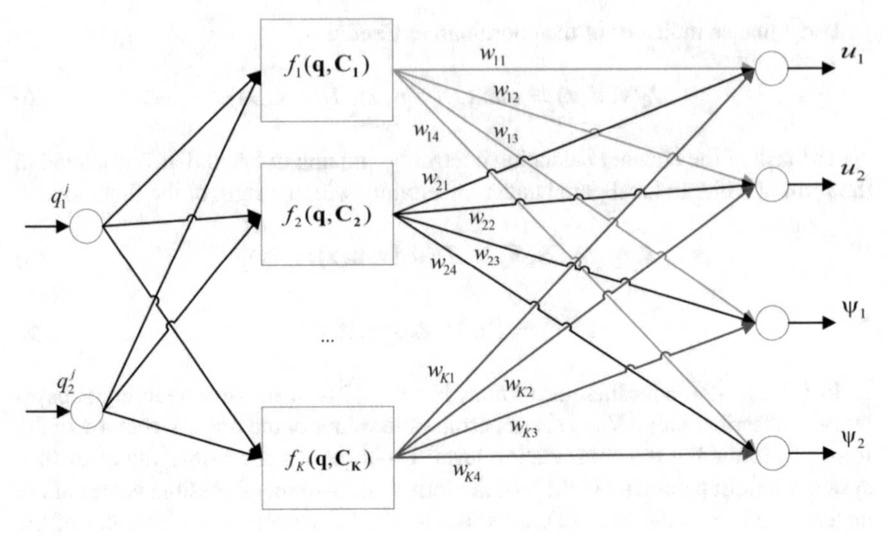

Fig. 3 RBF-network architecture used to the synthesis of the optimal control law $\mathbf{B}^{opt}(q_1, q_2)$ in real time

The number of neurons of the hidden layer is equal to the power of the training set: $K = |\mathbf{L}|$. The activation functions of neurons of the hidden layer are presented in the form

$$f_i(\mathbf{q}, \mathbf{C}_i) = \exp\left(-\frac{\|\mathbf{q} - \mathbf{C}_i\|}{\sigma_i^2}\right). \tag{10}$$

In (10) $\mathbf{C}_i$—centers of activation functions of neurons of the hidden layer, located in the space of input signals at the points of the training set $\mathbf{L}$: $\mathbf{C}_i = \mathbf{q}^i, i = \overline{1, |\mathbf{L}|}$; σ_i—width of the "window" of the activation function.

4 Computational Experiment

The trained RBF network synthesizes the control law $\mathbf{B}^{opt}(q_1, q_2)$ in real time in the form of interpolation surfaces $\psi_1^{opt}(q_1, q_2)$, $\psi_2^{opt}(q_1, q_2)$, $u_1^{opt}(q_1, q_2)$, $u_2^{opt}(q_1, q_2)$ (Figs. 4, 5, 6 and 7) on a set $\mathbf{Z}$ of species

$$\mathbf{Z} = \left\{\mathbf{q} = [q_1, q_2]^T \,|0 \le q_1 \le 3; 0 \le q_2 \le 3\right\}, \tag{11}$$

Set $\mathbf{Z}$ of the form (11) allows the occurrence of overload mode RSM channels. The results of the trained RBF network functioning are shown in Figs. 4, 5, 6, 7.

A study of the RSM resource balancing efficiency on all range $\mathbf{Z}$ of change of the input queues is made. Figure 8 shows examples of diagrams of changes of input data

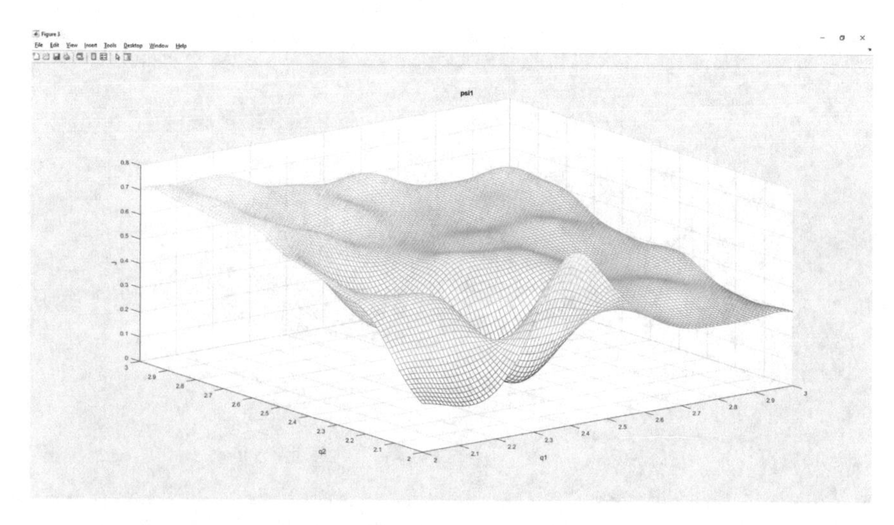

Fig. 4 Optimal balance control law $\psi_1^{opt}(q_1, q_2)$

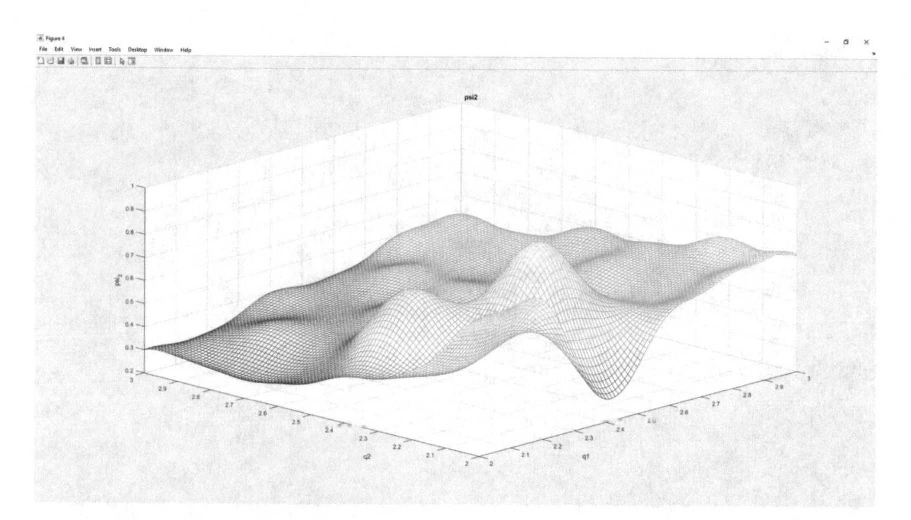

Fig. 5 Optimal balance control law $\psi_2^{opt}(q_1, q_2)$

queues in subsystems 1 and 2 from the range (11) are presented. Figures 9, 10 show the graphs of time delay changes in the system when using the laws of balancing resources on the basis of COSTEC and STEC, respectively. The dotted line indicates the acceptable level of the system delay time $\tau_{\max} = 0.314$, ensuring adequate perception of the situation by the operators and the possibility of prediction.

A comparative analysis of these balancing modes shows that in conditions of overload, starting from the level $q_1 + q_2 \geq 5$, the use of STEC does not provide

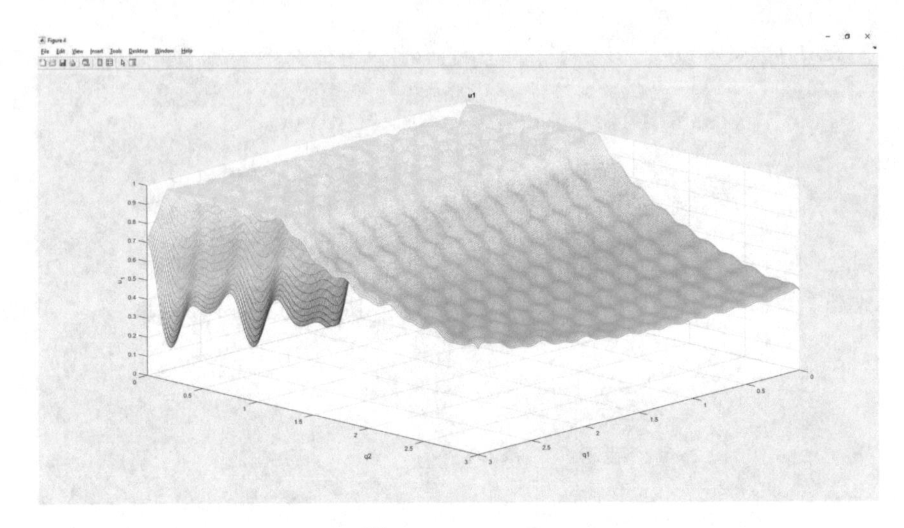

Fig. 6 Optimal balance control law $u_1^{opt}(q_1, q_2)$

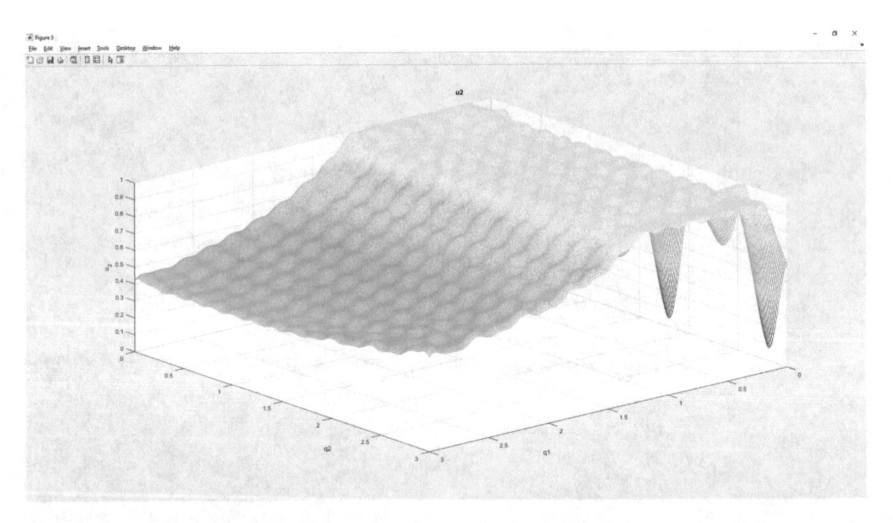

Fig. 7 Optimal balance control law $u_2^{opt}(q_1, q_2)$

the required level of efficiency. The use of COSTEC provides a robust quality that guarantees a given level of efficiency of the system over the entire range **Z** of uncertain factor.

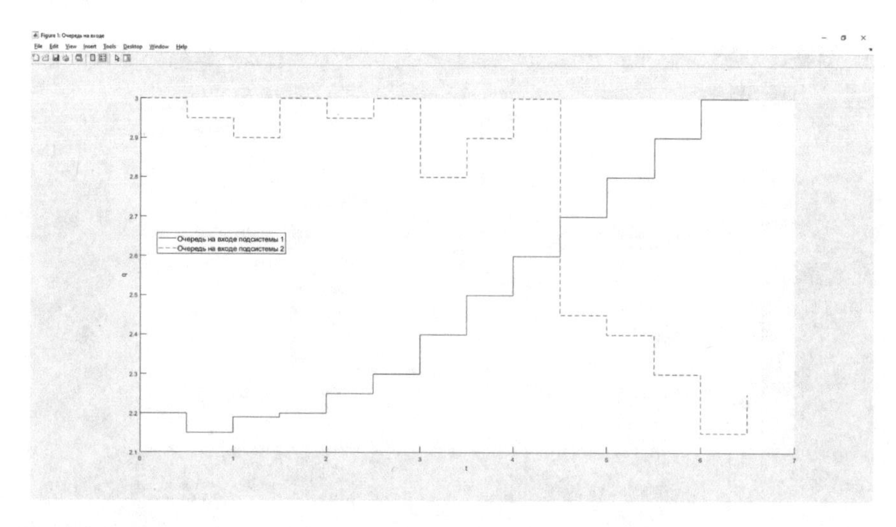

Fig. 8 Diagrams of changes in input data queues in subsystems 1 and 2

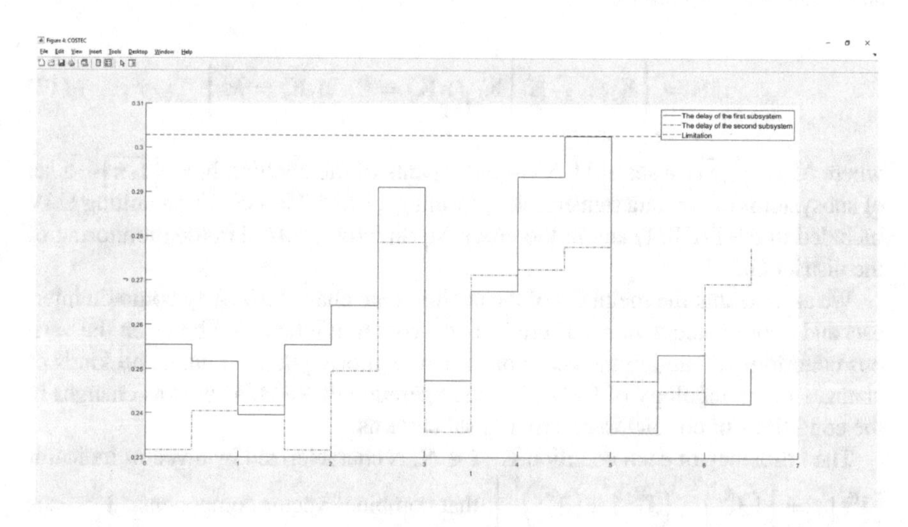

Fig. 9 Diagram of changes in time delays in the system, using the resources balancing law based on COSTEC

5 Situational Model of Remote Monitoring

It is proposed to form a situational model of remote monitoring in the form of a coalition structure. The coalition structure **K** is a breakdown of the set of UAVs involved in the monitoring process into coalitions:

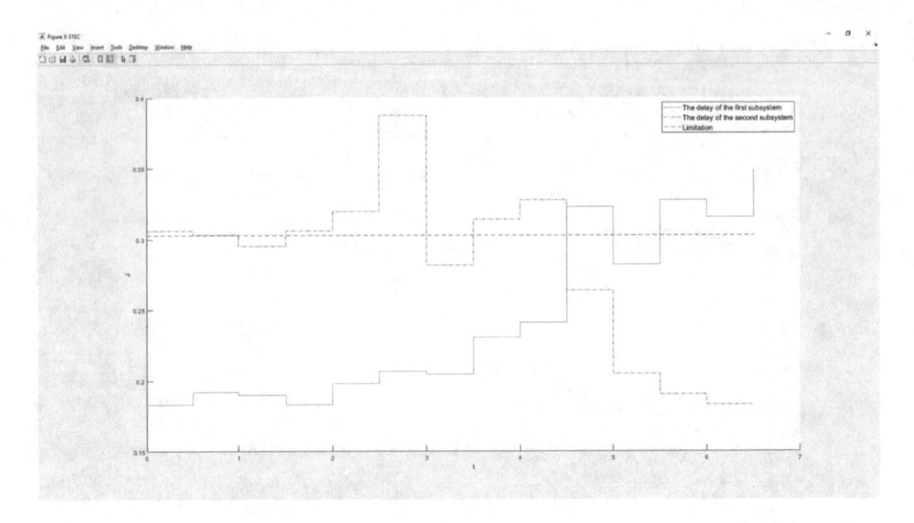

Fig. 10 Diagram of changes in time delays in the system, using the resources balancing law based on STEC without coordination

$$\mathbf{K} = \left\{ \mathbf{K}_1, \ldots, \mathbf{K}_n \middle| \mathbf{K}_i \underset{i \neq j}{\cap} \mathbf{K}_j = \emptyset;\ \underset{i \in \mathbf{N}}{\cup} \mathbf{K}_i = \mathbf{M} \right\} \tag{12}$$

where $\mathbf{M} = \left\{\overline{1, m}\right\}$ a set of UAVs—participants of the conflict; $\mathbf{N} = \left\{\overline{1, n}\right\}$—a set of subsystems of the data transmission, forming a RSM; $\mathbf{K}_i$—coalition uniting UAV included in the FANET_i and in the group M_i directly involved in the monitoring of the district G_i.

We assume that the members of the coalition are characterized by common interests and coordinated actions. There is a non-coalition interaction between the various coalitions. Changing the coalition structure allows you to simulate all kinds of changes in the topology of FANET in the operation of RSM, as well as changes in the conditions of conflict interaction of subsystems.

The efficiency of each coalition $\mathbf{K}_i$, $i \in \mathbf{N}$, is characterized by a vector indicator $\left[\mathbf{J}^{\mathbf{K}_i}\right]^T = \left[\left(\mathbf{J}_1^{\mathbf{K}_i}\right)^T, \left(\mathbf{J}_2^{\mathbf{K}_i}\right)^T, \left(\mathbf{J}_3^{\mathbf{K}_i}\right)^T\right]$, that combines vector components: $\mathbf{J}_1^{\mathbf{K}_i}$—the efficiency of information transfer; $\mathbf{J}_2^{\mathbf{K}_i}$—the efficiency of monitoring of the G_i area; $\mathbf{J}_3^{\mathbf{K}_i}$—the efficiency of the configuration of FANET_i and grouping M_i.

6 Conclusion

The task of the resources balancing of the remote monitoring system based on FANET is formalized in the form of a hierarchical game with the right of the first move under uncertainty, for which the COSTEC principle is used.

The problem of RSM channels balancing optimal law synthesis depending on the input load on the basis of the COSTEC principle is solved. It is shown that the use of a hierarchical balancing model based on COSTEC provides a significant improvement in the quality of RSM functioning over the entire range of operating conditions.

The optimal law of balancing RSM resources is implemented in real time on the basis of the RBF-network.

A situational monitoring model in the form of a coalition structure is proposed, which allows to carry out a comparative multicriteria analysis of the effeciency of various options of RSM infrastructure in case of changes in the conditions of functioning.

References

1. Bekmezci, I., Sahingos, O., Temel, S.: Flying Ad Hoc networks (FANETs): a survey. Ad Hoc Netw. **11**(3), 1254–1270 (2013)
2. Wang, J., Jiang, C., Han, Z., Ren, Y., Maunder, R., Hanzo, L.: Taking drones to the next level: cooperative distributed unmanned-aerial-vehicular networks for small and mini drones. IEEE Veh. Technol. Mag. **12**(3), 73–82 (2017)
3. Gupta, L., Jain, R., Vaszkun, G.: Survey of important issues in UAV communication networks. IEEE Commun. Surveys Tuts. **18**(2), 1123–1152, Second Quart (2016)
4. Sahingoz, O.K.: Networking models in flying Ad Hoc networks (FANETs): concepts and challenges. J. Intell. Robotic Syst. **74**(1), 513–527
5. Kim, D., Lee, J.: Topology construction for flying Ad Hoc networks (FANETs). In: Proceedings of International Conference on ICT Convergence (ICTC), Jeju, South Korea, hh. 153–157 (Oct 2017)
6. Leonov, A.V., Chaplygin, V.A.: Network FANET. Omsk Scientif. Bull. **3**(143) (2015)
7. Bekmezci, I., Ermis, M., Kaplan, S.: Connected multi UAV task planning for flying Ad Hoc networks. In: Communications and Networking (BlackSeaCom), 2014 IEEE International Black Sea Conference on IEEE, 2014, C. 28–32.1270 (May 2013)
8. Khan, M.A. et al.: Flying Ad Hoc networks (FANETs): a review of communication architectures, and routing protocols. In: Electrical Engineering and Computing Technologies (INTELLECT), 2017 First International Conference on Latest trends in IEEE, pp. 1–9 (2017)
9. Akkarajitsakul, K., Hossain, E., Niyato, D.: Coalition-based cooperative packet delivery under uncertainty: a dynamic bayesian coalitional game. IEEE Trans. Mobile Comput. **12**(2), 371–385 (2013)
10. Altman, E., Kumar, A., Singh, C., Sundaresan, R.: Spatial SINR games of base station placement and mobile association. IEEE/ACM Trans. Netw. (TON) **20**(6), 1856–1869 (2012)
11. Chuyko, J., Polishchuk, T., Mazalov, V., Gurtov, A.: Wardrop equilibria and price of anarchy in multipath routing games with elastic traffic. Game Theory Appl., pp. 9–19 (2012)
12. Eidenbenz, S., Kumar, A., Zust, S.: Equilibria in topology control games for Ad Hoc networks and generalizations. Mobile Netw. Appl. **11**(2), 143–159 (2006)
13. Han, Z., Niyato, D., Saad, W., Basar, T., Hjerungnes, A.: Game Theory in Wireless and Communication Networks: Theory, Models, and Applications, p. 536. Cambridge University Press (2012)
14. Jaramillo, J.J., Srikant, R.: A game theory based reputation mechanism to incentivizc cooperation in wireless Ad Hoc networks. Ad Hoc Netw. **8**(4), 416–429 (2011)
15. Long, C., Zhang, Q., Li, B., Yang, H., Guan, X.: Non-cooperative power control for wireless Ad Hoc networks with repeated games. IEEE J. Selected Areas Commun. **25**(6), 1101–1112 (2007)

16. Ren, H., Meng, M.Q.-H.: Game-theoretic modeling of joint topology control and power scheduling for wireless heterogeneous sensor networks. IEEE Trans. Autom. Sci. Eng. **6**(4), 610–625 (2009)
17. Saad, W., Han, Z., Ba§ar, T., Debbah, M., Hjarungnes, A.: Network formation games among relay stations in next generation wireless networks. IEEE Trans. Commun. **49**(9), 2528–2542 (2011)
18. Shi, H.-Y., Wang, W.-L., Kwok, N.-M., Chen, S.-Y.: Game theory for wireless sensor networks: a survey. Sensors **12**(7), 9055–9097 (2012)
19. Voronov, E.M.: Methods of Optimization of Multiobject Multicriteria Systems Control on the Basis of Stable-Effective Gaming Solutions, 576c. BMSTU, Moscow (2001)
20. Serov, V.A.: Stable-equilibrium control in a hierarchical game model of a structurally complex system under uncertainty. In: Popkov, Y.s.M. (eds.) Proceedings of the Institute of System Analysis RAS. Dynamics of Heterogeneous Systems, vol. 10, 1st edn, pp. 64–76. Komkniga (2006)
21. Vanin, A.V., Voronov, E.M., Serov, V.A.: Development of the method of multicriteria optimization of hierarchical control system based on coordinated stably effective compromises. Vestnik BMSTU. Ser. Nat. Sci. vol. 6, CC. 31–47 (2018). https://doi.org/10.18698/1812-3368-2018-6-31-47
22. Voronov, E.M., Serov, V.A.: A coordinated stable-effective compromises based methodology of design and control in multi-object systems. In: Gorodetskiy, A., Tarasova, I. (eds.) Smart Electromechanical Systems. Studies in Systems, Decision and Control, vol. 174, pp. 147–157. Springer, Cham
23. Serov, V.A.: Genetic computational procedure for finding vector equilibrium by Stackelberg in a hierarchical game model of functioning of a structurally complex system. Intelligent systems (INTELS'2006): Proceedings of the VII International Symposium (Russia, Krasnodar, 26–30 June 2006), pp. 73–74. RUSAKI, Moscow (2006)
24. Serov, V.A: Adaptive fitness functions in evolutionary game control optimization models in structure complicated systems. Vestnik BMSTU. Ser. Instrum. Making **2**(113), 111–122 (2017)
25. Serov, V.A.: Genetic algorithms of conflict equilibriums-based multicriteria systems control optimization under uncertainty. Vestnik BMSTU. Ser. Instrum. Making **4**(69), 70–80 (2007)
26. Serov, V.A., Voronov, E.M.: Evolutionary algorithms of stable-effective compromises search in multi-object control problems. In: Gorodetskiy, A., Tarasova, I. (eds.) Smart Electromechanical Systems. Studies in Systems, Decision and Control, vol. 174, pp. 19–29. Springer, Cham (2019)
27. Serov, V.A., Babintsev, J.N., Chichurin, A.V.: Neuroevolution technology multi-objective optimization of data flowcontrol in an automated monitoring system in situations of conflict and uncertainty. Neurocomputers: Development and Application **9**, 65–71 (2008)
28. Serov, V.A., Babintsev, Yu.N., Kondakov, N.S.: Neurocontrol of multicriteria conflict systems. Monograph, p. 136. MosGU, Moscow (2011)
29. Serov, V.A., Chechurin, A.V.: Control of multilevel systems in real time on the basis of hierarchical neural ensembles. Ind. ACS Controll. **7**, 37–41 (2011)
30. Sigov, A.S., Andrianova, E.G., Zhukov, D.O., Zykov, S.V., Tarasov, I.E.: Quantum informatics: review of the main achievements. Russian Technol. J. **7**(1), 5–37. https://doi.org/10.32362/2500-316x-2019-7-1-5-37 (2019)

Switching Operation Modes Algorithm for the Exoskeleton Device

Valery G. Gradetsky, Ivan L. Ermolov, Maxim M. Knyazkov, Eugeny A. Semenov and Artem N. Sukhanov

Abstract *Problem statement*: For the organization of feedback in the control system of the exoskeleton, information-measuring modules, sensors are essential elements. They allow monitoring the state of the system, to obtain information about the environment, objects of manipulation. Sensors are necessary elements of the master devices that allow evaluating the control signals of the operator. In this paper we propose our algorithm based on mathematical model of human skeletal muscle. *Purpose*: The movement of the exoskeleton with low speed is necessary to perform any technological operations and it requires increased accuracy of the desired movement. This is especially important when holding or moving heavy cargo or fragile objects. The mode of movement at high speeds allows operator to move quickly the links of the exoskeleton to the desired position in space. This mode is essential for application in situations where frequent changes of direction of movement are taking place and high speeds are required. A feature of this mode is the requirement for a short transition time of the position and speed of the links of the exoskeleton. *Results*: The algorithm for recognizing the desired action of the operator and selecting the desired mode of operation of the exoskeleton with the adjustment of the characteristics of the task generator was proposed. To obtain reliable experimental data SEMS system was used. The simulation shows that this algorithm improves the quality of control of the exoskeleton drive, using different control formation laws for the corresponding tasks. *Practical significance*: The field of exoskeleton devices application is determined by the scientific and technical tasks assigned to such systems. The use of exoskeletons is relevant in emergency situations where they are performing tasks related to the movement of heavy loads, ammunition suspension and the implementation of power support while debris removing, repair of agricultural machinery. The carried out researches allows revealing requirements to quality of drive system control of the

V. G. Gradetsky · I. L. Ermolov (✉) · M. M. Knyazkov (✉) · E. A. Semenov · A. N. Sukhanov
Ishlinsky Institute for Problems in Mechanics of the Russian Academy of Sciences, Prospect Vernadscogo 101-1, Moscow 119526, Russia
e-mail: ermolov@ipmnet.ru

M. M. Knyazkov
e-mail: Ipm_labrobotics@mail.ru

A. N. Sukhanov
e-mail: sukhanov-artyom@yandex.ru

© Springer Nature Switzerland AG 2020

A. E. Gorodetskiy and I. L. Tarasova (eds.), *Smart Electromechanical Systems*, Studies in Systems, Decision and Control 261, https://doi.org/10.1007/978-3-030-32710-1_10

exoskeleton device applied for any human activity. Various modes of operation of the exoskeleton device were proposed. Each mode should meet the requirements of operations that the operator should perform in the exoskeleton.

Keywords Control algorithms · Exoskeleton · Biopotentials · Situational control · EMG · Biocontrol · SEMS

1 Introduction

For the organization of feedback in the control system of the exoskeleton, information-measuring modules, sensors are essential elements. They allow monitoring the state of the system, to obtain information about the environment, objects of manipulation. Sensors are necessary elements of the master devices that allow evaluating the control signals of the operator. There are three types of data received from the information system.

The Strategic level of the exoskeleton requires information about the planned movements of a person based on his current actions. The Tactical level uses information about the objects of manipulation and the state of the environment. The Executive level uses information about the state of the motor system, movements in the drives, the state of the brake elements, the state of the limit switches. In exoskeleton systems, information and measurement systems are most often position sensors, strain gauges, sensors of muscle activity, as well as inertial sensors.

2 Data Acquisition

In our previous work [1, 2] we proposed a control technique for the designed [3, 4] exoskeleton system. The main goal for the current research is to improve efficiency of control by using torque control in the exoskeleton system to provide information about interaction of exoskeleton parts with operator. As a source of input signals the operator controls the result of the actuators motion via visual and torque information.

The input signal for the control system of the designed exoskeleton motion is obtained from the EMG sensors. The main advantage of using biopotentials in control is the high speed of obtaining that signal from the operator. In the traditional approach, based on the use of strain sensors or an interface based on the force action on the control elements, the nervous system of the operator sends a signal for muscle contraction. When excited, muscle fibers begin to contract, driving the human bone [5]. The consequence of this movement is the impact on the control element of the exoskeleton system.

Biopotential sensors allow receiving a signal at the beginning stage of muscle fiber contraction, which gives the opportunity to start processing this signal before contraction ends [6, 7]. The complete time of skeletal muscle contraction varies from

50 to 60 ms. This time period can be used to process the signal of muscle activity and generate a control signal for the exoskeleton motors. There are several control algorithms suitable for the exoskeleton device based on the muscles activity. These algorithms include:

- Binary Algorithm, which provides control signal for movement when reaching the predetermined amplitude of the EMG signal. In this case, the reverse motion of actuated parts of the exoskeleton is often carried out by springs or elastic elements [8, 9]. This algorithm is easy to implement, but it cannot provide control the speed of movement of the exoskeleton link. This algorithm is mainly used in prostheses.
- Variable Control Algorithm, which allows to change the speed of movement of the exoskeleton link in the threshold or continuous mode, as well as to control the magnitude of the torque implemented by the drive system of the joint participating in the movement [10]. To implement this algorithm, the control system uses a proportional link that connects the value of the digitized filtered biopotential signal with the output values of the control signal. This approach requires a low-pass bandpass filter and smoothing of the amplitude function of the EMG signal from time to time to avoid jitter or sudden movements of structural elements.
- The algorithm based of natural movements (Natural Reach and Pinch Algorithm). This algorithm includes the use of a database of signal patterns of biopotentials removed from human muscles [11]. The operator takes a course of training, performing certain actions. EMG sensors attached to the skin of the operator collect information about muscle activity and transmit it to a preliminary database for further processing.

In this paper we propose our algorithm based on mathematical model of human skeletal muscle [7].

Every skeletal muscle has its own maximal force contraction parameter which depends on plenty of factors. It depends mainly on the physiological characteristics of the operator. The technique of determining this parameter is necessary for the operation of the control system. To do this, the exoskeleton should not perceive the control from the EMG sensor, and the force action of the operator on the structural elements should be perceived as an external disturbance. To obtain the EMG signal, the non-invasive surface electrodes should be applied to the skin of the operator. The operator needs to apply maximum force to change the position of the links of the exoskeleton. The force sensor located on the handle will record the maximum value, and the position sensors will record the current values of the generalized coordinates. Thus, the control system will identify the maximum force that the operator can develop for the current configuration, and find the maximum weight of the load $m_{\max}$, which the operator are able to hold.

The force developed by the muscle is expressed by the following way:

$$F_{des}(t) = a(t)F_{\max} f_{FV}(V_a) f_{Fl}(l_a) \tag{1}$$

Here $f_{FV}(V_a)$ is functional dependence of contraction force from speed of muscle contraction, $f_{Fl}(l_a)$ is dependence of contraction force from the current length of the muscle [12]. Mathematical models of these dependencies were obtained by the authors [13, 14]. The activation level $a(t)$ is the value determined by the ratio of the desired force $F_{des}(t)$, to maximum possible force $F_{\max}$, developed under the current parameters of the length $l_a(t)$ and the reduction rate $V_a(t)$.

Muscle fibers generate electric signals when muscles are contracted. The intensity of muscle innervation means a fraction of the maximum effort. Sending impulses, the brain generates tension in the muscle. Pulse frequency (4–900 Hz) and the amplitude of the voltage (±0.0012 V) [15] received by the sensors of the action potential and the data after filtering and averaging, are supplied to the controller. Filters and the smoothing algorithm are implemented in the block of the interaction of human and exoskeleton. It is a link between the duty cycle of the operator's nervous system and the motor control system.

Control of the performed actions, as well as evaluation of the environment state goes through a visual channel, and tactile sensations allow the operator to evaluate the strength of reactions when interacting with objects in the environment. If there is an intermediary system between a human and the environment like exoskeleton device, it is necessary to understand that the introduction of such a link into the information channel of interaction between a person and an object leads to a violation of the naturalness of this interaction.

To restore the naturalness of human interaction with objects, many developers of exoskeleton systems use force feedback, which allows getting an idea of the presence of an object or its shape without direct contact with the object.

3 Muscle Model

In free movement the vector of generalized torques $M_{i.a}$ for exoskeleton motors will be formed by the muscle activity of the operator. Taking into account the delay in the formation of the control, the vector of generalized torques $M_{i.e}$ of the exoskeleton system will be formed with its own delay, which can lead to position error for the operator's arm links and exoskeleton links both. Thus, there is $|q_{i.e}(t) - q_{i.a}(t)| \leq \xi_i$, where ξ_i is the maximum position error. Attachment devices of the exoskeleton impose restrictions on the movement of the operator's limbs [16]. In this case, there is a gap $\varepsilon_i > 0$, while $\varepsilon_i < \xi_i$, which is the possible constructive error of the arm link of the operator from the corresponding link of the exoskeleton. In this case, the movement of the arm will be subject to a one-way constraint. For the exoskeleton device, this constraint turns to a disturbing effect [17].

Within a particular designed gap ε_i the operator's hand can be moved freely setting the movement of the links of the exoskeleton by the force of muscles. When the arm contacts with the link of the exoskeleton the not-holding joint appears, generating reactive efforts, making changes in the dynamics of the nominal system and in the dynamics of the desired motion of the exoskeleton.

The mathematical description of the movement of the exoskeleton-human system will be determined as follows:

$$\begin{cases} \frac{d}{dt}\left(\frac{\partial T_a}{\partial \dot{q}_{i.a}}\right) - \frac{\partial T_a}{\partial q_{i.a}} + \frac{\partial W_a}{\partial q_{i.a}} = M_{i.a} + R_i l_{i.a} \\ \frac{d}{dt}\left(\frac{\partial T_e}{\partial \dot{q}_{i.e}}\right) - \frac{\partial T_e}{\partial q_{i.e}} + \frac{\partial W_e}{\partial q_{i.e}} = M_{i.e} - M_{i.ext} - R_i l_{i.e} \\ R_i = \begin{cases} 0, & if\, |q_{i.e}(t) - q_{i.a}(t)| \le \xi_i \\ -k_i(q_{i.e}(t) - q_{i.a}(t) - \varepsilon_i), & if\; q_{i.e}(t) - q_{i.a}(t) > \varepsilon_i \\ k_i(q_{i.e}(t) - q_{i.a}(t) + \varepsilon_i), & if\; q_{i.e}(t) - q_{i.a}(t) < -\varepsilon_i \end{cases} \end{cases} \quad (2)$$

where k_i is a proportional factor, $l_{i.a}$ is arm for the force in the i-th joint of the operator's arm, $l_{i.e}$ is arm for the force in the i-th joint of the exoskeleton. T_a is the kinetic energy of the arm, W_a is the potential energy of the arm, $q_{i.a}$ are joint coordinates associated with the operator's arm, $M_{i.a}$—torques corresponding to the their coordinates $q_{i.a}$. The kinetic energy of the considered arm can be found as the sum of the kinetic energies of its components: $T_a = \sum_{i=1}^{n} T_{i.a}$, where n is the number of considered links. T_e is kinetic energy of the exoskeleton, W_e is potential energy of the exoskeleton, $q_{i.e}$ ere joint coordinates associated with the parts of the system of exoskeleton, $M_{i.e}$—generalized forces (torques) acting on an exoskeleton system, $M_{i.ext}$ is vector of external disturbing torques resulting from the interaction of links of the exoskeleton with the objects of the external environment.

As for the muscle its excitation-to-activation dynamics is described like the next equation:

$$\dot{a}(t) = f(u(t), a(t)) \quad (3)$$

Here *u(t)* is muscle fibers excitation (EMG), $a(t)$ is activation signal. The activation-to-force (contraction) dynamics can be described as follows:

$$\dot{f}^{MT} = g\big(a(t), l^{MT}, v^{MT}, f^{MT}\big) \quad (4)$$

where l^{MT} is current muscle length, v^{MT} is contraction velocity, f^{MT} is real force applied by the muscle. From the first Eq. (1) $f_{FV}(V_a)$ and $f_{Fl}(l_a)$ elements are nonlinear functions. The muscle force-length dependency can be described as:

$$f_{Fl}(l_a) = \begin{cases} 0, & if\, \frac{l_a}{l_o} \le 1 \\ e^{\frac{-(\frac{l_a}{l_o}-1)^2}{\gamma}}, & if\, \frac{l_a}{l_o} > 1 \end{cases} \quad (5)$$

Here l_a is the current muscle length, l_o is optimal length of the current muscle. $f_{FV}(V_a)$ depends on the contraction velocity:

$$f_{FV}(V_a) = \begin{cases} 0, & if\, \frac{V_a \tau_c}{l_o} \le -1 \\ \frac{l_o k_{CE1} + V_a \tau_c k_{CE1}}{k_{CE1} l_o - V_a \tau_c}, & if\, -1 < \frac{V_a \tau_c}{l_o} \le 0 \\ \frac{l_o k_{CE2} + f_V^{\max} V_a \tau_c}{k_{CE2} l_o + V_a \tau_c}, & if\, \frac{V_a \tau_c}{l_o} > 0 \end{cases} \quad (6)$$

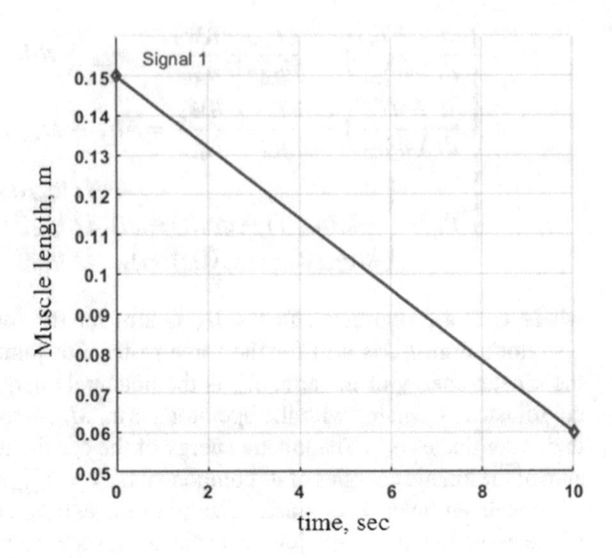

Fig. 1 Muscle length change

Here τ_c is muscle reaction time, $\gamma,\ k_{CE1},\ k_{CE2}$ and $f_V^{\max}$ are experimental coefficients. In our work we used the next numerical parameters [18]:

$$\tau_c = 0.1, \gamma = 0.45, k_{CE1} = 0.25, k_{CE2} = 0.06,\ f_V^{\max} = 1.6$$

With current parameters we can obtain contraction force for the skeletal muscle under certain length. Figure 1 shows length change for the skeletal muscle.

Thus we can obtain the virtual model of the muscle. Besides active elements in the model there are passive elements that are described with subsystem F^{PE} (Fig. 2). In our work we used the next expression for the passive elements:

$$F^{PE} = F_{\max}(e^{k_{PE}} + 1) \tag{7}$$

Here $k_{PE} = 3$ is experimental coefficient.

"fl(LM)"-subsystem is presented on Fig. 3. It contains force-length dependency from (5).

"fV(VM)"-subsystem is presented on Fig. 4. It contains force-velocity dependency from (6).

The scheduler is equivalent to the target task, represented in a graphical form. In this case, the desired law of motion is represented in the form of a linear change of the generalized coordinate from 0° corresponding to the situation of lowered down of the operator's arm to 135°, which corresponds to the arm bent in the elbow joint for example and changing the biceps brachii length from 150 to 60 mm. The model response for the length changing is developing force, presented in Fig. 5.

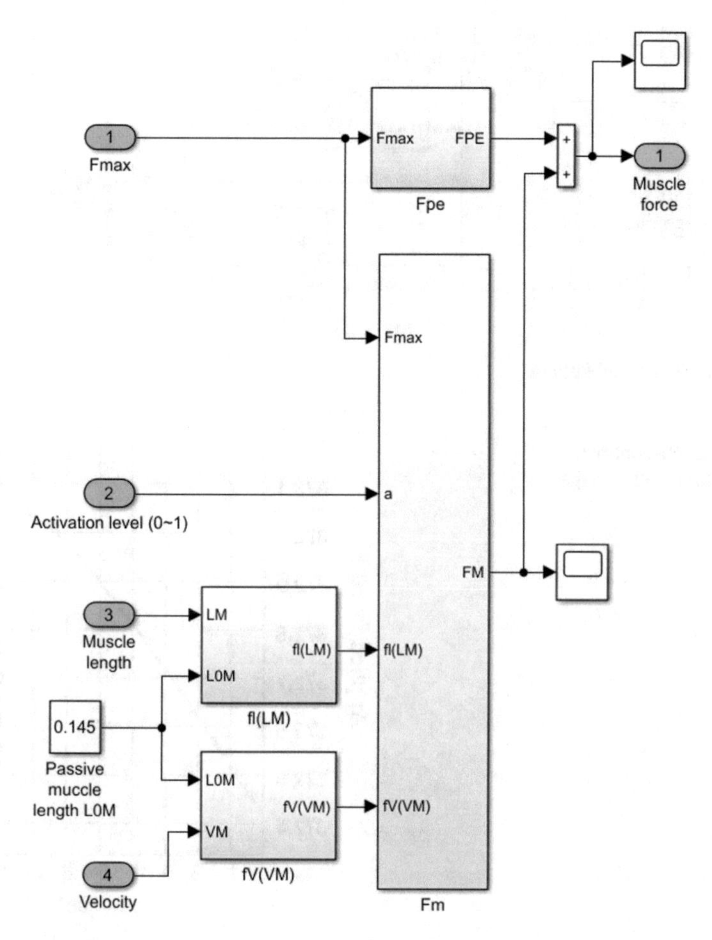

Fig. 2 Virtual muscle model

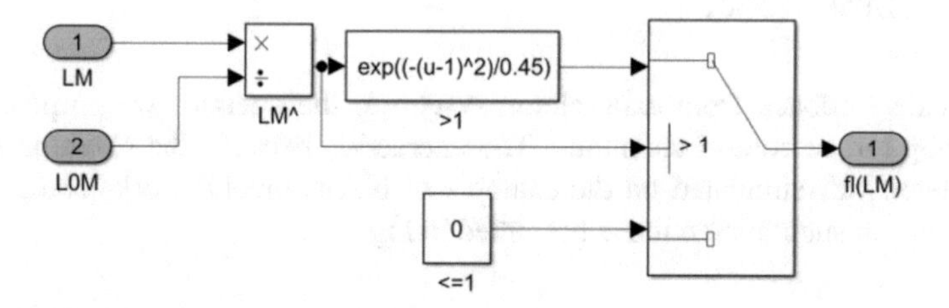

Fig. 3 "fl(LM)"-subsystem

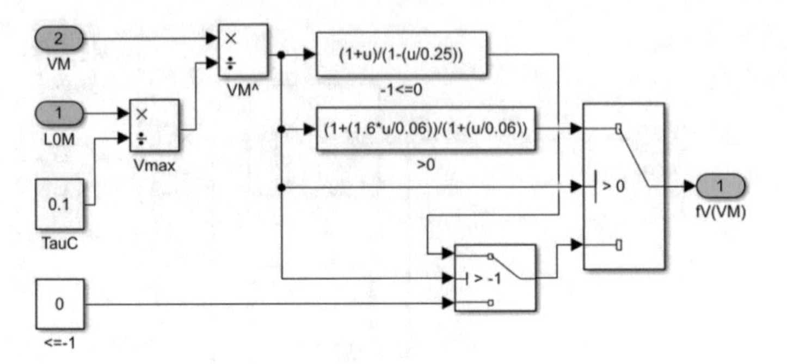

Fig. 4 "fV(VM)"-subsystem

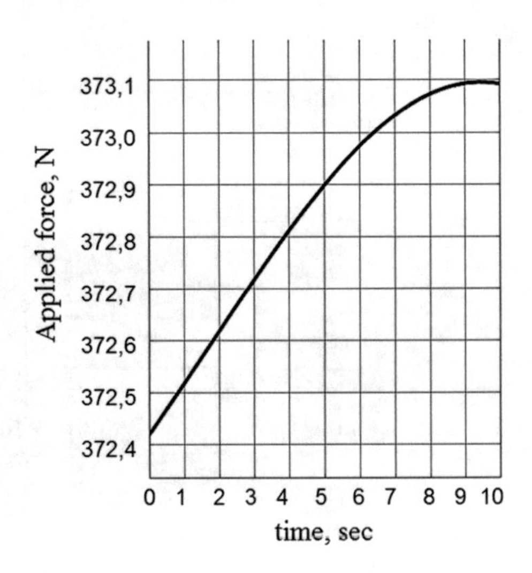

Fig. 5 Muscle contraction force under length change

4 Interface Model

To provide feedback from exoskeleton system to the operator we propose using close loop torque control technique. The interaction between the operator and the exoskeleton was simulated on the example of biceps brachii work. The interface subsystem for such interaction is presented in Fig. 6.

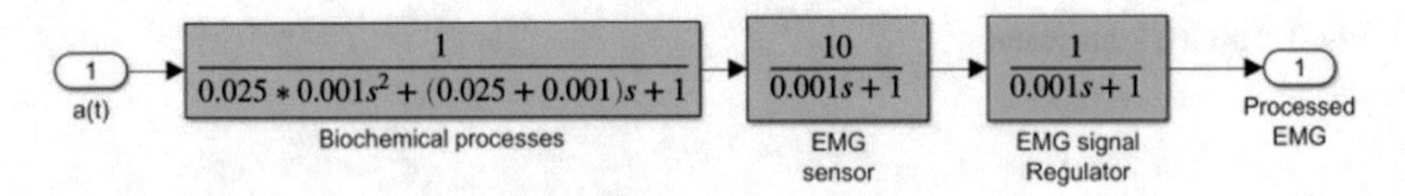

Fig. 6 Operator-Exoskeleton interaction block

The activation level $a(t)$ is the control signal for the Operator-Exoskeleton interaction block. It is an important information parameter for the muscle contraction model. This parameter is a source of information for the biopotential sensor (EMG sensor). In this block (Fig. 6) the time delay of the reaction of the synapse, the excitation of the muscle fiber, as well as the reaction of calcium exchange in the cells of the muscle were taken into account.

Delay in processing the signal of nerve impulses for the synapse and the process of calcium ion exchange in cells formed as aperiodic links. The transfer function of the biopotential sensor corresponds to the integrating unit with a sufficient gain ratio, which corresponds to the actual amplification of the EMG sensor. Also it has a PI controller that generates a control signal to the actuator input of the exoskeleton. It has customizable settings for integration time, the T_{reg} and transfer coefficient K_{reg}. This block and its parameters were described in our previous paper [1, 2].

The actuator of the exoskeleton is a multi-circuit electromechanical system, closed via torque. In addition to torque feedback, this system provides current feedback (Fig. 7). The torque feedback is essential for torque regulation in the system.

The result of the regulation is shown in the Fig. 8.

Exoskeleton drive is intended to compensate any external torque. Therefore, the operator may control speed of exoskeleton's links movement via biopotential signal

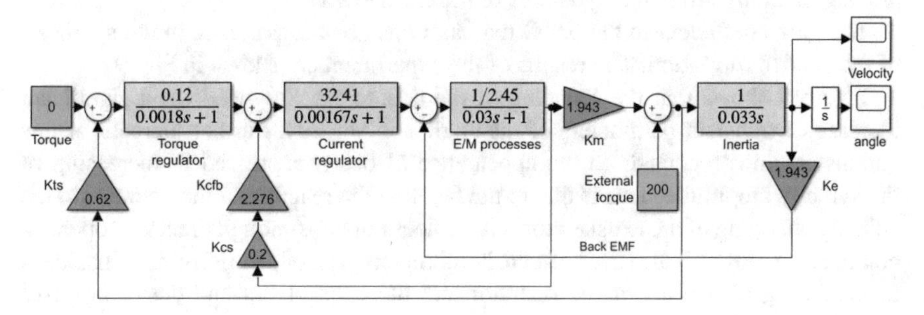

Fig. 7 Exoskeleton's drive with torque feedback

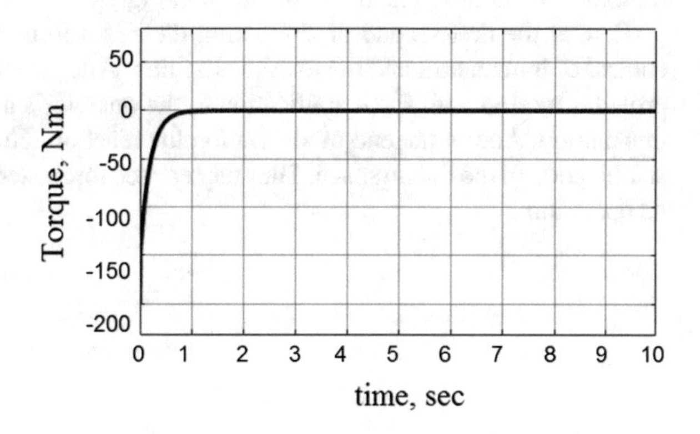

Fig. 8 Exoskeleton's drive with torque feedback

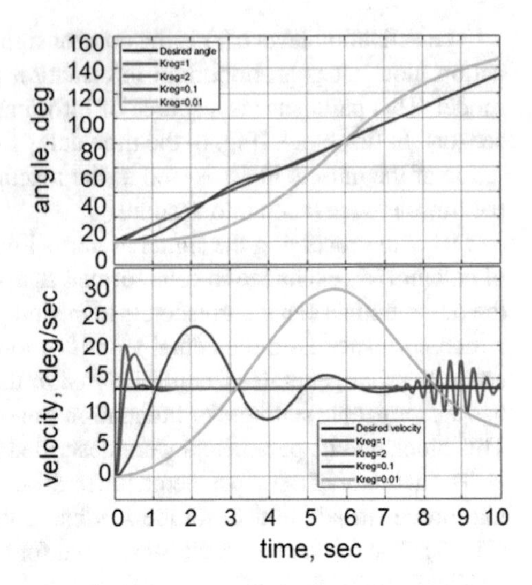

Fig. 9 The influence of K_{reg} variation on the behavior of the drive system

from his muscles. The exoskeleton drive control depends on the parameters of the controller and biopotential sensor. The choice of the parameters of these elements will significantly affect the dynamics of the exoskeleton.

The gain coefficient in the EMG regulator has great importance in the synthesis of the control formation. The results of the experiment are shown in Fig. 9.

Figure 9 shows that the reducing of the coefficient K_{reg} leads to a significant increase of transient parameters of the angle and velocity. Further increase of this parameter, however, leads to the appearance of beats, expressed in increasing of the velocity amplitude. This is due to the fact that the reaction of the operator to the velocity changing of the exoskeleton's links does not have enough time. The operator continues to move his arm even after drive compensation of position error. That leads to mistiming of actions of the operator and the exoskeleton and desynchronizes movement.

In the simulation we tried to obtain the reaction force between the operator and the exoskeleton link. The result is shown in Fig. 10.

Here at the first second of simulation there is a force leap that means physical contact of human arm and the exoskeleton link. After that leap the exoskeleton link provides motion with force application to the operator's arm, trying to help him in this motion. And in the end of simulation the reaction returns to zero which means stabilization of the link in space. Thus the realized torque feedback control performed its operation.

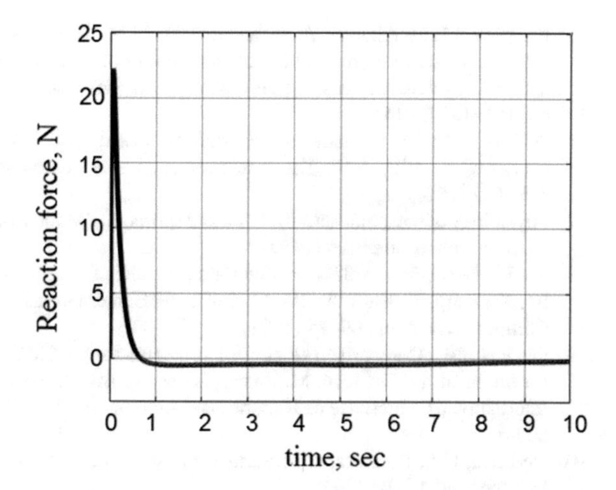

Fig. 10 Reaction force in handle

5 Conclusions

To implement the desired movement of the exoskeleton links it is necessary to process information coming from various sources (operator, reactive forces from the objects). Processing that data is essential for generation of control signal for the exoskeleton system. In this paper we propose our algorithm based on mathematical model of human skeletal muscle. To provide stability in control of the exoskeleton device its drive system should have torque feedback.

Acknowledgements The present work was supported by the Ministry of Science and Higher Education within the framework of the Russian State Assignment under contract No. AAAA-A17-117021310384-9.

References

1. Gradetsky, V., Ermolov, I., Knyazkov, M., Semenov, E., Sukhanov, A.: The influence of various factors on the control of the exoskeleton. In Interactive Collaborative Robotics, proc. of the 3rd International Conference ICR 2018, pp. 60–70. Springer, Switzerland, AG 2018 Printforce, The Netherlands (2018)
2. Gradetsky, V., Ermolov, I., Knyazkov, M., Semenov, E., Sukhanov, A: The dynamic model of operator-exoskeleton interaction. In Interactive Collaborative Robotics, proc. of the 3rd International Conference ICR 2018, pp. 52–59. Springer, Switzerland, AG 2018 Printforce, The Netherlands (2018)
3. Gradetsky, V., Ermolov, I., Knyazkov ,M., Sukhanov, A.: Generalized approach to bilateral control for EMG driven exoskeleton. In Conference Volume **113**, 2017, 12th International Scientific-Technical Conference on Electromechanics and Robotics, volume 113 of Zavalishin's Readings, pp. 1–5 (2017)

4. Ermolov, I.L., Sukhanov, A.N., Knyaz'kov, M.M., Kryukova, A.A., Kryuchkov, B.I., Usov, V.M.: A sensory control and orientation system of an exoskeleton.In: Saint Petersburg International Conference on Integrated Navigation Systems, icons 2015—Proceedings, vol. 22, pp. 181–185 (2015)
5. Wilkie, D.R.: The mechanical properties of muscle. British Med. Bull. vol. 12 (1956)
6. Gasser, H.S., Hill, A.V.: The dynamics of muscle contraction. Proc. R. Soc. Lond. B **96**, 398–437 (1924)
7. Physiology of muscular activity, labor and sports, L., (Manual of physiology); Hill, A.: Mechanics of muscle contraction (1969)
8. Ito, K.: EMG pattern classification for a prosthetic forearm with three degrees of freedom. In: Ito, K., Tsuji, T., Kato, A., Ito, M. (eds.) IEEE International Workshop on Robot and Human Communication, pp. 69–74 (1992)
9. DiCicco, M.: Comparison of control strategies for an EMG controlled orthotic exoskeleton for the hand. In: DiCicco, M., Lucas, L., Matsuoka, Y. (eds.) Proceedings of the 2004 IEEE International Conference on Robotics and Automation, New Orleans, LA, pp. 1622–1627 (Apr 2004)
10. De Luca, C.J., De Luca, C.J.: Surface Electromyography: Detection and Recording. DelSys Incorporated, p. 10 (2002)
11. Kuribayashi, K.: A discrimination system using neural networks for EMG-control prostheses-Integral type of EMG signal processing. In: Kuribayashi, K., Shimizu, S., Okimura, K., Taniguchi, T.: Proceedings of the 1993 IEEE/RSJ International Conference on Intelligent Robots and Systems, pp. 1750–1755 (1993)
12. Abbott, V.S., Wilkie, D.R.: The relation between velocity of shortening and the tension-length curve of skeletal muscle. J. Physiol. vol. 120 (1953)
13. Haeufle, D.F.B., Grimmer, S., Seyfarth, A.: A 2010 the role of intrinsic muscle properties for stable hopping—stability is achieved by the force–velocity relation. Bioinspir. Biomim. **5**, 016004. https://doi.org/10.1088/1748-3182/5/1/016004
14. Zajac, F.E., Gordon, M.E.: Determining muscle's force and action in multi-articular movement. Exerc. Sport Sci. Rev. **17**, 187–230 (1989); Zajac, F.E.: Muscle and tendon: properties, models, scaling, and application to biomechanics and motor control. CRC Critical Rev. Biomed. Eng. **17**, 359–411 (1989)
15. Vladimir, R.: Neuromodulation: action potential modeling. Master of Science Thesis in Biomedical Engineering, p. 130 (2014)
16. Samuel, A.: Active muscle control in human body model simulations. Master's Thesis in Automotive Engineering, CHALMERS, Applied Mechanics, Master's Thesis 2013, **62**, 64 (2013)
17. Novoselov, V.S.: On mathematical models of molecular contraction of skeletal muscles. Vestnik SPbGU. Ser. 10. **3**, 88–96 (in Russian) (2016)
18. Sancho-Bru, J.L., Pérez-González, A., Mora, M.C., León, B.E., Vergara, M., Iserte, J.L., Rodríguez-Cervantes, P.J., Morales, A.: Towards a realistic and self-contained biomechanical model of the hand (2011)

Design and Control for Vacuum Contact Devices of Mobile Wall Climbing Robot Application in Complex Environment

Valery G. Gradetsky, Maxim M. Knyazkov, Eugeny A. Semenov and Artem N. Sukhanov

Abstract *Problem statement*: The structure of vacuum grippers includes a system of sensitivity of the vacuum level, the angle of inclination of the surface and the trajectory of the robot. The vacuum level in the vacuum contact devices may vary in a wide range of values and allows the efficient adjustment of the flow level and pressure in the supply system of the mobile robot. Strategy based on decision making as example of control was developed taking into account the situational changes in the complex environment in which the robot moves. *Purpose*: To test the developed control algorithms, a series of experiments with vacuum grippers in air and water environments at different depths of underwater application was carried out. Based on the results of the work, recommendations were given for the design of adaptive vacuum grippers, which can be used on robots of vertical movement during their operation in complex environments. To obtain reliable experimental data SEMS system was used. *Results*: The design and control of the vacuum contact device of the mobile wall climbing robot implemented for vertical movement and operating in complex environments are considered. Variable design of vacuum contact devices capable for adaptation to changes in the dynamic environment on surfaces arbitrarily oriented in space was proposed. *Practical significance*: Various control modes for vacuum creation devices are proposed. It depends on the quality of the attachment surface and the operating environment of the wall climbing mobile robot.

Keywords Control algorithms · Wall climbing robots · Adaptive contact devices · Situational control · Complex environments · SEMS

V. G. Gradetsky · M. M. Knyazkov · E. A. Semenov (✉) · A. N. Sukhanov (✉)
Ishlinsky Institute for Problems in Mechanics of the Russian Academy of Sciences, Prospect Vernadscogo 101-1, 119526 Moscow, Russia
e-mail: sim1165@mail.ru

A. N. Sukhanov
e-mail: sukhanov-artyom@yandex.ru

M. M. Knyazkov
e-mail: Ipm_labrobotics@mail.ru

© Springer Nature Switzerland AG 2020

A. E. Gorodetskiy and I. L. Tarasova (eds.), *Smart Electromechanical Systems*, Studies in Systems, Decision and Control 261, https://doi.org/10.1007/978-3-030-32710-1_11

1 Introduction

We consider those vacuum contact devices (VCD) that are intended for satisfy contact between wall climbing robot and surfaces placed under different angles to horizon. The control of the such types of vacuum of vacuum contact devices was under consideration in some previous paper, but no adaptation has taken into account [1]. Climbing machines for underwater applications have a lot of innovations in design, construction, materials and components for mechanical, sensory, control and tool systems [2–4]. In common case climbing underwater machine is a mobile underwater robot (UCR) intended for realization technological prescribed tasks. The robot's mechanical and control systems, include vacuum contact devices, drives, sensors with special design that permitted to work in underwater conditions up to 10 m depth [5]. Vacuum contact devices (VCD) are one of the main components required special attention.

The paper presents discussion related with prescribed tasks, information parameters needed for design and control of UCR, experimental characteristics, demands for basic schemes, recommendations for vacuum contact device design, information data for simulation and modeling. Control motion information and design are realized by using information parameters such as force, pressure, vacuum, proximity, position, angle, velocity and flow. In our study we suggested to consider as example of control—decision making on the base of fuzzy logic system.

Previous UCR had supervision control [6]. In our paper the attention was made to increase autonomy and reliability of UCR by means fuzzy control and decision making.

Former much research in robotics deals with different problems of the motion of wheeled mobile robots but not UCR and the motion control of wheeled mobile robots in unstructured environments. Fuzzy logic approaches to mobile robot navigation and obstacle avoidance have been investigated by several researches. Many application works of fuzzy logic in the mobile robot field have given promising results.

Strategy was presented in paper [7] for the autonomous navigation of field mobile robots on hazardous natural terrain using logic approach and a novel measure of terrain traversability. The navigation strategy is comprised of three simple, independent behaviors: seek-goal, traverse-terrain, and avoid obstacles. This navigation strategy requires no a priori information about the environment.

The sensor-based navigation of a mobile robot in an indoor environment is very well presented in [8]. The paper deals with the problem of the navigation of a mobile robot either in an unknown indoor environment or in a partially-known one. Fuzzy controllers are created for the navigation of the real robot. The good results obtained illustrate the robustness of a fuzzy logic approach with regard to sensor imperfections.

Intelligent mobile robot motion in unknown environments reflected in many papers including [9–25] but not applied previous to UCR.

In this paper the following tasks are solving:

- Design of UCR with VCD;
- Vacuum contact devices study by use information characteristics;

- Control algorithms;
- Decision making and strategy of UCR as example of control.

2 Design of Underwater Wall Climbing Robot with Vacuum Contact Devices

Besides general features for all type underwater vehicles such as increased pressure, fluidity environment, etc., a surface quality is important of a UCR reliable motion. As a rule, the surfaces, located under water are covered by the obstacles in the form of different kinds of cracks, sea-weeds and deposits. Therefore for motion along such surfaces it is necessary special vacuum contact devices and control with decision making possibility to have reliable contact with surface or to avoid such kind of obstacles.

Wall climbing robot has to reliable move along vertical or slope surfaces characterized not predicted quality. In such complex conditions robot has to estimate and decide itself what to do, how to move, what direction of the motion is preferable to avoid cracks or holes, and to decide about value of detaching force depends on situation.

The designed UCR consists of two platforms (Fig. 1). Here 1 is internal platform with leg group 3 and vacuum contact devices (VCD) on every leg; 2—leg group with VCD of external platform; 4—transport pneumatic drive; 5—rotating unit; 6—external platform; 7—piston-drive; 8—technological platform; 9—technological equipment; 10–11—hermetic units of control system; 12–15—platforms for sensory equipment. All VCD equipped with proximity and force, vacuum sensors, pneumatic drives have proximity sensors for final position fixations.

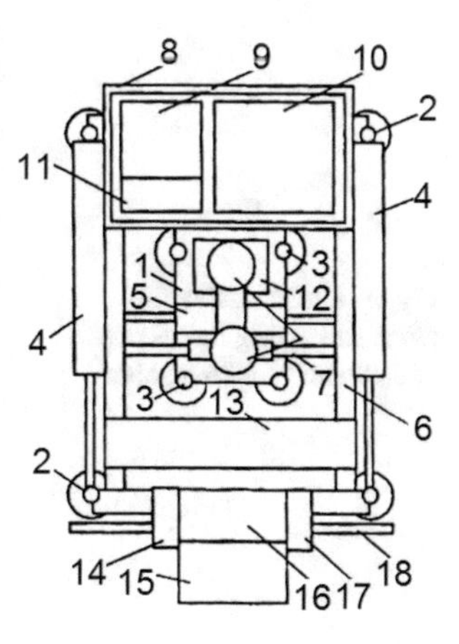

Fig. 1 Scheme of underwater climbing robot

A technological module is installed on the platform. A video camera is used for navigation and orientation of the robot. A control unit organizes automatic control of the whole system.

The URC moves as follows. When the internal group of the VCD is connected to motion surface by means of the grippers, the external one has possibility to move easily with a piston-rod relative to the platform or the platform can rotate to change the direction of motion by rotation unit. When the external group of the VCD is fixed on the surface, the internal one can move with the platform, and so on.

The view of one prototype version is shown in Fig. 2. The following forces are acting on UCR in underwater position (Fig. 3):

F_1—technological force, F_2—acting force of water column, F_3, F_N and F_T—friction, normal and tangential forces, acting on every vacuum contact device, M_c—moment acting on every vacuum contact device, M_1—overturn moment applied at center of mass of UCR.

Fig. 2 View of underwater climbing robot with VCD

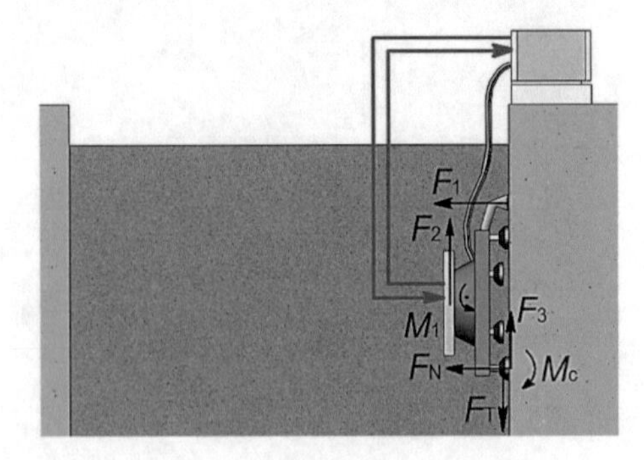

Fig. 3 Underwater robot place and acting forces on underwater climbing robot

Acting vibrations on UCR as result of technological processes may be harmonic type, $F_1 = A \sin \omega t$.

Motion equation of the pneumatic drive piston of UCR with technological equipment mass m may be written as follows:

$$\sum F_i = 0 \text{ or } \ddot{x} = \frac{1}{m}(F_0 + F_1 + F_2 + F_3 - F_4),$$

where

F_1—technological force, F_2—acting force of water level, F_3—dry friction force, $F_4 = D\ddot{x}$—viscous friction in pneumodrive, $D = 2.5 * 10^{-2}$—coefficient of viscous friction. F_0—additional payload of UCR, $P_0 = m_0 g \cdot \sin \alpha$, m_0—total UCR mass, g—acceleration due to gravity, α—angle between surface and VCD ends.

Reduction gear sets the desired value of the output drive velocity. The microprocessor connected with regulator and rack carries out a control of the PC working longitudinal motion.

This drive design provides both the continuous low speed mode to fulfill technological operations by the robot, and the discrete high speed mode. The discrete high speed mode is convenient for the fast transportation of the technological equipment by the UCR to working zone. The discrete high speed mode is switched on by a control muff disconnection. In the continuous low speed mode a control muff connects the engine block with PC. In this case, the drive output velocity depends on the engine velocity.

Prescribed prospective tasks for underwater climbing robots may be as follows:

- inspection and repair of the surfaces of the pools in nuclear power station;
- technological processes producing such as cleaning, cutting, welding, along surfaces in such pools;
- survey and cleaning of ships surfaces, during parking in ports (marine growth on the hull reduces a speed of a ship and increases a fuel consumption, it takes a long time, and the toxic and expensive agents to destroy a marine growth in dry dock);
- survey and cleaning of channels and dams walls;
- survey and repair of underwater parts of petroleum platforms;
- inspection and repair of underwater pipelines;
- inspection and underwater cutting of sink ships;
- help for divers.

For execution of such work it is necessary to fix the robot on the surface and then to perform the robot's movement along the surface. For these cases it is recommended to take into attention the experience of use the underwater climbing robots (UCR) special design. UCR are capable to move along vertical and slope surfaces, located under water with the help of vacuum grippers. Advantages of such robots are ability to move along the surfaces of various materials (metal, concrete, etc.) and reliable fastening on surface.

3 Vacuum Contact Devices Study

Vacuum contact device is intended for realize gripping function of UCR with a surface in underwater position. The standard friction gripper (FG) have a normal gripping force of over 1000 N, but they need an initial press force about 150 N to seal the possible roughness and cracks of a motion surface. There are standard sealing grippers (SG), which can seal the roughness up to 4 mm and more by an elastic edge, but they have no sufficient gripping force. The combination of these two kinds of grippers gives a possibility of having a high gripping force with a sealing effect on the rough surfaces.

This two-staged VCD is shown in Fig. 4, where 1—elastic sealing, 2—pad is realize function of force sealing and is connected with piston drive 5, 4—ejector, 6—position sensor, P_{S1}, P_{S2}—supply pressure for piston drive. At the initial moment only the SGs are actuated. They produce an attaching force enough to seal the roughness for the FGs. This is the first "sealing" stage. After the sealing response time is finished in about 1.7 s. the power ejector is switched on. As result the FGs carry out a force fixation of the UCR by means VCD on the motion surface in about 2 s response time. Time is the second "force" stage of the gripping. The used FGs with diameter 160 mm provide an attaching force up to 1100 N. The two-staged gripping takes approximately 4 s and then it is possible to begin a working stage to fulfill a stride by moving one group of the VCD relative to another one, which is fixed to a motion surface.

The gas ejector 4 (Fig. 4) of VCD is intended for produce vacuum P_2 in the chamber 3. Air and water mass flow through the every of channels 1, 2, 3 of the ejector 4 (Figs. 4, 5) can calculated as

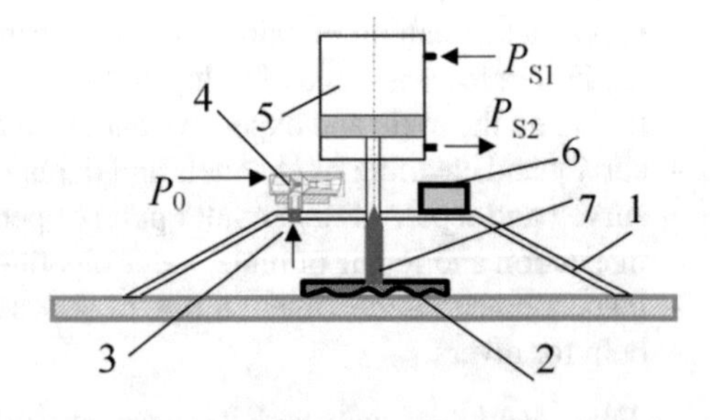

Fig. 4 Scheme of vacuum contact device with a piston drive

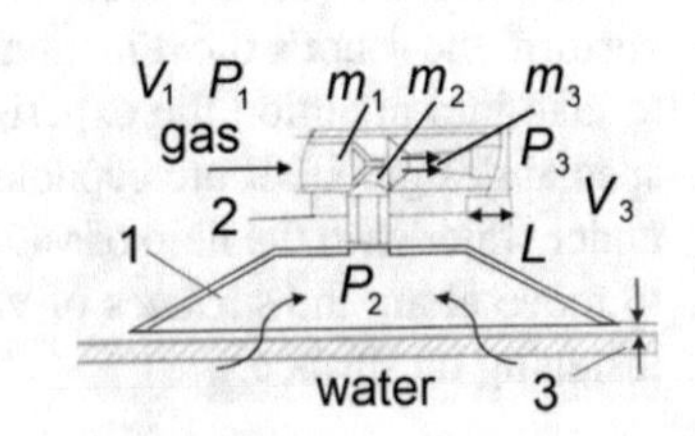

Fig. 5 Scheme of vacuum contact device with ejector

$m_i = \rho_i V_i S_i$, where ρ_i is density, V_i is velocity, S_i is space of cross section of i-flow. For gas stream $S_1 = \frac{\pi d_1^2}{4}$, for water stream $S_2 = \frac{\pi d_2^2}{4}$, for mixture flow $S_3 = \frac{\pi D^2}{4}$.

The following balance equations are valid for gas-water ejector [5, 6]: motion balance: $m_1 V_1 + P_1 S_1 + m_2 V_2 + P_2 S_2 = (m_1 + m_2) V_3 + P_3 S_3$; energy balance: $m_1 \frac{V_1^2}{2} + m_2 \frac{V_2^2}{2} = (m_1 + m_2) \frac{V_3^2}{2}$. Using those equation, it is possible to find vacuum P_2, if other parameters are known [20]:

$$\begin{cases} P_2 = \frac{1}{S_2} \left[\begin{array}{l} m_1 V_1 + P_1 S_1 + m_2 V_2 - P_3 S_3 - (m_1 + m_2)\sqrt{K} \\ -\frac{K}{2D}(\lambda P_3 (L - L_T) \Phi^2) \end{array} \right] \\ K = \frac{m_1 V_1^2 + m_2 V_2^2}{m_1 + m_2} \end{cases} \quad (1)$$

where P_1, P_2, P_3—pressure in the cross Sects. 1, 2, 3, coefficient $\lambda = \frac{0.3164}{(\mathrm{Re})^{1/4}}$, Re—Reynolds number, Φ^2—Martinelly parameter [8].

We supposed that pressure P_2 is in chamber 1 (Fig. 5), x—variable pneumatic restriction where pressure change from atmospheric P_a up to vacuum P_2, 3—constant pneumatic restriction, and $P_a - P_2 = kP$—linear approximation.

The force characteristics depends on the depth are presented in Fig. 6.

The number of the VCD for such a purpose is calculated beforehand depending on a total weight of the UCR with technological equipment. A diameter of one VCD is got from the next formula: $d = \sqrt{\frac{4F_n}{\pi \Delta \rho}}$, where

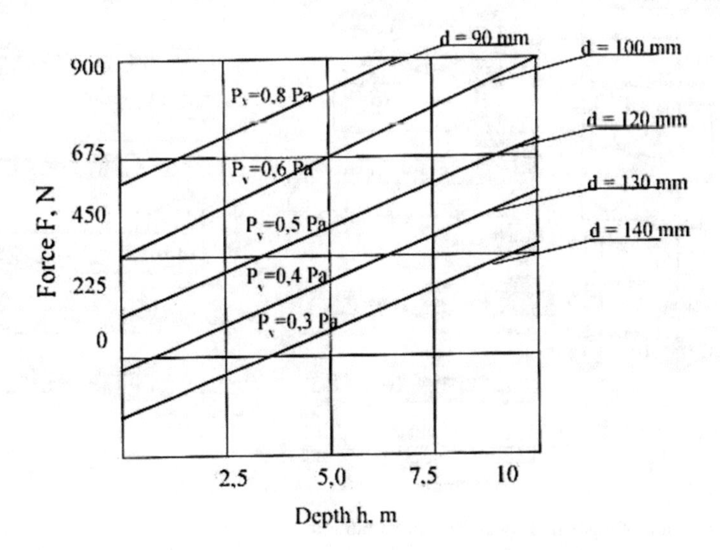

Fig. 6 Acting forces of vacuum control devices depend of water depth

F_n—normal detaching force, $\Delta\rho$—pressure difference between VCD vacuum volume and environment.

The normal detaching force is calculated as $F_n \geq \frac{F_t}{\mu} + F_e$ where

F_t—tangential detaching force,

F_e—normal detaching force from technological equipment,

μ—friction coefficient.

4 Algorithms of the Control Robot Motion

To satisfy control UCR with VCD motion the conforming algorithms were developed. Appropriate algorithms intended for satisfy as automatic control for typical kinds of robot motion.

Program software was realized on the base developed algorithms that satisfy translation, rotation transport motions and VCD motion for robot connecting to the walls.

As examples, some block-diagrams of algorithms are presented on illustrations (Fig. 7). Structure of algorithm for supervision control of robot motion includes level of man-operator and level of robot.

Where VCDC—vacuum contact device central, VCDO—vacuum contact device outside, EJ—ejection, PD—piston drive, PR—pressure reducer, JK—joystick.

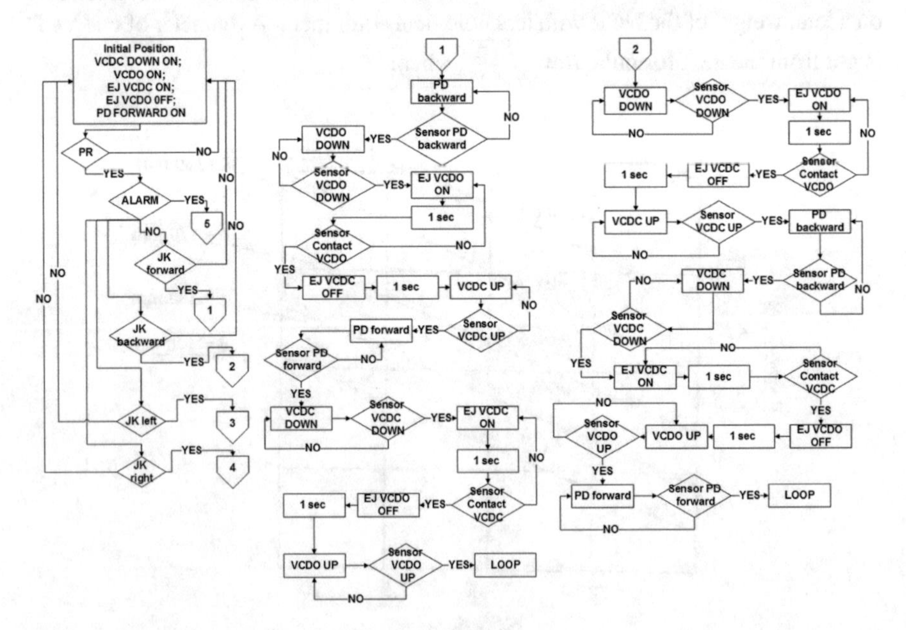

Fig. 7 Structure of algorithm for control robot motion

5 Decision Making and Strategy of Underwater Wall Climbing Robot Control

The structure includes two levels hierarchical system of the software for remote control of UCR vertical moving in a supervisory mode. The main processor is at the top level of a system and serves supervisory mode of control.

Operator of this system submits commands to the central computer, observing on distance or on a screen of monitor the conditions at place of action of a robot, according to which robot will execute determined movements action. The software of a supervisory control level provides remote data acquisition from system of valuation gauges about operating conditions and from microprocessors of mechatronic drives, also executes a reception of commands of operator and putting out the information on a screen of a monitor and an operator console through interface of user.

The general operating system of all hierarchical structure is stored in a memory of top level computer.

The fulfillment by a robot of determined actions is executed in an automatic mode after acceptance of decisions at the help of algorithms and the software of fuzzy logic system.

The logic analyses of the information, arriving from gauges installed in drives, and received on expert valuations is executed at this second level of a system.

The analysis of the fuzzy logic processor provides acceptance of decisions on the basic of mathematical model of the world, of preliminary expert valuations and of knowledge about the purpose.

Algorithms of a reception of the information concerning with operating programs realizing them and also storage of this information can be directly at a fuzzy logic level, or this task is decided with the help of algorithms and of software by common exchange with of supervisory control level.

The bottom level of a software system is intended for fulfillment of work by mechatronic drives. This function is provided with the help of built-in microprocessors and the programs developed as drivers in codes.

In this section fuzzy control is applied to the autonomous wall climbing robot reliable motion over unstructured surfaces with such types of obstacles like cracks, deposits, hills, slopes. It is supposed that robot has structure as presented in Fig. 1. Force, vacuum, proximity sensors perform the measuring applied variable forces and angles of the slope VCD end. To reliable move towards the target and avoiding obstacles, the VCD of UCR has to changes orientation or/and direction of motion depends on sign of the force.

When an obstacle in an unknown surface is very close, VCD of UCR rapidly changes orientation. The navigation goal is reliable UCR motion to the target position while avoiding the obstacles in unstructured surfaces.

The intelligent UCR behavior is formulated by means fuzzy rules. Fuzzy-logic-based control is applied to realize UCR reliable motion with simple decision making over unknown surfaces.

Inputs of the fuzzy controller are the following:

- reference force F inside VCD as a result of vacuum value,
- variable applied force ΔF inside VCD as a result of vacuum change depending of obstacle,
- angle orientation α VCD end of slope,
- derivative angle orientation $\Delta\alpha$ VCD end of slope.

Outputs of the fuzzy controller are the following:

- motion direction x of VCD end;
- motion position correction Δx of VCD end;
- direction y of VCD legs motion or
- direction y of UCR legs motion with VCD.

Reference force $F = F_2$ and output force $F = F_{out}$ inside VCD and variable applied force ΔF inside VCD are determined by means of vacuum sensors. Variable applied force is equal to $\Delta F = F_2 - F_{out}$.

Reference motion position $x = x_2$ and output position $x = x_{out}$ of UCR are determined by means of proximity gerkon sensor. Variable or derivative position orientation $\Delta x = x_2 - x_{out}$.

Direction y of VCD legs motion is realized when robot decide that are no possibility to have reliable contact VCD with surfaces and it is necessary to change position and produce next step to avoid the obstacle. Then robot can decide about the direction of motion taking into account the final motion target.

The block diagram of the fuzzy inference system is presented in Fig. 8.

The rule-base for UCR fuzzy control are the following:

R1: If F is positive and α is positive then Δx is zero; R2: If F is positive and α is negative then Δx is zero; R3: If F is negative and α is negative then x is positive (motion down); R4: If F is negative and α is positive then y is positive (new position); R5: If F is negative and $\Delta\alpha$ is positive then y is positive (new position); R6: If ΔF is negative and α is positive then y is positive (new position); R7: If ΔF is negative and α is positive then x is positive (motion down); R8: If ΔF is positive and $\Delta\alpha$ is positive then Δx is zero; R9: If F is positive and $\Delta\alpha$ is positive then Δx is zero.

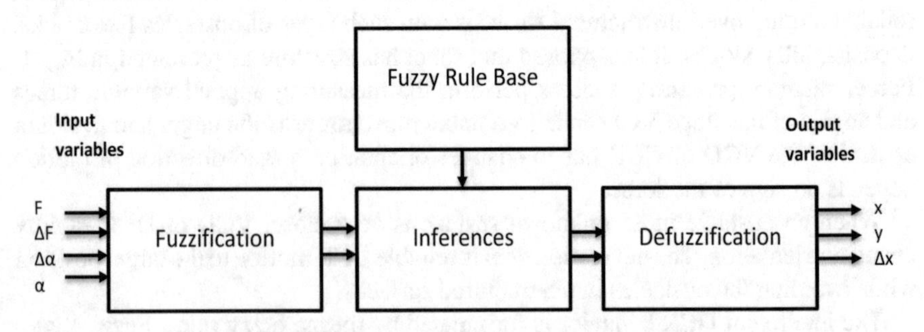

Fig. 8 The block diagram of fuzzy inferences system

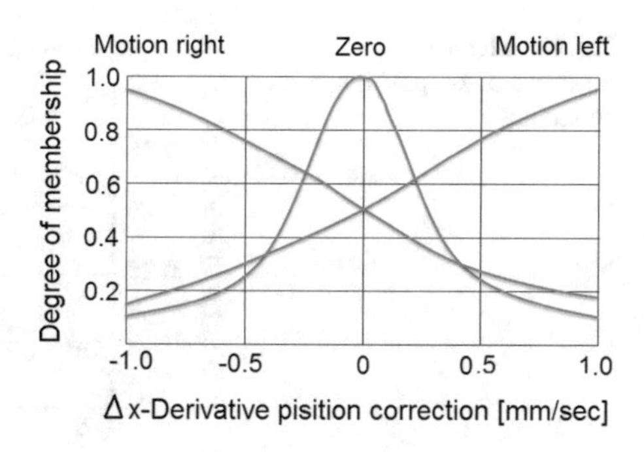

Fig. 9 Membership functions of the direction of motion

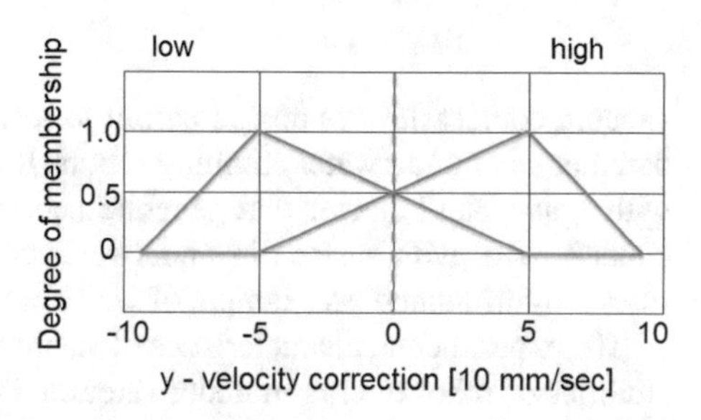

Fig. 10 Membership functions of velocity

For proposed fuzzy controller input variables for angle α of UCR (obstacle orientation) are expressed using two linguistic labels: Gaussian membership functions—sloped left and sloped right. Input variables for the reference force inside VCD are expressed two linguistic labels: Gaussian membership functions – low and high.

Input variable for the derivative force ΔF inside VCD are expressed by two linguistic labels: Gaussian membership functions—low and high. Input variables for the derivation $\Delta\alpha$ of UCR are expressed using two linguistic labels: Gaussian membership functions—left and right.

The fuzzy sets for the output variables of the VCD derivative direction of motion is presented in Fig. 9. Output variables are normalized between Δx [-1.0; $+1.0$ mm/sec].

The fuzzy sets of the output variable—new piston position y correction by velocity changing is presented in Fig. 10. Output variables are normalized between [-10; $+10$ mm/sec].

The fuzzy sets of the output variabl—VCD position x correction is presented in Fig. 11. Output variables are normalized between [-0.5; $+0.5$ mm].

6 Conclusions

Main particularities and information parameters wall climbing robot design and motion on underwater conditions are under consideration. Recommendations for

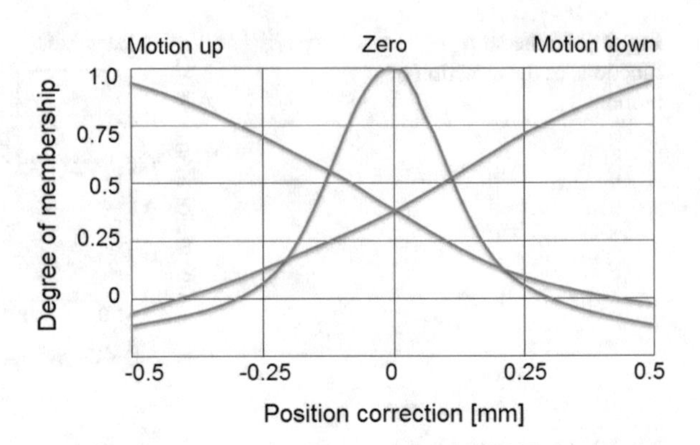

Fig. 11 Membership functions of x—position correction

vacuum contact devices design permit to select needed force and hydromechanical parameters of underwater climbing robots. It is supposed that every robots equipped with "gas-water" ejector that generate necessary vacuum for satisfy reliable force contact robot with surface in underwater conditions. Decision making and fuzzy rules are formulated, as example of wall climbing robot control.

The experimental characteristics of vacuum contact devices are illustrate the possibilities of robot motion in underwater environments. Suggested hydromechanical parameters are used to prepare realization of underwater technologies by means wall climbing robots.

Acknowledgements This study was supported by the Russian Foundation for Basic Research Grant 18-08-00357.

References

1. Sattar, T., Chen, S., Bridge, B., Shang, J.: Design of a climbing robot for inspection aircraft wings and fuselage. Int. J. **34**, 445–502 (2007)
2. Gradetsky, V.G., Rachkov, M.Y.: Wall Climbing Robots, pp. 223. The Institute for Problems in Mechanics RAS, Moscow (1997)
3. Sattar, T.P., Hilton, P., Howlader, M.F.: Deployment of laser cutting head with wall climbing robot for nuclear decommissioning. In: Proceedings of CLAWAR 2016 International Conference, Advance in Cooperative Robotics, pp. 725–732, London, UK (Sept 2016)
4. Zhao, Z., Shirkoohi, G.: Climbing robot design for NDT inspection. In: Proceedings of CLAWAR 2017 International Conference, Human—Centric Robotics, pp. 259–266, Porto, Portugal (Sept 2017)
5. Kozlov, D., Pavlov, A., Yaschuk, V.: The Mathematical Model of the Gas-Liquid Ejector with the Straight Mixing Chamber, pp. 8–11. Izvestia of Volgograd State Technical University (2010)
6. Gradetsky, V., Knyazkov, M., Sukhanov, A., Semenov, E., Chashchukhin, V., Kryukova, A.: Possibility of using wall climbing robots for underwater application. In: Proceedings of CLAWAR-2017 International Conference, Human-Centric Robotics, Porto, Portugal, pp. 239–246 (Sept 2017)
7. Mester, G.: Intelligent mobile robot motion control in unstructured environments. Acta Polytechnica Hungarica **7**(4), 153–162 (2010)

8. Mester, G.: Obstacle—slope avoidance and velocity control of wheeled mobile robots using fuzzy reasoning. In: Proceedings of the IEEE 13th International Conference on Intelligent Engineering Systems, pp. 245–249, INES 2009, Barbados, Library of Congress: 2009901330, https://doi.org/10.1109/ines.2009.4924770, ISBN 978-1-4244-4113-62009
9. Seraji, H., Howard, A., Tunstel, E.: Terrain-based navigation of planetary rovers: a fuzzy logic approach. In: Proceedings of the 6th International Symposium on Artificial Intelligence and Robotics and Automation in Space: i-SAIRAS 2001, pp. 1–6, Quebec, Canada (2001)
10. Maaref, H., Barret, C.: Sensor-based navigation of a mobile robot in an indoor environment. Robotics Autonom. Syst. **38**, 1–18 (2002)
11. Xu, W.L., Tso, S.K., Fung, Y.H.: Fuzzy reactive control of a mobile robot incorporating a real/virtual target switching strategy. Robotics Autonom. Syst. **23**, 171–186 (1998)
12. Kozlowski, K., Kowalczik, W.: Motion control for formation of mobile robots in environment with obstacles, studies in computational intelligence, towards intelligent. Eng. Inf. Technol. **243**, 203–219, Library of Congress: 2009933683, Springer, https://doi.org/10.1007/978-3-642-03737-5_15 (2009). ISBN 978-1-642-03736-8
13. Mester, G.: Improving the mobile robot control in unknown environments. In: Proceedings of the YUINFO'2007, pp. 1–5, Kopaonik, Serbia (2007)
14. Mester, G.: Obstacle avoidance of mobile robots in unknown environments. In: Proceedings of the IEEE SISY 2007, pp. 123–128, Subotica, Serbia (2007)
15. Mester, G.: Obstacle and slope avoidance of mobile robots in unknown environments. In: Proceedings of the XXV. Science in Practice, pp. 27–33, Schweinfurt, Germany (2007)
16. Szepe, T.: Sensor-based control of an autonomous wheeled mobile robot. In: Proceedings from PROSENSE 3rd Seminar Presentations, pp. 34–37, Institute Jozef Stefan, Ljubljana, Slovenia (Jan 2010)
17. Szepe, T.: Robotiranyitas tamogatasa tavoli erzekelorendszerrel, *VMTT konferencia, konferenciakiadvany*, pp. 527–532, Ujvidek, Szerbia (2010)
18. Pozna, C., Prahoveanu, V., Precup, R-E.: A new pattern of knowledge based on experimenting the causal relation. In: Proceedings of the 14th IEEE International Conference on Intelligent Engineering Systems INES 2010, pp. 61–66, Las Palmas of Gran Canaria, Spain (2010)
19. Pozna, C., Precup, R-E., Minculete, N., Antonia, C.: Properties of classes, subclasses and objects in an abstraction model. In Proceedings of the 19th International Workshop on Robotics in Alpe-Adria-Danube Region, RAAD, 23–25 June 2010, pp. 291–296, Budapest, Hungary (2010)
20. Robotics in Nuclear Facilities.: Special Issue for Exhibition of the 11th International Conference on Structural Mechanics in Reactor Technology (SmiRT 11), Tokyo, pp. 25–28 (Aug 1991)
21. Rashed, M.Al., Kimball, M., Vega, L., Vera, D., Soler, J., Correa, M., Garcia, A., Virk, G., Sattar, T.: Climbing robot for NDT applications. In: Proceedings of Clawar 2017 International Conference, Porto-Portugal, pp. 285–292 (2017)
22. Gradetsky, V., Veshnikov, V., Kalinichenko, S.: Multilink walking robot. In: Proceedings of the 7th International Conference on Advanced Robotics, ICAR'95, vol. 1, pp. 401–405, Sant Feliu de Guixols (Sept 1995)
23. Mester, G.: Intelligent mobile robot controller design. In: Proceedings of the Intelligent Engineering System, INES 2006, pp. 282–286, London, United Kingdom (2006)
24. Mester, G.: Motion control of wheeled mobile robots. In: Proceedings of the IEEE SISY 2006, pp. 119–130, Subotica, Serbia (2006)
25. Mester, G.: Obstacle avoidance and velocity control of mobile robots. In: 6th International Symposium on Intelligent Systems and Informatics, Proceedings of the IEEE SISY 2008, pp. 1–5, IEEE Catalogue (2008)

Information and Measuring Support of Situational Control

Robotized Imaging System Based on Sipm and Image Fusion for Monitoring Radiation Emergencies

A. V. Vasileva, A. S. Vasilev, A. K. Akhmerov and Victoria A. Ryzhova

Abstract *Introduction*. Emergencies caused by ionizing radiation occur less frequently than natural, transport or domestic ones. However, they are often global in nature and have long-term negative consequences. The problem of radiation emergencies is that they are often difficult to reveal quickly. Radiation monitoring means must also be designed to minimize people involvement, since radiation causes irreparable harm to them. *Purpose*: development a robotic system for imaging of ionizing and radioactive radiation. *Methods*. A system scheme is proposed which is based on sensor fusion to increase the demonstrativeness and efficiency of radiation monitoring. *Results*. A principle of ionizing radiation imaging using a coded aperture and mathematical processing is presented. Processing is implemented on the basis of periodic correlation carried out in frequency domain. A method of heterogeneous channels calibration and a technique of their images fusion performed using exponential transformation are shown. A special feature of the system is an ionizing radiation imaging channel, which is implemented on SiPM and thereby has a high sensitivity and a wide range of detected energies. The design features of the system provide high protection of its components from penetrating ionizing radiation, allowing monitoring in a dangerous radiation environment. *Practical relevance*. The main application is the automated radiation monitoring of SEMS group operating environment, which is carried out by robotized systems and allows prompt detecting of an increased level of ionizing radiation, preventing high damage from radiation accidents.

Keywords Situation control · Radiation monitoring · Robotic automation systems · Visualization · Image fusion · Ionizing radiation · SiPM

A. V. Vasileva · A. S. Vasilev · A. K. Akhmerov · V. A. Ryzhova (✉)
ITMO University, St. Petersburg, Russia
e-mail: victoria_ryz@corp.ifmo.ru

A. V. Vasileva
e-mail: avasileva@itmo.ru

A. S. Vasilev
e-mail: a_s_vasilev@itmo.ru

© Springer Nature Switzerland AG 2020

A. E. Gorodetskiy and I. L. Tarasova (eds.), *Smart Electromechanical Systems*, Studies in Systems, Decision and Control 261, https://doi.org/10.1007/978-3-030-32710-1_12

1 Introduction

Ionizing radiation emergencies occur less frequently than natural, transport, industrial or domestic ones. However, they are often global in nature and have long-term negative consequences affecting the population and the environment. During the most famous and catastrophic radiation disasters thousands of people died in a short period of time due to a lethal dose of radiation, and even more died within a few years after that from radiation induced diseases, in particular cancer. Increased doses of ionizing radiation may occur not only because of global catastrophes, but also because of local acts of radiation terrorism, as were the cases in Mexico and Iraq in 2017 and 2016 by abduction of radioactive iridium-192.

Radiation emergencies are often difficult to reveal quickly, which increases the risks of negative consequences. Monitoring means that provide information on radiation situation in a visual way are very useful and therefore are being researched and developed by many research groups around the world [1–4]. Visual monitoring systems provide an opportunity not only to see an ionizing radiation source, but also to understand exactly where it is located [5]. This is achieved by using two separate channels to form images of optical and ionizing radiation and by combining them by one of image fusion methods [6–8]. So, sensor fusion provides the highest demonstrativeness of radiation monitoring.

Performance, sensitivity and operational features of the imaging system are directly determined by a sensor used in the ionizing radiation imaging channel. There are many possible solutions, such as radiation counters [9], photomultiplier tubes [10–12], CCD and CMOS arrays [1, 13], Medipix tracking detectors [3, 15]. Compared to these detectors, a silicone photomultiplier (SiPM) seems to be more preferable due to a significantly higher sensitivity and better operational characteristics, so nowadays it is actively investigated [15, 16]. In our development, we propose to implement the ionizing radiation imaging channel based on SiPM.

The design of radiation monitoring means must also minimize the involvement of people in order to eliminate the negative radiation impact on them [17]. For this, a monitoring system should be as automated as possible and able to be remotely controlled. This leads to the main objective of this work, namely development of a robotized system for imaging of ionizing and radioactive radiation. In this paper, we propose a scheme of radiation imaging system, its design and algorithms for information obtaining and processing.

2 Structure and Principle of Radiation Monitoring System

The developed imaging system must provide information about the radiation situation in a visual form. It is supposed to be used in a harsh radiation environment, so it must operate without direct involvement of people and as autonomously as possible. At the same time, opportunities for remote monitoring and control should be provided.

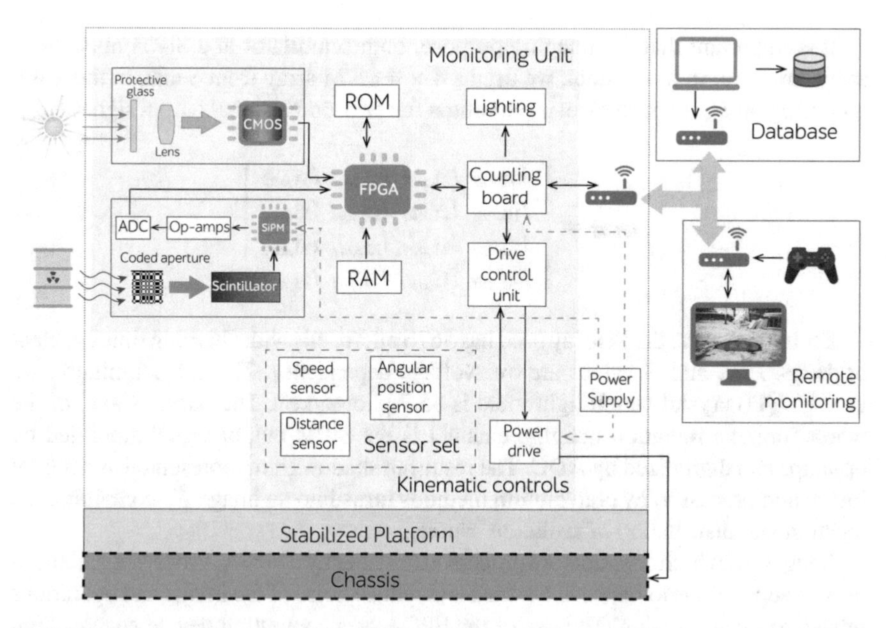

Fig. 1 Structure of the robotized imaging system for radioactive monitoring

In sum, a block diagram of the system that takes into account the requirements stated has the form shown in Fig. 1.

The main functions of the whole developed system are performed by the Monitoring Unit. It contains separate imaging channels for both visible and ionizing radiation. In the visible spectral range, radiation is focused by a lens; an optical image F_λ is formed by a CMOS array.

In the ionizing radiation imaging channel, a coded aperture is used instead of the lens to form the image. The coded aperture works as a radiation modulator and generally requires mathematical processing by convolution methods to obtain an actual image of the source [18, 19]. In our case, the coded aperture forms a shadowgram of ionizing radiation on a scintillator, which converts it into a shadowgram of optical radiation. The latter is detected by SiPM, output signal of which is a photocurrent formed as follows [20, 21]:

$$I_{out} = \int_{t_e} \left\{ \frac{G \cdot q}{10^6 \cdot \tau} \left[1 - \exp\left(\frac{-PDE \cdot E_\gamma \cdot \chi \cdot n_\gamma}{N_p} \right) \right] \right\} \quad (1)$$

where t_e is signal integration time, ms, χ is the scintillator light yield, photon/keV, E_γ is the energy of ionizing particle, keV, n_γ is the number of ionizing particles, q is the elementary charge, $q = 1.6 \times 10^{-19}$ C. There are the following SiPM parameters in the Eq. (1): τ is the microcell charging time constant, ns, N_p is the total number of microcells, PDE is photon detection efficiency, G is SiPM gain.

It is important that for imaging purpose, both scintillator and SiPM must be of array structure. In our system, we used a 4×4 SiPM array from Sensl. In this case, output signal I_{out} of each pixel is part of the full two-dimensional output signal I_{shad}:

$$I_{shad} = \begin{bmatrix} I_{00,out} & I_{01,out} & I_{02,out} & I_{03,out} \\ I_{10,out} & I_{11,out} & I_{12,out} & I_{13,out} \\ I_{20,out} & I_{21,out} & I_{22,out} & I_{23,out} \\ I_{30,out} & I_{31,out} & I_{32,out} & I_{33,out} \end{bmatrix} \quad (2)$$

Each pixel of the SiPM array used has 5676 microcells with charging time constant of 48 ns. *PDE* and *G* values are overvoltage dependent [22]. As a scintillator, we used CsI(Tl) crystal which light yield is 55.7 photon/keV. The output signal of the whole ionizing radiation imaging channel is the SiPM output signal amplified by op-amps and digitalized by ADC. The resulting shadowgram represented as a digital image and processed by convolution methods turns into an image F_γ containing the location and distribution of radiation source.

Images from both channels are processed together by a FPGA. Processing includes primary operations to improve contrast and reduce noise of the image and operations related to images fusion. The use of the FPGA as a computing device ensures high speed of processing due to its inherent parallelism. A coupling board is used to connect the Monitoring Unit with external devices, such as remote monitoring and a database server containing data of the radiation environment. For data transmission, Wi-Fi is available, but any other wired or wireless interface may be used.

For spatial orientation of the robotized system, a sensor set is provided. In a common case, a speed, a distance, and an angular position (direction) sensors are required. If necessary, this set can be extended by other sensors (wheel rotation, acceleration, etc.). The information from sensors is analyzed by a drive control unit and sent to a power drive to move the system to the target position and orientation set by people in remote monitoring.

Chassis are used to move the system. Depending on the application specifics, different types of chassis can be used, such as tracked, wheeled or walking ones [23]. When the system is moving, vibrations and shaking are possible, which can cause distortions during images fusion and errors in the sensors output data. To avoid this, the entire system including monitoring unit and kinematic controls is on a stabilized platform.

3 Image Processing by Fusion

The image displayed on the operator display contains fused information from optical and ionizing radiation channels, which significantly increases the demonstrativeness and efficiency of radiation monitoring [8, 24]. The main fusion stages are pixel intensities matching and result presentation. However, since the image F_γ has a low spatial resolution and a limited field of view due to the features of ionizing radiation visualization, the first stage of fusion should be the spatial alignment and calibration of visible and ionizing radiation channels.

3.1 Spatial Alignment and Calibration

Calibration is required to minimize spatial misalignments, which are inevitable when designing the two-channel imaging system. The calibration method described below has been specifically developed to align the visible and ionizing radiation channels; in both channels, calibration objects must be observed simultaneously.

As the calibration objects, we used white and UV LEDs for the visible and ionizing radiation channels, respectively. The radiation energy of the UV LED is sufficient to initiate the scintillation process [25]. LEDs must be located at predetermined points, at a distance $\left(B_x, B_y\right)$ from each other (Fig. 2). It is advisable the LEDs to be placed in one plane, i.e. on a single basis perpendicular to the sight axis of the channels, in order to reduce the number of computational operations at further stages.

In Fig. 2, $X_\gamma Y_\gamma$ and $X_\lambda Y_\lambda$ are coordinate systems of the ionizing radiation and visible radiation channel, respectively; *XY* is the global Euclidean coordinate system. The objective of calibration is to find the relationship between $X_\gamma Y_\gamma$ and $X_\lambda Y_\lambda$, while

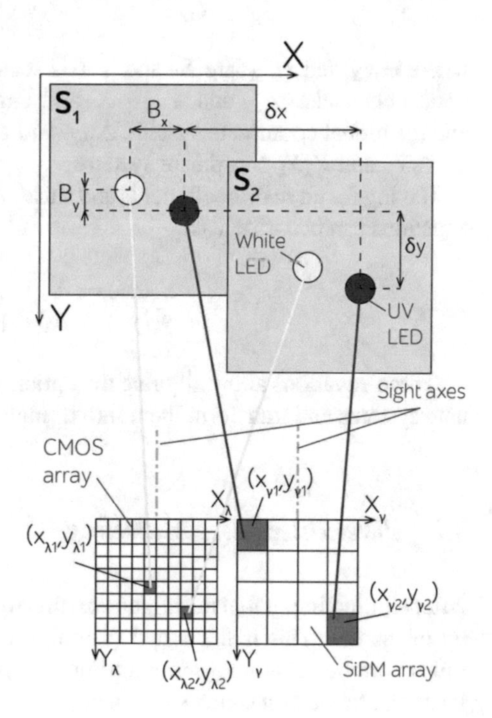

Fig. 2 Calibration geometry

$X_\lambda Y_\lambda$ is considered as the target coordinate system. Thus, the following sequence of actions must be performed.

1. Calculating the image coordinates of calibration objects placed in the first position S_1. The coordinates of UV LED are $(x_{\gamma 1}, y_{\gamma 1})$ and $(x_1 + B_x, y_1 + B_y)$ in ionizing radiation channel and global coordinates systems, respectively. The coordinates of white LED are $(x_{\lambda 1}, y_{\lambda 1})$ and (x_1, y_1) in optical channel and global coordinates systems, respectively. The coordinates are calculated using weighted summation [26].
2. Placing calibration objects in the second position S_2 and calculating the coordinates $(x_{\gamma 2}, y_{\gamma 2})$, $(x_2 + B_x, y_2 + B_y)$, $(x_{\lambda 2}, y_{\lambda 2})$, and (x_2, y_2) the same way.
3. Calculating the displacements: $(\delta x_\gamma, \delta y_\gamma) = (x_{\gamma 2} - x_{\gamma 1}, y_{\gamma 2} - y_{\gamma 1})$, $(\delta x_\lambda, \delta y_\lambda) = (x_{\lambda 2} - x_{\lambda 1}, y_{\lambda 2} - y_{\lambda 1})$, $(\delta x, \delta y) = (x_2 - x_1, y_2 - y_1)$.
4. Finding the scale coefficients and shifts by the equations

$$m_{\gamma,X} = \frac{\delta x_\gamma}{\delta x_\lambda}; m_{\gamma,Y} = \frac{\delta y_\gamma}{\delta y_\lambda}; m_{\lambda,X} = \frac{\delta x_\lambda}{\delta x}; m_{\lambda,Y} = \frac{\delta y_\lambda}{\delta y} \tag{3}$$

and

$$\begin{array}{l} \Delta_{\gamma,X} = x_\lambda - B_x m_{\lambda,X} - x_\gamma m_{\gamma,X} \\ \Delta_{\gamma,Y} = y_\lambda - B_y m_{\lambda,Y} - y_\gamma m_{\gamma,Y} \end{array}, \tag{4}$$

where $m_{\gamma,X}$ and $m_{\gamma,Y}$ are X- and Y-axis scales between the ionizing radiation and optical channels, $m_{\lambda,X}$ and $m_{\lambda,Y}$—X- and Y-axis scales between the optical channel and the global coordinate system, $\Delta_{\gamma,X}$ and $\Delta_{\gamma,Y}$ are shifts between X- and Y-axes of $X_\gamma Y_\gamma$ and $X_\lambda Y_\lambda$ coordinate systems.

Having found scale coefficients and shifts, the resulting coordinate transformation equations can be written as

$$\begin{array}{l} x_\lambda(x_\gamma) = \Delta_{\gamma,X} + x_\gamma m_{\gamma,X} \\ y_\lambda(y_\gamma) = \Delta_{\gamma,Y} + y_\gamma m_{\gamma,Y} \end{array}. \tag{5}$$

These equations allow aligning the optical and ionizing radiation channel coordinate systems and transform them into a single global coordinate system.

3.2 Pixel Intensities Matching

After eliminating spatial misalignment, the two multi-sensor images can be combined by intensities. This is the actual fusion process which must be carried out in real time to provide continuous monitoring and updating information about the level and location of radiation sources with a high frequency. Therefore, in order to reduce the time between robot actions and information displaying, it is advisable to perform the fusion in one computational flow without buffering. This imposes a limit on the computational operations number.

We implemented fusion by the exponential transformation method which consists in pixel-by-pixel raising of the visible-channel image F_λ to the normalized intensity value of the inverted gamma-channel image F_γ [25, 27]:

$$Q(x, y) = F_\lambda(x, y)^{1 - F_\gamma(x,y)/2^n}, \quad (6)$$

where n is the bit depth of the F_γ image.

Since the F_γ image is normalized, the exponent is in the range of [0; 1]. After inverting F_γ, minimum intensity values of its pixels become close to 1, which results in closeness in intensities of images F_λ and Q. In the opposite case, the intensities of the F_λ pixels undergo significant changes. In the end, a narrow range of pixel intensities of the image F_λ expands resulting in an increase in the Q image contrast.

3.3 *Presenting the Fused Image*

The image Q should be presented in a form convenient to be analyzed and interpreted by the operator. Since humans distinguish between thousands of shades of color and only about two dozen shades of gray, the most effective presentation is in pseudo-colors [28, 29]. Pseudo-color images are useful for searching and detecting objects on a complex background, since they accentuate key information. To form the pseudo-color image, a color assignment rule is used, the mathematical description of which is given below.

The first step is to find the difference between the fused image Q and the original optical-channel image F_λ:

$$D(x, y) = Q(x, y) - F_\lambda(x, y), \quad (7)$$

The pixels of the image D have values in the $0 \ldots 2^n - 1$ range, where n is the bit depth of the F_λ image. This range is to be quantized to k intervals, each of which represents a certain color:

$$X_s = \frac{2^n - 1}{k}, \quad (8)$$

where X_s is the range of image D pixel values that will correspond to one color.

For each of k colors, an array C is formed, in which 1 indicates that the value is within the specified range while shows 0 the opposite:

$$C_j(x, y) = \begin{cases} 1, & D(x, y) \in [j \cdot X_s; (j + 1) \cdot X_s] \\ 0, & D(x, y) \notin [j \cdot X_s; (j + 1) \cdot X_s] \end{cases}, \quad (9)$$

where $j = 0 \ldots k - 1$. Using the arrays obtained, (R_Q, G_Q, B_Q) values for the target pseudo-color image are calculated according to the equations:

$$\begin{aligned} R_Q(x, y) &= C_1(x, y) \cdot R'_1 + C_2(x, y) \cdot R'_2 + \cdots + C_j(x, y) \cdot R'_j \\ G_Q(x, y) &= C_1(x, y) \cdot G'_1 + C_2(x, y) \cdot G'_2 + \cdots + C_j(x, y) \cdot G'_j, \\ B_Q(x, y) &= C_1(x, y) \cdot B'_1 + C_2(x, y) \cdot B'_2 + \cdots + C_j(x, y) \cdot B'_j \end{aligned} \quad (10)$$

where R'_j, G'_j, B'_j are red, green and blue values of each j-th pseudo-color in the RGB color system.

Elements (R_Q, G_Q, B_Q) are used to be placed on the corresponding positions of the fused image. For the red component, this condition looks like

$$Q_R(x, y) = \begin{cases} Q(x, y), R_Q(x, y) = 0 \\ R_Q(x, y), R_Q(x, y) \neq 0 \end{cases}. \quad (11)$$

A similar condition applies to Q_G and Q_B with values G_Q and B_Q, respectively. Finally, each element of the fused image Q in pseudo-color representation becomes a three-dimensional vector Q_{psd}:

$$Q_{psd}(x, y, 3) = \begin{bmatrix} Q_R(x, y) \\ Q_G(x, y) \\ Q_B(x, y) \end{bmatrix}. \quad (12)$$

An example of the fused image in pseudo-color representation is shown in Fig. 3. Figure 3 shows that pseudo-color representation significantly simplifies the radiation monitoring process for the operator and thereby reduces the probability of type I and type II errors.

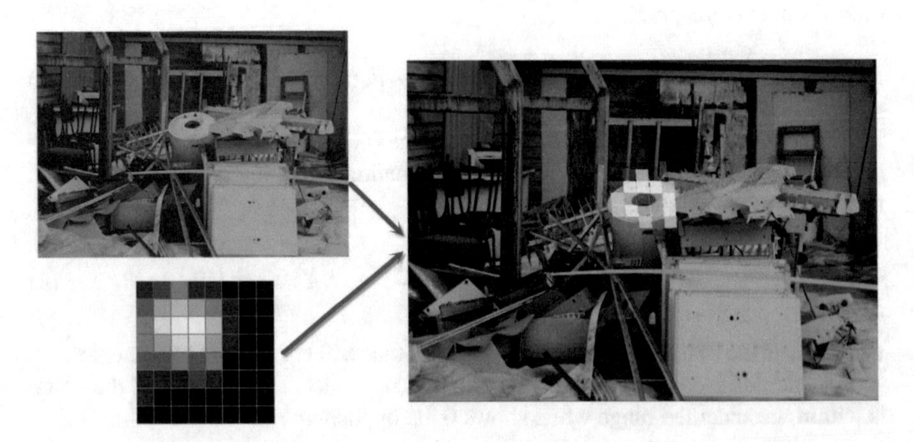

Fig. 3 Fused image represented in pseudo-colors

4 Design Features for Functional Components Protection from Radiation Effects

The intended use of the developed imaging system in harsh radiation conditions makes special demands on its design. To ensure uninterrupted operation, the main functional units must be protected from penetrating ionizing radiation. This primarily refers to digital electronic (FPGA, DSP, ADC, memory, etc.) and optical components. For electronics, exposure to radiation can lead to temporary failures or to irreversible changes in parameters of transistors that are part of integrated circuits [30, 31]. For optics, radiation causes staining of glass and leads to darkening of optical surfaces and decrease in transmittance [32]. To avoid these effects, three approaches can be applied:

- technological methods of radiation attenuation (special radiation-resistant element base);
- constructive methods of radiation attenuation (lead screens, special optical materials, etc.);
- redundancy of main system components.

Radiation-resistant elements are significantly more expensive than similar elements of a wide purpose. The same is true for redundancy of main elements. Moreover, redundancy increases the probability of fail-safe functioning, but does not fully protect against failures. Considering these factors, we applied constructive methods to reduce the influence of radiation on electronic components (Fig. 4).

To protect the electronic components including CMOS array and processing unit, we developed a special double enclosure. Its inner part is made of lead, while the outer part is aluminum (Fig. 4a). The lead enclosure made of 5 mm thick sheet material attenuates gamma radiation of 1.5 MeV almost two times and soft gamma radiation (tens of keV) up to 50 times. The lead enclosure does not protect against beta radiation, because Bremsstrahlung may occur [33]. Therefore, the outer 4 mm thick aluminum enclosure is provided. This design provides effective protection, but

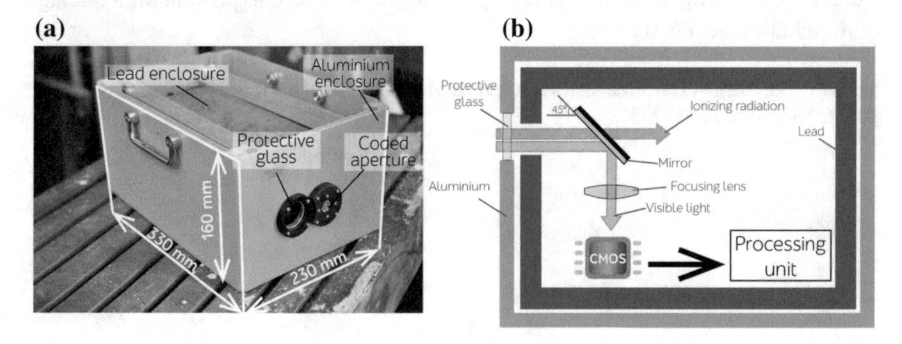

Fig. 4 Protecting double enclosure (**a**) and its internal layout (**b**)

has a large weight and size (Fig. 4b). The total mass of the double enclosure shown in Figure is 16 kg.

Direct impact of ionizing radiation on the focusing lens and CMOS sensor is eliminated by light reflection. Visible light (green arrow in Fig. 4a) hits the optical components after being deviated, while ionizing radiation (purple arrow in Fig. 4a) passes in its original direction. The optical axis is deviated by an inclined flat mirror mounted at 45° relative to the light propagation direction. In addition, ionizing radiation entering through the input window of the optical channel is attenuated by protective heavy flint glass. In our design, the 10 mm thick glass provides attenuation equivalent to a lead plate 2.5 mm thick.

Thus, the cumulative design features for protection of electronic and optical components allow maintaining their characteristics for a long period of time and thereby ensure the failure-free operation of the developed imaging system.

5 Conclusion

The paper describes the robotized system for imaging of ionizing and radioactive radiation. The system scheme is based on sensor fusion. The ionizing radiation channel uses a scintillator coupled with a silicon photomultiplier, which ensures high sensitivity and efficiency of radiation detection. To obtain an ionizing radiation image, a coded aperture is used. To increase demonstrativeness and efficiency of radiation monitoring, information from optical and ionizing radiation channels is fused. Fusing method developed consist in spatial alignment of images, their pixels matching and presenting the result in pseudo-color palette. The paper also shows the design features of the developed system which ensures protection of the main functional units from penetrating ionizing radiation and thereby from their failures.

The developed system operates autonomously and is controlled remotely. This eliminates the negative radiation impact on people and allows performing monitoring in a harsh radiation environment. Therefore, the system can be used for automated radiation monitoring of SEMS group operating environment to prevent high damage from radiation accidents.

Acknowledgements The research was carried out at the expense of the Russian Science Foundation grant (project No. 18-79-00048).

References

1. Mirion Technologies, Inc.: IPIX Ultra Portable Gamma-Ray Imaging System. http://www.gammadata.se/assets/Uploads/iPIX-SS-C47705.pdf. Accessed 26 Apr 2018
2. Okada, K., Tadokoro, T., Ueno, Y., Nukaga, J., Ishitsu, T., Takahashi, Nagashima, K.: Development of a gamma camera to image radiation fields. Prog. Nucl. Sci. Tech, **4**, 14–17 (2014). http://doi.org/10.3131/jvsj2.57.51
3. Wahl, C.G., Kaye, W.R., Wang, W., Zhang, F., Jaworski, J.M., King, A., He, Z.: The polaris-H imaging spectrometer. Nucl. Instrum. Methods Phys. Res., Sect. A **784**, 377–381 (2015). https://doi.org/10.1016/j.nima.2014.12.110
4. Wang, Y., Shuai, L., Li, D., Hu, T., Zhang, Z., Wei, C., Wei, L.: Development of a portable gamma imager based on SiPM and coded aperture technology. In: 2015 IEEE Nuclear Science Symposium and Medical Imaging Conference (NSS/MIC), pp. 1–3. San Diego, October 2015. IEEE. http://dx.doi.org/10.1109/NSSMIC.2015.7581764
5. Vlasenko, A.I., Lapin, O.E., Pervishko, A.F., Demchenkov, V.P., Arkadyev, V. B., Lupal, S.D.: Portable combined system of fusing video and gamma images of radiation sources. Issues of defense technology. Tech. Means Terrorism Countering. **16**(7–8), 90–94 (2012). (in Russian)
6. Ghassemian, H.: A review of remote sensing image fusion methods. Inf. Fusion **32**, 75–89 (2016). https://doi.org/10.1016/j.inffus.2016.03.003
7. Liu, Y., Liu, S., Wang, Z.: A general framework for image fusion based on multi-scale transform and sparse representation. Inf. Fusion **24**, 147–164 (2015). https://doi.org/10.1016/j.inffus.2014.09.004
8. Vasilev, A.S., Korotaev, V.V.: Research of the fusion methods of the multispectral optoelectronic systems images. In: Automated Visual Inspection and Machine Vision, vol. 9530, p. 953007. Munich, June 2015. International Society for Optics and Photonics. https://doi.org/10.1117/12.2184554
9. Vlasenko, A.N.: Portable gamma source imaging system. Instrum. Radiat. Meas. News IRMN **1**(84), 34–44 (2016). (in Russian)
10. Santo, J.T. et al.: Application of remote gamma imaging surveys at the Turkey point PWR reactor facility. In: Proceedings of the Institute of Nuclear Materials Management (INMM). pp. 454–487 (2006)
11. Israelashvili, I., Coimbra, A.E.C., Vartsky, D., Arazi, L., Shchemelinin, S., Caspi, E.N., Breskin, A.: Fast-neutron and gamma-ray imaging with a capillary liquid xenon converter coupled to a gaseous photomultiplier. J. Instrum. **12**(09), P09029 (2017). https://doi.org/10.1088/1748-0221/12/09/P09029
12. Zeraatkar, N., Sajedi, S., Fard, B.T., Kaviani, S., Akbarzadeh, A., Farahani, M.H., Ay, M.R.: Development and calibration of a new gamma camera detector using large square photomultiplier tubes. J. Instrum. **12**(09), P09008 (2017). https://doi.org/10.1088/1748-0221/12/09/P09008
13. Okada, K., Tadokoro, T., Ueno, Y., Nukaga, J., Ishitsu, T., Takahashi, I., Nagashima, K.: Development of a gamma camera to image radiation fields. Prog. Nucl. Sci. Tech. **4**, 14–17 (2014). http://dx.doi.org/10.15669/pnst.4.14
14. Martynyuk, Y.N., Vishnevsky, I.B.: Industrial prototype of a portable gamma camera for use at nuclear facilities. Instrumentation and Radiation Measurement News IRMN. **4**, 13–23 (2016). (in Russian) http://www.doza.ru/docs/pub/2016/13-23-4-2016.pdf
15. Cozzi, G., Busca, P., Carminati, M., Fiorini, C., Gola, A., Piemonte, C., Regazzoni, V.: Development of a SiPM-based detection module for prompt gamma imaging in proton therapy. In: 2016 IEEE Nuclear Science Symposium, Medical Imaging Conference and Room-Temperature Semiconductor Detector Workshop (NSS/MIC/RTSD), pp. 1–5. Strasbourg, October 2016. IEEE. https://doi.org/10.1109/NSSMIC.2016.8069393
16. Xie, Y., Bentefour, E., Janssens, G., Smeets, J., Dolney, D., Yin, L., Prieels, D.: MO-FG-CAMPUS-JeP1-02: proton range verification of scanned pencil beams using prompt gamma imaging. Med. Phys. **43**(6Part31), 3717–3717. (2016). https://doi.org/10.1118/1.4957339

17. Arkad'ev, V.B., Vlasenko, A.N., Golubeva, O.A., Lapin, O.E., Pervishko, A.F.: Radiation monitoring devices for equipping mobile robotic systems. Issues of defense technology. Tech. Means Terrorism Countering. **5–6**, 27–33 (2011). (in Russian)
18. Caroli, E., Stephen, J.B., Di Cocco, G., Natalucci, L., Spizzichino, A.: Coded aperture imaging in X-and gamma-ray astronomy. Space Sci. Rev. **45**(3–4), 349–403 (1987). https://doi.org/10.1007/BF00171998
19. Gottesman, S.R., Fenimore, E.E.: New family of binary arrays for coded aperture imaging. Appl. Opt. **28**(20), 4344–4352 (1989)
20. Rosado, J.: Performance of SiPMs in the nonlinear region. Nucl. Instrum. Methods Phys. Res. Sect. A **912**, 39–42 (2017)
21. Sensl: An Introduction to the Silicon Photomultiplier System. https://www.sensl.com/downloads/ds/TN%20-%20Intro%20to%20SPM%20Tech.pdf. Accessed 14 Jul 2018
22. Sensl: High PDE and Timing Resolution SiPM Sensors in a TSV Package. http://sensl.com/downloads/ds/DS-MicroJseries.pdf. Accessed 14 Jul 2018
23. Vasilev, A.V.: Development and classification principles of ground mobile robot's and planet rover's chassis. St. Petersburg State Polytech. Univ. J. Comput. Sci. Telecommun. Control Syst. **1**(164) (2013)
24. Pohl, C., Van Genderen, J.L.: Review article multisensor image fusion in remote sensing: concepts, methods and applications. Int. J. Remote Sens. **19**(5), 823–854 (1998)
25. Vasileva, A.V., Vasilev, A.S.: Research and development of a high-energy radiation imaging system based on SiPM and coding aperture. In: Electro-Optical Remote Sensing XII, vol. 10796, p. 107960 N. International Society for Optics and Photonics (2018)
26. Jiang, J., Xiong, K., Yu, W., Yan, J., Zhang, G.: Star centroiding error compensation for intensified star sensors. Opt. Express **24**(26), 29830–29842 (2016)
27. Liu, Z., Laganiere, R.: Context enhancement through infrared vision: a modified fusion scheme. SIViP **1**(4), 293–301 (2007)
28. Gonzalez, R., Woods, R.: Digital image processing, 3rd edn. Prentice Hall, New York (2008)
29. Alpar, O., Krejcar, O.: Quantization and equalization of pseudocolor images in hand thermography. In: International Conference on Bioinformatics and Biomedical Engineering, pp. 397–407. Springer, Cham (2017)
30. Ostler, P.S., Caffrey, M.P., Gibelyou, D.S., Graham, P.S., Morgan, K.S., Pratt, B.H. et al.: SRAM FPGA reliability analysis for harsh radiation environments. IEEE Trans. Nucl. Sci. **56**(6), 3519–3526 (2009)
31. Quinn, H., Robinson, W.H., Rech, P., Aguirre, M., Barnard, A., Desogus, M. et al.: Using benchmarks for radiation testing of microprocessors and FPGAs. IEEE Trans. Nucl. Sci. **62**(6), 2547–2554 (2015)
32. Gusarov, A.I., Doyle, D.B.: Modeling of gamma-radiation impact on transmission characteristics of optical glasses. In: Photonics for Space and radiation Environments II, vol. 4547, pp. 78–86. International Society for Optics and Photonics (2002)
33. Van Pelt, W.R., Drzyzga, M.: Beta radiation shielding with lead and plastic: effect on bremsstrahlung radiation when switching the shielding order. Health Phys. **92**(2), S13–S17 (2007)

Solid-State Optical Radiation Matrix Receivers in Robots' Vision Systems

Anastasiya Y. Lobanova, Victoria A. Ryzhova, Valery V. Korotaev and Daria A. Drozdova

Abstract *Problem statement*: Video sensors based on matrix optical receivers are often being used as a part of binocular and multi-angle portable robot's vision systems. Their main aim is to provide robot a possibility to orient in any environment. Due to complex architecture of matrix receiver's surface, instrumental errors may occur, which can have a significant impact the measuring result in robot's orientation system. Thus, systematic research of matrix photodetector's parameters, which affect the value of electric signal, is significant for the measuring video information schemes of portable robots' video sensors accuracy analysis. *Purpose of research*: development of an algorithm for measuring the variation of optical parameters of the photodetector surface. *Results*: basing on ellipsometry method, algorithm for passing optical rays through the multilayer matrix structure and software for its implementation were developed, the variation of parameters of the solid-state matrix receiver's surface were calculated. *Practical significance:* the ellipsometry method can be applied to control the quality of matrix optical receivers. The program, developed during this research, gives a possibility to automate the calculation of the optical parameters of video sensors matrices.

Keywords Ellipsometry · Portable robot · Optical matrix receivers · Inverse ellipsometry problem · Polarization · Surface control

A. Y. Lobanova · V. A. Ryzhova (✉) · V. V. Korotaev · D. A. Drozdova (✉)
ITMO University, Saint Petersburg, Russia
e-mail: victoria_ryz@corp.ifmo.ru

D. A. Drozdova
e-mail: dar-drozdova@yandex.ru

A. Y. Lobanova
e-mail: alobanova2144@gmail.com

V. V. Korotaev
e-mail: korotaev@grv.ifmo.ru

© Springer Nature Switzerland AG 2020
A. E. Gorodetskiy and I. L. Tarasova (eds.), *Smart Electromechanical Systems*, Studies in Systems, Decision and Control 261, https://doi.org/10.1007/978-3-030-32710-1_13

1 Introduction

Video sensors are an essential part of robots' sensor system, which provides initial information to the executive system of the mobile robot. Video sensors perform orientation of the robot in the surrounding space and form the initial information for its executive and manipulation functions. Binocular and multi-angle vision systems of a mobile robot based on video sensors are used in space, aviation, ground, surface and underwater devices, that is, where automatic image analysis is required to form solutions and generate control signals [1–4]. General technical requirements for the robots determine basic requirements for parameters and characteristics of their video sensors: range, accuracy, speed. The most widely used these days are video sensors based on solid-state photovoltaic receivers: CCD and CMOS lines and matrices [5–8].

Vision systems based on solid-state multi-element photodetectors have high sensitivity, high spatial resolution and serve as indispensable sources of information when creating mobile systems with supervisory control [9–11]. However, complex structure of the matrix surface with spatially distributed sensing elements (pixels) leads to instrumental errors, which have a significant impact on the measurement result in the robot orientation system.

Pixels can be characterized by individual optical, geometric and electrical characteristics, their own noise levels [12–15]. The heterogeneity of the characteristics of individual pixels of the matrix leads to the variation of its absolute sensitivity across the field, creates a deterministic spatial structure of the surface, which affects the ideal image and misrepresents it. This type of distortion is called spatial or geometric noise, which is typical for all multi-element photodetectors, regardless of the manufacturing technology and architecture of devices, and is absolutely absent in systems based on single-element receivers [16–19].

Distortions created by spatial noise are an important source of distance measurement errors, as well as errors in detecting objects when moving and orientation in mobile robot's space. Therefore, the measurement of optical parameters' spreading on the surface structure of the matrix, affecting the distribution of its sensitivity and the value of the electrical signal, considered to be appropriate for its further accounting and compensation in the development of video cameras for mobile robot's vision systems.

2 Principles of Signal and Noise Generation in Optical Sensors Based on Matrix Structures

The image read from the camera contains geometric noise due to limitations associated with physical processes occurring in multi-element photodetectors. Since the technological possibilities of reducing intrinsic noise and spread of pixel parameters of multi-element receivers are limited [18, 20, 21]. Therefore, detection conditions can be improved by enlarging the signal/noise ratio by increasing the size of the matrix photocells and the reading circuit's operating voltages [22–24], as well as the implementation of algorithms for automatic correction of inhomogeneity of the images read by software [25–27].

When implementing high-precision devices and systems, additive and multiplicative components of geometric noise are considered, which, being determined by noise, are amenable to algorithmic correction at the stage of digital signal processing [25–27]. The additive noise is being caused by the unevenness of the average density of dark accumulation currents in the crystal and appears itself as a standard deviation of the number of "thermal" charges in various storage cells of the matrix from the average value of the crystal. The multiplicative component of the spatial noise is due to the uneven sensitivity within the active area of the matrix, and appears itself as the dispersion of the fluctuations in the number of charges formed in pixels under the action of background radiation, and the number of signal charges accumulated in the cells of the matrix.

These days the simplest and the most effective is the hardware [7, 24, 28], and software linear correction of unevenness of sensitivity on the basis of signal conversion from each receiving element according to the linear law with individually selected coefficients loaded into the processor memory after starting the calibration program [27, 29].

Normally, it is not possible to fully compensate for the influence of spatial noise by this method. There can be several reasons for that: irradiation's of the photosensitive area of the receiver heterogenety due to the imperfection of the source of the background radiation, change of the values of the receiver's parameters for the time elapsed since the last calibration and the main reason, the deviation of actual receivers from the linear model [28, 30, 31].

If at the design stage of the video sensor, the distribution of technological parameters of optical systems as a part of individual pixels of a specific matrix is known, it will be possible to obtain a sensitivity distribution over the sensor's working area, simplify the algorithm for correcting geometric distortions, improve the video sensor calibration scheme, and reduce the load on the digital processor.

Consider the relations of the pixel output signal with its optical properties. The value of the signal charge taken from a single pixel can be described by the following expression [12, 22]:

$$q = i A_{el} T_c / e, \tag{1}$$

where A_{el}—element's area, T_{c}—charge accumulation time, e—electron's charge; i—photocurrent at the output of matrix's sensor element, which can be determined by the following expression:

$$i = \frac{e\eta\beta\lambda}{hc}\Phi'' = \frac{e\eta\beta\lambda}{hc}T(\varepsilon, n)\Phi \tag{2}$$

where η—photodetector's quantum efficiency; β—the coefficient of separation of pairs of charge carriers; h—Planck's constant; c—the speed of light in a vacuum; λ—wavelength of radiation; ε—angle of incidence at the interface between the media; n—refractive index; T—the energy transmittance coefficient of the boundaries of the media, which describes the ratio between the energy of the incident Φ and the passing Φ″ waves.

If the radiation at the input of the optical receiver is polarized, the polarization sensitivity of the isotropic photodetector occurs for inclined beams. In this case, the media section boundary transmittance is described by the following expression [32–34]:

$$T = \frac{n_2 \cos s'}{n_1 \cos s}\left(t_s \cos^2 \alpha_{in} + t_p \sin^2 \alpha_{in}\right) = T_s \cos^2 \alpha_{in} + T_p \sin^2 \alpha_{in}, \quad (3)$$

where ε''—refraction angle, n_1 and n_2 refractive indices of optical media, T_s and T_p are energy transmittance coefficients for radiation that is polarized perpendicularly and parallel to the plane radiation's incidence, α_{in}—incident radiation's azimuth of polarization, t_s and t_p—amplitude transmission coefficients for mutually orthogonal radiation components.

The main parameter of the optical receiver is the effective polarization sensitivity ΔS, which is determined as the difference between the p—and s—components [32]:

$$\Delta S(\lambda, \varepsilon) = S_p(\lambda, \varepsilon) - S_s(\lambda, \varepsilon) = \frac{\eta e \beta \lambda}{hc}\left[T_p(\varepsilon, n) - T_s(\varepsilon, n)\right]. \quad (4)$$

This formula relates the sensitivity of the photodetector to the transmittance or refractive indices of the optical media forming it.

Thus, in order to research the distribution of sensitivity on the sensor site, it is necessary to know the distribution of refractive indices of the media through which the radiation passes.

3 Materials and Methods of Research

The variation of refractive indices on the surface of an isotropic matrix photodetector was studied by measurements of polarization-optical parameters at various points of the matrix performed on an ellipsometer. The generalized ellipsometry scheme is based on the functional relationship between the selective values of refractive indices of a multilayer system and ellipsometric angles that characterize its reflective properties [35, 36].

The quantitative measure of the change in the state of polarization of the radiation at the sample output is ellipsometric angles that characterize the relative reflectance of the system and are determined as [35]:

$$r_s/r_p = \mathrm{tg}\Psi\ \exp(i\Delta), \tag{5}$$

where r_P, r_S—amplitude reflection coefficients for the complex components of the amplitude of the reflected wave, polarized in the plane of its incidence on the reflecting boundary of the media (*p*), and perpendicular to the plane of incidence (*s*). These values are functions of the optical constants of the reflecting system, the thickness of plane-parallel layers on the reflecting surfaces, as well as the angle of incidence and wavelength of radiation.

The method used is based on the zero method, which consists in establishing a connection between the ellipsometric angles and the orientation of the elements of the optical scheme of the device, which corresponds to the minimum intensity (attenuation) of radiation at its output.

The theoretical basis of the experimental method are the relations that relate ellipsometric angles ψ_0, Δ_0 and ψ_1, Δ_1,characterizing the system to the angles, and that describe the state of polarization of the incident and reflected waves, respectively. These relations have the following form [35–37]:

$$\begin{gathered}\Delta = \Delta_1 - \Delta_0 \Rightarrow \Delta_0 = \Delta_1 - \Delta, \\ \mathrm{tg}\Psi = \mathrm{tg}\Psi_0/\mathrm{tg}\Psi_1 \Rightarrow \mathrm{tg}\Psi_0 = \mathrm{tg}\Psi \cdot \mathrm{tg}\Psi_1.\end{gathered} \tag{6}$$

In order to determine the generalized refractive indices of the structural elements of the matrix, it is necessary to solve the inverse problem of ellipsometry, for which the ellipsometric angles Ψ and Δ, measured at different angles of incidence of radiation on the sample are taken as initial data. Figure 1 shows the main stages of solving the inverse ellipsometry problem.

The algorithm used for solving the inverse problem for the model of a multilayer reflecting system consistently produces an iterative procedure for finding the refractive indices of non-absorbing isotropic layers at a given search range and calculating their thicknesses based on the measurement information [34, 35]. The first stage of

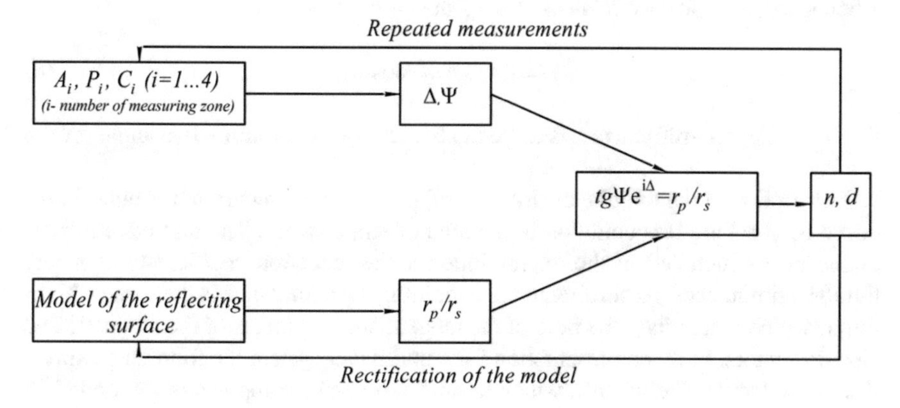

Fig. 1 Main stages of solving the inverse ellipsometry problem

the presented algorithm is the measurement of ellipsometric angles Ψ and Δ, using an ellipsometer. At the second stage, the formation of a mathematical model of the reflecting surface is carried out.

To solve the inverse problem of ellipsometry, numerical nonlinear methods are used. They are based on the repeated solution of the direct problem with the selected according to some rules specifying the system parameters and the comparison of the calculated values Ψ and Δ, with the experimental ones. This procedure continues until the difference between the calculated and experimental values of Ψ and Δ is less than a given value. In order to clarify the model, repeated measurements are carried out under different conditions that do not affect the studied values. The final stage of the calculation is the determination of surface parameters on the measured values Ψ and Δ, at one or several angles of incidence of the radiation.

3.1 Mathematical Model

Various methods of forming a model of the reflecting structure are known, which differ in the way of mathematical description of the interaction of polarized radiation with matter. Traditionally formulas of Scandone-Ballerini for generalized Fresnel coefficients are used [34, 35, 38, 39].

For multilayer coatings containing j layers, recurrence relations are used. Each next layer is expressed through the previous one. Generalized Fresnel coefficients $R_p(j, 0)$ and $R_s(j, 0)$ acquire a general expression [32, 35]:

$$R_{j,0} = \frac{r_{j+1,j} + R_{j-1,0}e^{-2i\delta_j}}{1 + r_{j+1,j}R_{j-1,0}e^{-2i\delta_j}}, \tag{7}$$

where $r_{j+1,j}$—Fresnel coefficients for the boundary «j+1st media and j media» , $R_{j-1,0}$—generalized Fresnel coefficients describing the whole system below the $j-1$ boundary; δ_j—phase thickness of the j-th layer:

$$\delta_j = 2\pi/\lambda n_j d_j \cos \varepsilon_j, \tag{8}$$

where n_j, d_j, ε_j—refractive index, geometric thickness and refractive angle for the j-th layer.

Description of the multilayer structure of an element by recurrent formulas introduced by Abel are less common in the area of ellipsometry. The refractive indices of media are included in the expressions for the reflection coefficients of media through admittances. According to the definition, the admittance is the surface high-frequency conductivity—the ratio of the tangential components of the magnetic and electric vectors. In this representation for a multilayer system the following expression for reflection coefficients, which is valid for p-and s-components are used [35]:

$$R_{j,0} = \mp \frac{u_{j+1} - Y_j}{u_{j+1} + Y_j}, \quad (9)$$

Here the sign "–" is taken for the p-component, the sign "+"—for the s-component; Y_j—generalized admittances are calculated by formulae:

$$Y_j = u_j \frac{Y_{j-1} + u_j itg\delta_j}{u_j + Y_j itg\delta_j}. \quad (10)$$

Values u_{jp} and u_{js} are surface admittances for the *j*-th media. They can be expressed as:

$$u_{jp} = \frac{n_j}{\cos \varepsilon_j}; u_{js} = n_j \cos \varepsilon_j, \quad (11)$$

where n_j—refractive index of the *j*-th media; ε_j—is the refractive angle in this media.

Despite the apparent difference in the formulae of Scandone-Ballerini and Abele, both methods give the same numerical results as a consequence of the application of Maxwell's equations to consider reflection of light [35, 39, 40]. The advantage of this Abele's method is that the reflection coefficients depend on the characteristics of the layer itself, while in Scandone and Ballerini's method they also depend on the characteristics of the previous layer [36]. That is, generalized admittances are an absolute characteristic of the reflecting system, regardless of the environment in which it is located.

Therefore, for to solve the inverse problem of ellipsometry for the calculation of the generalized refractive index of the medium, a model of the pixel optical system based on Abele admittances was chosen, a solution algorithm and a calculation program in C++ were formed. For non-absorbing layers of matrices on a silicon substrate, the program adopted the following optical constants for a working wavelength of laser radiation of 0.623 microns: $n_{Si} = 3.865$ and $\kappa_{Si} = 0.023$.

3.2 Experiment

Experimental measurements were carried out on two samples of CMOS–matrices manufactured by OmniVision Tech., which are used as part of the video sensors: color sensor, matrix OV5620 (Color CMOS QSXGA (5.17 MPixel)); monochrome sensor, matrix OV9121 (B&W CMOS SXGA (1.3 MPixel)). Since a pixel is conceived as it is shown on Fig. 2. The first layer is microlens, the second is a light filter (the monochrome sensor does not have these layers), the third is a transparent electrode made of polycrystalline silicon or an alloy of indium and tin oxide, the fourth is silicon oxide.

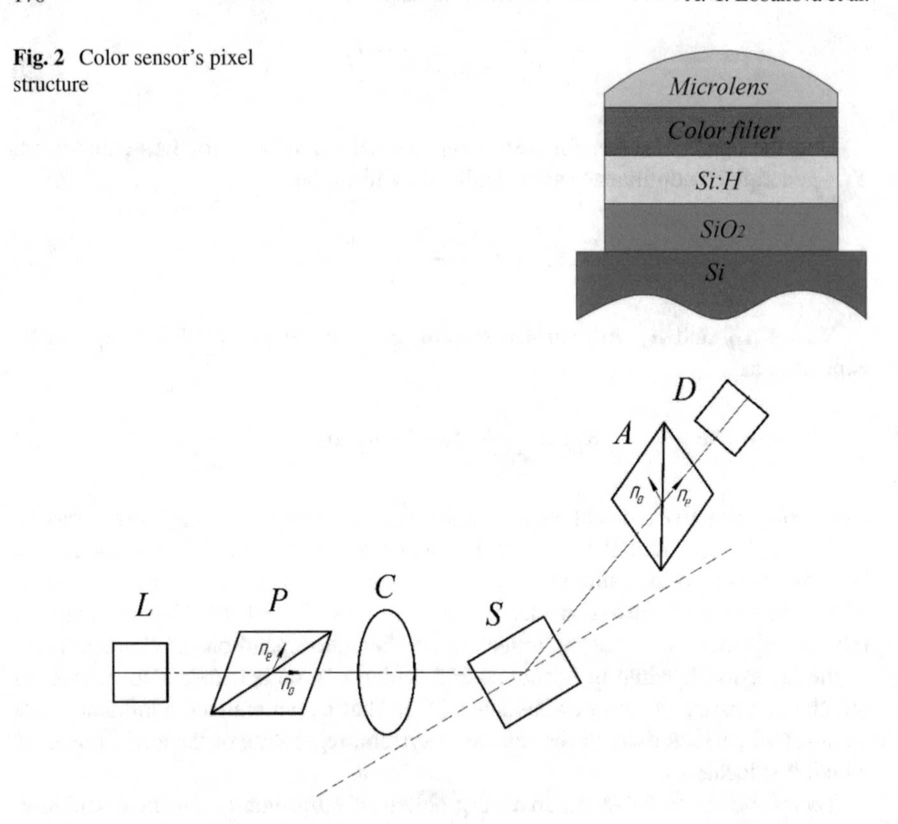

Fig. 2 Color sensor's pixel structure

Fig. 3 The arrangement of anisotropic polarization elements in the scheme of an ellipsometer

Ellipsometric angles measured by a four-zone method using an ellipsometer are taken as input data for the solution of the inverse problem. The ellipsometer is built according to the zero scheme *PCSA*, which is also shown in Fig. 3

The implementation of the zero optical measurement method consists in achieving the minimum intensity of the light beam at the output of the analyzer *A*, by alternating turns of the polarizer *P* and the analyzer *A*. The angular position of the compensator *C* is fixed so that its fast axis is at angles of +45° and −45° to the plane of incidence of the beam of light on the sample *S*.

This provides the highest accuracy of the method and gives the simplification of the procedure of calculation of values of ellipsometric angles Ψ and Δ at the final stage of measurement [35, 37]. For each operating position of the compensator, two measuring situations are formed, which correspond to certain positions of the polarizer and the analyzer, ensuring complete suppression of radiation at the output of the optical scheme of the device. The calculations are performed in the approximation of an isotropic medium based on the results of measurements in four measuring zones. Each measuring zone corresponds to a certain formula for calculating ellipsometric angles Ψ and Δ:

$$\Psi_1 = \Psi_A^{(1)},\ \Delta_1 = -2\gamma_P^{(1)} - \frac{\pi}{2} + 2\pi l \quad \text{for zone 1;}$$
$$\Psi_2 = \Psi_A^{(2)},\ \Delta_2 = -2\gamma_P^{(2)} + \frac{\pi}{2} + 2\pi l \quad \text{for zone 2;}$$
$$\Psi_3 = \Psi_A^{(3)},\ \Delta_3 = 2\gamma_P^{(3)} + \frac{\pi}{2} + 2\pi l \quad \text{for zone 3;}$$
$$\Psi_4 = \Psi_A^{(4)},\ \Delta_4 = 2\gamma_P^{(1)} - \frac{\pi}{2} + 2\pi l \quad \text{for zone 4;}$$

$\gamma_p^{(i)} = P_i - P_0$, further values of polarization angle ψ get counted according to the expressions $\Psi_A^{(i)} = |A_i - (A_0 - 90°)|$. $l = 0, 1, 2 \ldots$, – l gets selected in a way that $0 \leq \Delta \leq 360°$. A_0, P_0 and C_0—adjusting settings. They are known from the passport of ellipsometer: $A_0 = 180°$, $P_0 = 90°$, $C_0 = 81°54'$.

It is known that the depth of the minimum intensity of the light beam at the output of the ellipsometer in the quenching situation depends on the quality of the surface and the degree of its structure. The higher the surface quality and the less structured it is, the greater the depth has a minimum [35, 37, 40].

One of the main features taken into consideration when choosing an angle of incidence of radiation is neighborhood of the Brewster angle. In this area, there is an increased sensitivity of ellipsometric angles to changes in the state of the surface. In the case of homogeneous isotropic non-absorbent substrate without the film, the angular dependence Ψ and Δ on the corner of Brewster are well-known features: the spearhead for ψ and a step height of 180° for Δ [35, 40]. Therefore, in order to illustrate the state of the surface for each sample, reflective characteristics were obtained at different points of the surface based on multi-angle measurements of ellipsometric angles. The angles of incidence were set from 45° to 60° every 5°.

Defining the term "measuring situation" by the totality of the provisions of the extinction of the analyzer and polarizer for each measurement zone on a single sample for a single angle of incidence in this case there were investigated 32 situation measuring.

In the analysis of the measured values, the systematic errors of the device and random errors of the measuring experiment were taken into account. The instrumental error of the ellipsometer is 30° s. To assess the reproducibility of measurements for each angle of incidence, tenfold measurements were carried out for each measuring zone. Then the average error for each angle of incidence was calculated, which depends on the operator's work, the properties of the system itself, the angle of incidence, the accuracy of the device:

$$\delta\Delta = t\sqrt{\frac{\sum_{i=1}^{N}\left(\Delta_i - \overline{\Delta}\right)^2}{N}}, \quad \overline{\Delta} = \frac{1}{N}\sum_{i=1}^{N}\Delta_i$$

$$\delta\Psi = t\sqrt{\frac{\sum_{i=1}^{N}\left(\Psi_i - \overline{\Psi}\right)^2}{N}}, \quad \overline{\Psi} = \frac{1}{N}\sum_{i=1}^{N}\Psi_i,$$

where $i = 1, 2 \ldots 10$; t –Student's coefficient; N—number of pairs of measured values; $\delta\Delta$ and $\delta\Psi$—standard deviation of ellipsometric angles.

3.3 Results of the Research and Analysis

Monochrome sample. Multi-angle measurements took place and relations between ellipsometric angles and angle of incidence were obtained (Fig. 4).

The Brewster angle was found at which the beam of light is extinguished at any position of the polarizer. In this case, the ellipsometric angle Δ is zero. By reducing the step when specifying an angle of incidence of the radiation in the vicinity of the Brewster angle was determined to be the value of 56°45′. The shape characteristics are close to a homogeneous isotropic non-absorbing substrate, indicating a relatively uniform surface structure at the point of incidence of the beam.

The errors of multi-angle measurements for ten series of measurements for each angle of incidence were determined. For one of the surface points, the results are shown in Table 1.

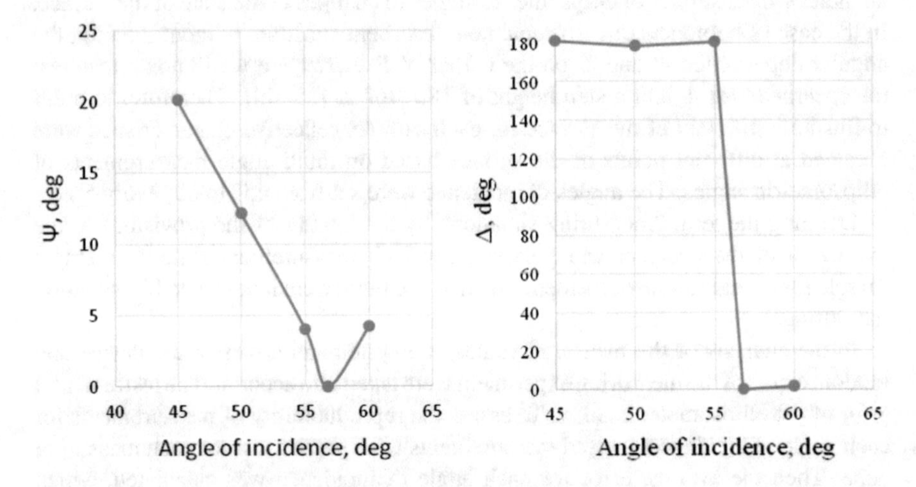

Fig. 4 Relations between ellipsometric angles and angle of incidence

Table 1 Measurement's results for monochrome sample

Angles, degree	45°		50°		55°		60°	
	Average	Standard deviation	Average	Standard deviation	Average	Standard deviation	Average	Standard deviation
Δ	181.013	0.458	178.440	1.226	180.44	2.837	181.762	2.111
Ψ	20.156	0.305	12.163	0.522	3.997	0.405	4.275	0.302

Further for to evaluate the homogeneity of polarization-optical parameters ellipsometric measurements were made at four points of monochrome matrix. Ellipsometric angles for 2 angles of incidence (60º and 70º) were measured at 4 points of matrix. Example of angles of polarizer and analyzer measured and counted based on 3 measurements (out of 10) are shown in Table 2:

Ellipsometric angles, errors of their measurements and averaged for two angles of incidence overall index of refraction for reflecting the structure of the pixels in each of the four points of the matrix were measured. The results are shown in Table 3.

The variation of the generalized refractive index and ellipsometric angle ψ, corresponding to the ratio of the mutual orthogonal components of the amplitude of the reflected radiation, over the site of the matrix was calculated. Variation of an Δ angle is difficult to estimate due to sharp phase changes in the vicinity of the Brewster angle. The results are presented in Table 4.

Thus, the maximum variation of the refractive index for a sample of a monochrome sensor was 3.8%.

Color sample. Multi-angle measurements were carried out and relations of ellipsometric angles Ψ and Δ on the angle of incidence of radiation were obtained (Fig. 5) for color CMOS sensor OV5620.

Brewster angle for the structure of the color sensor was not detected, as complete damping could not be achieved due to multiple re-reflection of the beam from the multilayer surface structure. The ellipsometric angle Δ did not approach zero in any measurement situation. This is an indicator of the complex heterogeneous structure of the optical system of pixels forming the matrix of the color sensor. At angles of incidence close to the angle of pseudo-Bruwster, the quenching area is very wide, so the error in determining ellipsometric angles at angles of incidence 55° and 60° is much higher than at 45°. The results are presented in Table 5.

Then ellipsometric measurements were carried out for to assess the relative homogeneity of the polarization-optical parameters at four points of the monochrome matrix. The ellipsometric angles measured for two incidence angles of radiation at four points of the matrix with the angle of incidence of 45°, which provided the best conditions for suppressing the radiation output of the analyzer. As an example, Table 6 shows the polarizer and analyzer orientation angles based on the results of three series of measurements (out of ten) for one of the points of the color sample surface.

Ellipsometric angles, errors of their measurements and averaged overall index of refraction for reflecting structure of the pixels in each of the four points of the matrix were measured. The results are shown in Table 7.

The generalized refractive index's and ellipsometric angles variations are shown in Table 8.

Maximum variation of refractive index for color sample calculated is 2.2%.

Thus, during the experiment, samples of CMOS sensors, which can be used as sensitive elements of the orientation system of a mobile robot, were investigated.

Table 2 Experimental data for a monochrome sample by points

No.	Angle of incidence	1 zone		2 zone		3 zone		4 zone	
		C = 126.9				C = 36.9			
		P	A	P	A	P	A	P	A
1	60°	46.88	83.67	136.89	94.18	43.2	83.68	133.2	95.15
		33.2	83.73	123.2	92.17	43.83	82.77	133.83	93.9
		34	83.6	124	93.63	50.58	84.25	140.58	93.5
	70°	49.33	66.5	139.33	111.67	41.83	67.17	131.83	112.57
		51.03	66.33	141.03	112.15	39.92	66.72	129.92	111.25
		49.92	67.25	139.92	110.15	40.52	68.1	130.52	112.33
2	60°	51.67	86.67	141.67	93.5	43.55	83.93	133.55	94.25
		41.57	85.93	131.57	95.05	44.9	86.27	134.9	94.58
		43.33	86.15	133.33	92.5	48.83	84.27	138.83	93.1
	70°	49.1	65.92	139.1	112.58	38.87	68.82	128.87	110.08
		51.25	67.72	141.25	109.47	39	68.57	129	109.58
		50.57	67.6	140.57	110	38.62	68.9	128.62	109.47
3	60°	43.18	83.22	133.18	93.67	50	83.5	140	92.1
		40.4	84.18	130.4	93.83	50.17	84.47	140.17	94.77
		38.15	83.15	128.15	92.65	49.77	83.3	139.77	93.85
	70°	48.92	66.58	138.92	111.37	35.25	69.28	125.25	112.02
		50.62	66.58	140.57	111.43	38.9	68.58	128.9	110.77
		48.93	66.75	138.93	111.15	37.67	66.07	127.67	111.7
4	60°	38.55	83.88	128.55	93.4	48.92	82.93	138.92	93.67

(continued)

Table 2 (continued)

No.	Angle of incidence	1 zone		2 zone		3 zone		4 zone	
		C = 126.9				C = 36.9			
		P	A	P	A	P	A	P	A
		40.67	82.5	130.67	94.05	44.57	83.5	134.57	94.4
		41.67	83.8	131.67	93.37	48.17	83.37	138.17	94.05
	70°	49.58	67.63	139.58	109.68	38.57	68.08	128.57	112.25
		47.75	66.67	137.75	111.63	38.93	66.8	128.93	110.48
		48.13	65.67	138.13	110.63	37.73	68.18	127.73	112.62

Table 3 Analysis of monochrome sample's surface heterogeneity

No.	Angle of incidence	$\overline{\Delta}$	$\overline{\Psi}$	$\delta\Delta$	$\delta\Psi$	n
1	60°	7.844	5.069	1.776	0.223	1.398
	70°	350.661	22.337	0.267	0.219	
2	60°	180.238	4.147	1.235	0.288	1.391
	70°	348.522	21.138	0.228	0.286	
3	60°	9.400	4.921	0.516	0.249	1.446
	70°	347.791	22.048	0.427	0.263	
4	60°	6.922	5.2458	0.575	0.142	1.426
	70°	349.922	22.022	0.234	0.282	

Table 4 Variation of polarization-optical parameters for monochrome sample

No. of an area	1–2	1–3	1–4	2–3	2–4	3–4
$\Delta_{\Psi,60}$, deg	0.922	0.148	0.174	0.774	1.095	0.325
$\Delta_{\Psi,60}$, %	18.2	2.9	3.4	15.7	20.8	6.2
$\Delta_{\Psi,70}$, deg	1.199	0.289	0.315	0.91	0.88	0.03
$\Delta_{\Psi,70}$, %	5.4	1.3	1.4	4.1	4.0	0.1
Δ_n, %	0.5	3.4	2.0	3.8	2.5	1.3

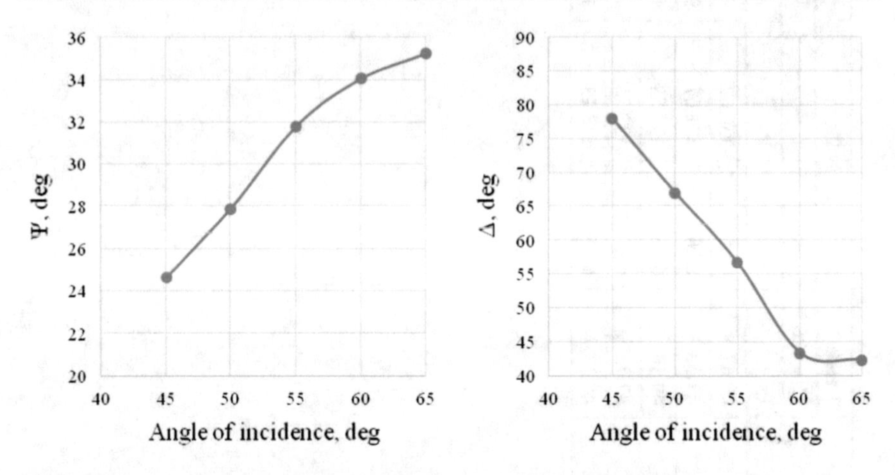

Fig. 5 Relations of ellipsometric angles Ψ and Δ on the angle of incidence of radiation

Table 5 Measurement's results for color sample

Angles	45°		50°		55°		60°	
	Average	Standard deviation	Average	Standard deviation	Average	Standard deviation	Average	Standard deviation
Δ	78.056	0.436	66.967	0.325	56.250	0.669	43.300	0.775
Ψ	24.622	0.170	27.844	0.284	31.267	0.539	34.041	0.642

Table 6 Experimental data for the color sample by points at 45°

No.	1 zone		2 zone		3 zone		4 zone	
	C = 126.9				C = 36.9			
	P	A	P	A	P	A	P	A
1	92.167	113.633	182.167	65.433	86.833	63.15	176.833	111.567
	94.233	115.633	184.233	65.633	85.1	61.8	175.1	111.7
	92.85	114.5	182.85	65.833	85.333	61.6	175.333	112.917
2	94.917	115.6	184.917	64.167	86.417	61.167	176.417	112.9
	94.033	113.75	184.033	64.533	85.683	62.933	175.683	112.417
	92	115.883	182	63.867	85.517	62.55	175.517	112.667
3	93.5	116.05	183.5	63.983	81.917	61.5	171.917	113
	92.2	114.6	182.2	64	84.45	62.083	174.45	113
	92.867	112.167	182.867	62.4	82.167	64.45	172.167	113.6
4	96.167	113.633	186.167	65.867	82.383	63.65	172.383	112.833
	92.283	114.467	182.283	64.25	83.167	64.333	173.167	113.767
	95.6	114.6	185.6	64.633	82.667	64.433	172.667	113.333

Table 7 Analysis of color sample's surface heterogeneity at 45°

No.	$\overline{\Delta}$	Ψ	$\delta\Delta$	$\delta\Psi$	n
1	82.672	24.7083	0.288	0.219	1.546
2	82.222	25.333	0.321	0.222	1.539
3	79.989	25.333	0.314	0.389	1.551
4	78.056	24.622	0.436	0.170	1.573

Table 8 Variation of polarization-optical parameters for color sample

No. of areas	1–2	1–3	1–4	2–3	2–4	3–4
Δ_{Ψ}, deg	0.625	0.625	0.086	0	0.711	0.711
Δ_{Ψ}, %	2.5	2.5	0.3	0	2.9	2.9
Δ_{Δ}, deg	0.45	2.683	4.616	2.233	4.166	1.933
Δ_{Δ}, %	0.5	3.3	5.9	2.8	5.3	2.5
Δ_{n}, %	0.5	0.3	1.7	0.8	2.2	1.4

4 Conclusion

The algorithm has been developed for calculating the variation of optical parameters of the surface of a matrix video sensor. The variations of ellipsometric angles and generalized refractive indices for the surface of two matrix receivers was calculated.

For the monochrome sensor matrix, the reflective characteristics correspond to a homogeneous isotropic non-absorbing structure. However, the maximum variation of polarization-optical parameters was revealed: 20.8% for the ellipsometric angle Ψ in the Brewster angle region and 5.4% outside it; the spread for the generalized refractive index of the surface structure of the matrix was 3.8%. This indicates a significant surface heterogeneity that must be taken into account when analyzing the geometric noise of the matrix.

For the color sensor, the interband dispersion of ellipsometric angles was observed in the vicinity of the Brewster angle, which belongs to the interval of incidence angles from 55° to 60°. This is due to the heterogeneity of ultrathin films forming the surface structure of the matrix. In this case, an important role is played by the heterogeneity of the surface within the light spot. It is confirmed that the most significant growth of interband dispersion in the neighborhood of the Brewster angle is observed for the angle Δ, which has much higher sensitivity for changes in the surface state in this neighborhood compared to the angle Ψ [35].

The smallest statistical error for the inhomogeneous surface of the color sensor is obtained at a minimum angle of incidence of 45°, which can be obtained on an ellipsometer. For the specified angle of incidence, the variation of polarization-optical properties for the surface of the color sensor matrix was revealed. The maximum variation counted are 2.9% for the ellipsometric angle Ψ (the main azimuth of the polarization ellipse reflected by the radiation surface); 5.9% for the ellipsometric angle Δ (the relative phase difference between the mutually orthogonal components of the reflected light); 1.7% for the generalized refractive index of the surface structure of the matrix.

Thus, despite the more complex and heterogeneous structure of the color optical system of pixels in comparison with the monochrome matrix, the color sensor has a more uniform distribution of the refractive index on the working surface.

In high-precision devices, where the accuracy of the refractive index is important up to the fourth decimal point, the revealed surface variation of refractive indices and the sensor sensitivity can lead to significant errors in the processing of information with the mobile robot control system.

Acknowledgements This work was financially supported by Government of the Russian Federation, Grant (08-08).

References

1. Wan, S., Zhang, X., Xu, M., Wang, W., Jiang, X.: Region-adaptive path planning for precision optical polishing with industrial robots. Opt. Express **26**(18), 23782–23795 (2018)
2. Teka, B., Raja, R., Dutta, A.: Learning based end effector tracking control of a mobile manipulator for performing tasks on an uneven terrain. Int. J. Intell. Robot. Appl. https://doi.org/10.1007/s41315-019-00081-8 (2019)
3. Xu, J., Chen, R., Liu, S., Guan, Y.: Self-recalibration of a robot-assisted structured-light-based measurement system. Appl. Opt. **56**(32), 8857–8865 (2017)

4. Jiang, G., Luo, M., Lu, L., Bai, K., Abdelaziz, O., Chen, S.: Vision solution for an assisted puncture robotics system positioning. Appl. Opt. **57**(28), 8385–8393 (2018)
5. Wang, D., Liang, H., Zhu, H., Zhang, S.: A Bionic camera-based polarization navigation sensor. Sensors **14**, 13006–13023 (2014). https://doi.org/10.3390/s140713006
6. Zhao, H., Xu, W.: A bionic polarization navigation sensor and its calibration method. Sensors. **16**, 1223 (2016). https://doi.org/10.3390/s16081223
7. Michael, F., Thorsten, G., Gene, P.: Very large area CMOS active-pixel sensor for digital radiography. IEEE Trans. Electron Devices. **56**(11) (2009)
8. Teyssieux, D., Euphrasie, S., Cretin, B.: MEMS in-plane motion/vibration measurement system based CCD camera. Measurement **44**(10), 2205–2216 (2011)
9. Hill, R.C., Lafortune, S.: Scaling the formal synthesis of supervisory control software for multiple robot systems. In: Conference: 2017 American Control Conference (ACC). p. 3840 (2017). https://doi.org/10.23919/acc.2017.7963543
10. Gonzalez, A.G., Alves, M.V., Viana, G.S., Carvalho, L.K., Basilio, J.C.: Supervisory control-based navigation architecture: a new framework for autonomous robots in industry environments. Ind. Inf. IEEE Trans. **14**(4), 1732–1743 (2018)
11. Macktoobian, M., Aliyari, M.Sh: Optimal distributed interconnectivity of multi-robot systems by spatially-constrained clustering. J. Adapt. Behav. **25**(2), 96 (2017). https://doi.org/10.1177/1059712317700500
12. Akimov, Y.K.: Silicon radiation detectors (Review). Instrum. Exp. Tech. **50**(1)1–28 (2007), ISSN 0020-4412
13. Zmuidzinas, J.: Thermal noise and correlations in photon detection. Appl. Opt. **42**(25), 4989 (2003)
14. Wang, D., Zhang, T., Kuang, H.G.: Relationship between the charge-coupled device signal-to-noise ratio and dynamic range with respect to the analog gain. Appl. Opt. **51**(29), 7103–7114 (2013)
15. Chen, L., Zhang, X., Lin, J., Sha, D.: Signal-to-noise ratio evaluation of a CCD camera. Opt. Laser Technol. **41**, 574–579 (2009)
16. Weiwei, F., Yanjun, J., Ligang, C.: The impact of signal–noise ratio on degree of linear polarization measurement. Optik—Int. J. Light Electron Opt. **124**(3), 192–194 (2013)
17. Davenport, J.J., Hodgkinson, J., Saffell, J.R., Tatam, R.P.: Noise analysis for CCD-based ultraviolet and visible spectrophotometry. Appl. Opt. **54**, 8135–8144 (2015)
18. Liu, H., Zhang, J.: Dark current and noise analyses of quantum dot infrared photodetectors. Appl. Opt. **51**, 2767–2771 (2012)
19. Yang D.X.D., Gamal, A.E.: Comparative analysis of SNR for image sensors with enhanced dynamic range (1999). http://isl.stanford.edu/groups/elgamal/abbas_publications/C068.pdf
20. Nam, H., Park, J.L., Choi, J.S., Lee, J.G.: The optimization of zero-spaced microlenses for 2.2um pixel CMOS image sensor. In: Proceeding of SPIE 6520, Optical Microlithography XX (2007)
21. Nussbaum, Ph, Völkel, R., Herzig, H.P., Eisner, M., Haselbeck, S.: Design, fabrication and testing of microlens arrays for sensors and microsystems. Pure Appl. Opt. **6**, 617–636 (1997)
22. Huber, M., Pauluhn, A., Culhane, J.: CCD and CMOS sensors. Observing Photons in Space ISSI Scientific Reports Series, ESA/ISSI, pp. 391–408 (2010)
23. Berger, C.R., Benlachtar, Y., Killey, R.I., Milder, P.A.: Theoretical and experimental evaluation of clipping and quantization noise for optical OFDM. Opt. Express **19**, 17713–17728 (2011)
24. Konnik, M.V., Manykin, E.A., Starikov, S.N.: Optical-digital correlator with increased dynamic range using spatially varying pixels exposure technique. Opt. Mem. Neural Net. Inf. Opt. **18**(2), 61–71 (2009)
25. Konnik, M.V., Welsh, J.S.: On numerical simulation of high-speed CCD/CMOS-based wavefront sensors in adaptive optics. SPIE Optical Engineering+Applications (2011)
26. Ding, R., Venetsanopoulos, A.N.: Generalized homomorphic and adaptive order statistic filters for the removal of impulsive and signal-dependent noise. IEEE Trans. Circuits Syst. **34**(8), 948–955 (1987)

27. Chen, Z., Wang, X., Liang, R.: Calibration method of microgrid polarimeters with image interpolation. Appl. Opt. **54**, 995–1001 (2015)
28. Meza, P., Machuca, G., Torres, S., Martin, C.S., Vera, E.: Simultaneous digital super-resolution and nonuniformity correction for infrared imaging systems. Appl. Opt. **54**, 6508–6515 (2015)
29. Liu, G., Tan, O., Gao, S.S., Pechauer, A.D., Lee, B., Lu, C.D., Fujimoto, J.G., Huang, D.: Postprocessing algorithms to minimize fixed-pattern artifact and reduce trigger jitter in swept source optical coherence tomography. Opt. Express **23**, 9824–9834 (2015)
30. Tyo, J.S.: Design of optimal polarimeters: maximization of signal-to-noise ratio and minimization of systematic error. J. Appl. Opt. **41**, 619–630 (2002)
31. Feng, W., Chen, L.: Relative orientation accuracy analysis of the polarizers in a polarization CCD camera. Optik **121**, 1401–1404 (2010)
32. Chipman, R.A.: Polarization ray tracing. In: Proceeding of SPIE 0766, Recent Trends in Optical Systems Design and Computer Lens Design Workshop, 10 June 1987. https://doi.org/10.1117/12.940204
33. Chipman, R.A.: Polarization analysis of optical systems. Opt. Eng. **28**(2) (1989)
34. Korotaev, V.V.: Computing the polarization of optical radiation crossing mirror and prism systems. [Article@METOD RASCHETA SOSTOYANIYA POLYARIZATSII OPTICHESKOGO IZLUCHENIYA PRI PROKHOZHDENII ZERKAL'NYKH I PRIZMENNYKH SISTEM.] (1979) Izvestia vyssih ucebnyh zavedenij. Priborostroenie, **22**(4), 77–82
35. Irene, E.A., Tompkins, H.G. (eds.): Handbook of ellipsometry. William Andrew Publisher (2005)
36. Fujiwara, H.: Spectroscopic ellipsometry: principles and applications. Wiley, London (2007)
37. Tikhiii, A.A., Gritskikh, V.A., Kara-Murza, S.V., Nikolaenko, YuM, Zhikharev, I.V.: Features of interpreting ellipsometric measurement results. Opt. Spectrosc. **112**(2), 300–304 (2012)
38. Azzam, R.M.A.: Photopolarimetric measurement of the Mueller matrix by Fourier analysis of a single detected signal. Opt. Lett. **2**(6), 148–150 (1978)
39. Chipman, R.A.: Polarization considerations for Optical Systems. Opt. Eng. **28**(2), 85 (1989)
40. Atkinson, Gary A., Ernst, Jürgen D.: High-sensitivity analysis of polarization by surface reflection. Mach. Vis. Appl. **29**(7), 1171–1189 (2018)

Peculiarities of Reducing the Impact of Air Tract on the Accuracy of Positioning Elements of Robotics at Analysis of a Diffraction Pattern of Air Tract Dispersion on a Photo Matrix Field

Ivan S. Nekrylov, Alexander N. Timofeev, Igor A. Konyakhin, Valery V. Korotaev and Tong Minh Hoa

Abstract The article deals with theoretical aspects of CMOS sensor cross-links effect as a main drawback of the dispersion method implementation with CMOS video camera and RGB optical radiation source. The paper is concerned with usage of diffraction grating to solve the problem of images overlapping on CMOS sensor with Bayer pattern. It is shown that the diffraction distribution allows to determine the energy centers of different images of the same RGB optical radiation source independently at the same time without overlapping. The text gives mathematical description of how diffraction image is formed and how to use it in dispersion method. The article is of interest to people who deals with optical radiation propagation through the air tract with vertical temperature gradient.

Keywords Temperature gradient · Dispersion method · Spatial position control · Positioning of robotics elements · Optical-electronic system · Diffraction pattern · Bayer pattern · Overlapping of images

1 Introduction

For positioning elements of robotic systems, more and more often means of automated non-contact monitoring of the spatial position of objects are used. One of such means, realizing the optical contactless method, is an optical-electronic system with active reference marks. Such a system generally contains a video camera, a reference mark and an information processing unit. The reference mark is located on the controlled object, the video camera registers a series of images of the reference mark, transfers them to the processing unit, where its coordinates in the space of objects

---

I. S. Nekrylov (✉) · A. N. Timofeev · I. A. Konyakhin · V. V. Korotaev · T. M. Hoa
ITMO University, St. Petersburg, Russia
e-mail: nekrylov@itmo.ru

A. N. Timofeev
e-mail: timofeev@itmo.ru

© Springer Nature Switzerland AG 2020

A. E. Gorodetskiy and I. L. Tarasova (eds.), *Smart Electromechanical Systems*, Studies in Systems, Decision and Control 261, https://doi.org/10.1007/978-3-030-32710-1_14

are calculated. Such a system is compact, requires a minimum of constructive intervention, has a large range of motion control and provides high positioning accuracy [1–3]. However, when controlled objects of robotic systems work in conditions of high temperature gradients of the air tract, for example, machine parts in steelmaking, the problem of considering the influence of temperature gradient on the error in determining the coordinates of the reference mark and, as a result, the positioning accuracy of the object being monitored arises [4]. Even at small control distances, the error in determining the coordinates of the reference mark can reach 20–30% with temperature gradients inherent in production.

In a few applications, this error is unacceptable, so the task is to minimize the influence of the temperature gradient of the air tract on the operation of the optical-electronic system for monitoring the spatial position of objects with active reference marks. The dispersion method for determining the vertical temperature gradient of the air tract using an active reference mark in the form of an RGB LED and a video camera with a color matrix photodetector is known. This method consists in estimating the linear difference of the energy centers of the images from the blue and red spectral component of the reference mark. The implementation of the dispersion method with the geometric course of the rays does not introduce constructive changes to the original system and requires only programmatic refinement [5]. However, the well-known implementation of the dispersion method, based on the geometric course of the rays, has a significant drawback—the presence of an error of cross-links between the elements of the matrix photodetector. This error arises because the images of different spectral components from the reference mark overlap in the plane of the photodetector, introducing an error in the determination of the energy center coordinates of each of the images.

To solve this problem, the authors propose to analyze the diffraction image of the reference mark, obtained by setting the diffraction grating in the plane of the entrance pupil of the lens of the video camera. In the diffraction pattern, the images of different spectral components of the reference mark will be separated so much that they will not overlap each other. Knowing the parameters of the diffraction grating, its location, the parameters of the lens and the source of optical radiation, it is possible to estimate the discrepancy between the energy centers of the images in the geometric course of the rays in the absence of cross links error between the elements of the matrix photodetector. The use of a diffraction grating made it possible to spread the images of the reference mark in the plane of the matrix photodetector to values equal and exceeding the size of the image of the reference mark, which makes it possible to analyze images in different physical areas of the matrix photodetector. Thus, the error of the influence of cross-links is eliminated.

2 Theory

Dispersion method. Figure 1 shows the geometric representation of the optical radiation passage through the air tract, considering the vertical temperature gradient along the path [6].

The vertical arrows in the diagram indicate the direction of the temperature gradient $\nabla_y T$.

The dependence of the refractive index on the wavelength of the incident light is determined as:

$$n = 1 + A + \frac{3 \cdot B}{\lambda^2} + \frac{5 \cdot C}{\lambda^4}, \tag{1}$$

where A = 2876,04 × 10^{-7}, B = 16,288 × 10^{-7}, C = 0136 × 10^{-7}—dispersion coefficients for the dry air including 0.03% of carbon dioxide [7].

Formula (1) shows that the refractive index decreases with temperature decreasing and wavelength increasing. Consequently, the radiation will change direction at an angle of $d\alpha_i$ at each point of the air tract, that leads to a linear displacement in a plane dy_i and the radiation direction will be shifted to a value of y_λ at the receiver plane [8, 9]:

$$\Delta y = \int_0^{x_I} \frac{x}{n \cdot \cos^2 \upsilon} \cdot \nabla_y n \, dx \tag{2}$$

where n—refractive index of air to emit a given wavelength, υ—the angle between base direction and the tangent line to the radiation direction at the observation point, x_I—abscissa of the beam exit point in the surface layer of the atmosphere, x—the abscissa of the point of observation.

The dependence of the refractive index vertical gradient of an air tract on the atmospheric parameters will generally take the form [10]:

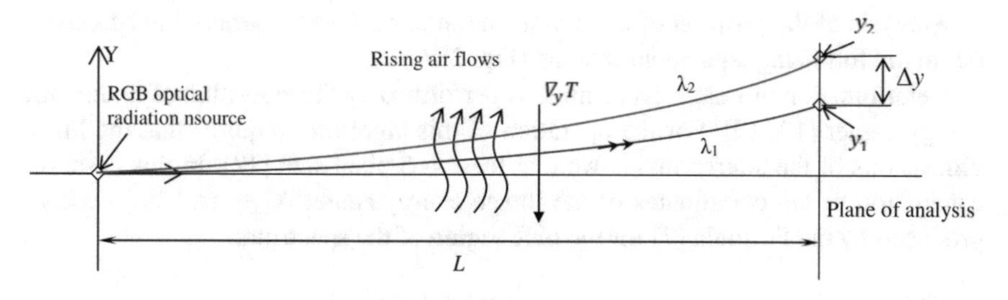

Fig. 1 Explanation of the radiation propagation through the atmosphere. L measurement distance, $\nabla_y T$—vertical temperature gradient, where ∇—nabla operator

$$\nabla_y n = -\frac{n-1}{T} \cdot \nabla_y T + \frac{n-1}{P} \cdot \nabla_y P - 0.155\frac{n-1}{P} \cdot \nabla_y P_p \tag{3}$$

where $\nabla_y T$, $\nabla_y P$, $\nabla_y P_p$ vertical gradient of temperature, pressure and partial pressure of water vapor.

Substituting (2) into (3) and neglecting components $\nabla_y P$ and $\nabla_y P_p$ because of their smallness, we obtain the formula [11]:

$$y_\lambda = -\int\limits_0^{x_l} \frac{n-1}{n \cdot T \cdot \cos^1 v} \cdot \nabla_y T \cdot x\,dx \tag{4}$$

Assume that the radiation beam propagates in a single "refraction unit" [10]. Then from Formula (4) we have:

$$y_\lambda = -\frac{n-1}{T} \cdot \nabla_y T \cdot \frac{L^2}{2} \tag{5}$$

In optoelectronic systems of the spatial position of objects control such displacement Δy will be made for changing the position of the reference mark, whereas in fact the displacement caused by the beam "deviation" from the base direction. Determination of the vertical temperature gradient using the radiation of one wavelength is difficult, because it is necessary to exact the same horizontal position of the radiation source and the receiver. It is proposed to use double-wavelength dispersion measuring method to determine $\nabla_y T$. This method implies to measure the shift y_{λ_1} at the wavelength λ_1 and the shift y_{λ_2} at the wavelength λ_2. Due to the phenomenon of dispersion, the displacements y_{λ_1} and y_{λ_2} will differ by a certain amount of H. Then the difference of displacements gives an opportunity to express the value of $\nabla_y T$ [7].

$$\nabla_y T = \frac{2\left(y_{\lambda_1} - y_{\lambda_2}\right) \cdot T}{L^2 \cdot (n_2 - n_1)} = \frac{2H \cdot T}{L^2 \cdot (n_2 - n_1)} \tag{6}$$

Analysis of the position of the image on the photodetector arrays can be carried out in the following equivalent scheme (Fig. 2):

Determination of image coordinates is performed by the algorithm of finding the energy center [12, 13]. For the operation of this algorithm requires that the linear dimensions of the source image were at least 4–5 pixel size [12]. In this case, the calculation of the coordinates of the image energy center X_{bEC} and Y_{bEC} can be produced by the Formula (7) for the blue region of the spectrum:

$$X_{bEC} = 2\frac{\sum_{j=0}^{j=M} \frac{\sum_{i=0}^{i=N} [Q_i(x_i)x_i]}{\sum_{i=0}^{i=N} Q_i(x_i)}}{M}$$

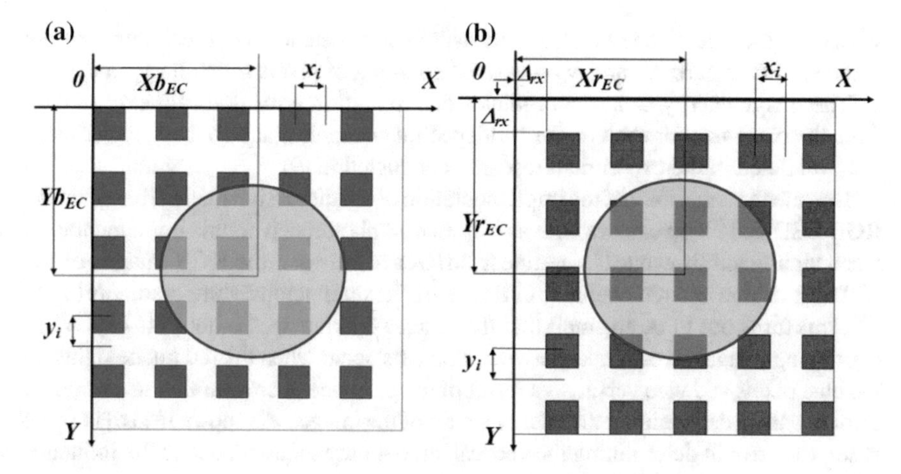

Fig. 2 Target image at the color sensor with Bayer filter (**a**—blue, **b**—red)

$$Y_{bEC} = 2\frac{\sum_{j=0}^{j=N} \frac{\sum_{i=0}^{i=M} [Q_i(y_i)y_i]}{\sum_{i=0}^{i=M} Q_i(y_i)}}{N} \tag{7}$$

where x_i, y_i—the coordinates of the elements included in the neighborhood of $M \times N$; M, N—the value of the neighborhood that contains an image of the object; Q_i—sum signal from elements with the j-th row belonging to the neighborhood of M, N.

The image energy center determination has a similar appearance for the red spectrum region, considering that each sensor with a red filter will be shifted from a blue element according to values Δ_{rx} and Δ_{ry} horizontally and vertically:

$$X_{rEC} = 2\frac{\sum_{j=0}^{j=M} \frac{\sum_{i=0}^{i=N} [Q_i(x_i)x_i]}{\sum_{i=0}^{i=N} Q_i(x_i)}}{M} + \Delta_{rx}$$
$$Y_{rEC} = 2\frac{\sum_{j=0}^{j=N} \frac{\sum_{i=0}^{i=M} [Q_i(y_i)y_i]}{\sum_{i=0}^{i=M} Q_i(y_i)}}{N} + \Delta_{ry} \tag{8}$$

It should be considered that the shifts y_{λ_1} and y_{λ_2} detecting on matrix will be associated with a linear shift in the object space, according to the geometrical optical laws:

$$y'_{\lambda_1} = \frac{y_{\lambda_1} \cdot L}{f} = \frac{p \cdot N_1}{f} \cdot L$$
$$y'_{\lambda_2} = \frac{y_{\lambda_2} \cdot L}{f} = \frac{p \cdot N_2}{f} \cdot L \tag{9}$$

where p—the size of one pixel; L—measurements distance; f—focal length of the receiver optical part; N_1 and N_2—vertical counting of pixels for shifts y'_{λ_1} and y'_{λ_2}.

The image energy center coordinates determination error does not exceed 0.05 from the pixel size for each of the multispectral channels that have been experimentally validated in the experiment research conducted in [9].

Despite the simplicity of the implementation of the dispersion method based on the RGB LED and video camera with a color matrix photoreceiver, this implementation has a significant drawback. The linear difference between the RGB images of the LED in the blue and red channels of the matrix at small temperature gradients (up to 1 °C/m) turns out to be so small that the images overlap each other [14, 15]. When overlaying images, cross-links between channels occur when the red image lights up the blue pixels and vice versa. As a result of the presence of cross-links, an additional error occurs in determining the energy center of the images. As shown in [16] in some cases, the error in determining the vertical temperature gradient, due to the influence of cross-links, can reach 100%.

To increase the linear difference between the images, it is advisable to increase the focal length of the lens of the optical system, which entails an increase in cost, size and mass of the system.

Diffraction pattern. The authors of this article propose to use the phenomenon of diffraction of optical radiation for independent analysis of images in the blue and red channel of the CMOS sensor. When placing the diffraction grating in front of the receiving part of the system of the matrix photodetector, a diffraction pattern will be observed for each of the spectral components of the two-color radiation source [17, 18]. In this case, you can calculate the difference between the centers of the images not by the central maxima as shown in Fig. 1, but by the m and $-m$ maxima of the order, respectively, since the difference between them will be significantly larger.

Figure 3a shows the path of the rays in the system for determining the vertical gradient of the airway temperature by the dispersion method without a diffraction grating and with a diffraction grating in the entrance pupil of the objective (Fig. 3b).

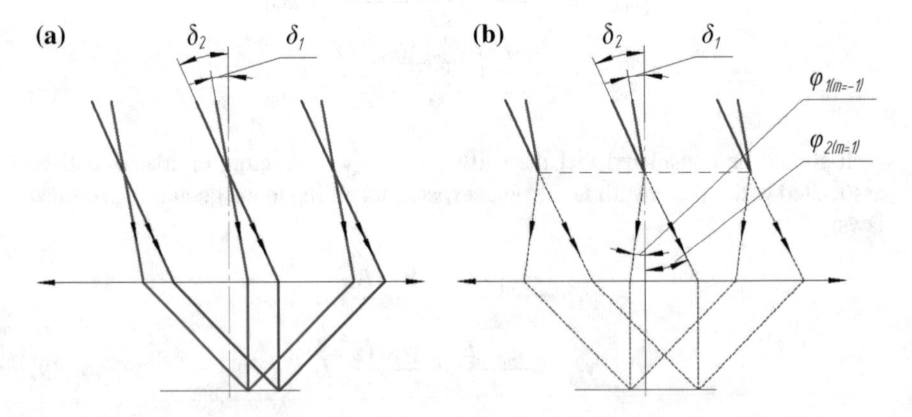

Fig. 3 Decomposition of optical radiation passing through a diffraction grating

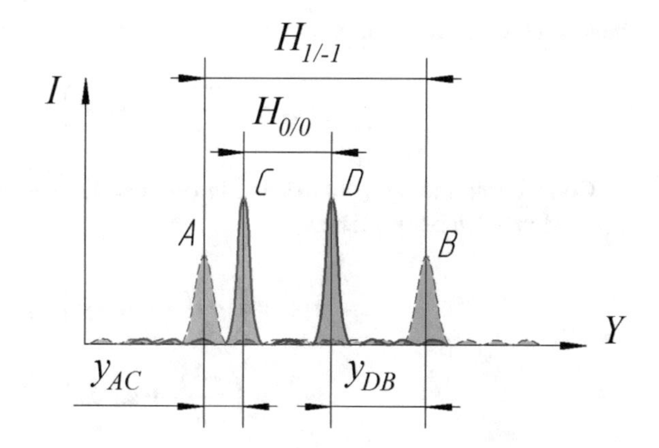

Fig. 4 Spatial distribution of the intensities 1/−1 and 0/0 orders of the diffraction pattern maxima

Figure 3b shows only the rays that form the maximum $-m$ order for the red spectral component and the maximum m order for the blue spectral component. The rays forming the central maxima do not deviate when they hit the diffraction grating, respectively, their course will coincide with the rays shown in Fig. 3a. Due to the large energy losses in high-order side maxima, we limit ourselves to considering the side maxima $m = 1$ and $m = -1$. Based on this, an intensity distribution will be observed in the analysis plane (Fig. 3).

In Fig. 4 $H_{1/-1}$—the difference between the coordinates of the maxima of the red and blue spectral component 1 and −1 of the order, respectively; $H_{0/-0}$—the difference between the coordinates of the central maxima.

As can be seen, $H_{1/-1}$ significantly exceeds the value of $H_{0/-0}$. The diffraction pattern in Fig. 3 is presented for a diffraction grating having 50 hatch/mm, radiation with a wavelength of 480 nm, incident at an angle of 25° and radiation with a wavelength of 680 nm, incident at an angle of 10°. The difference in the coordinates of the maxima of 1 and −1 is about two times greater than the difference between the coordinates of the central maxima.

In order to express the value of $H_{0/-0}$ in terms of $H_{1/-1}$, we use the well-known formula for determining the sine of the diffraction angle for the maximum of the m-order:

$$\sin \varphi = \frac{\pm m\lambda}{d} + \sin \delta \tag{10}$$

where $m = 0, 1, 2, \ldots$ is the order of the diffraction maximum, λ is the wavelength of the incident radiation, d is the period of the diffraction grating, δ is the angle of incidence of the radiation on the diffraction grating. Denote the 0-th order maximum center coordinate for the red spectral component by the letter A, and the C maximum for the −1st maximum. Similarly, we denote the coordinate of the *0*-th order maximum center for the blue spectral component by the letter D, the 1-st maximum by B. the fact that a parallel beam of rays falls on the optical system, and the analysis plane is in the back focal plane, from the Formula (10) we find that the distance $y_{0/m}$

between the central maximum and the m-order maximum will be

$$y_{0/m} = \left(\frac{\pm m\lambda}{d} + \sin \delta \right) \cdot f' \quad (11)$$

Considering that $(y_{AC}) = -1$ and $m\ (y_{DB}) = 1$, expression (11) for the values of y_{AC} and y_{DB} can be written as:

$$y_{AC} = \left(\frac{-\lambda_r}{d} + \sin \delta_K \right) \cdot f' \quad (12)$$

$$y_{DB} = \left(\frac{-\lambda_b}{d} + \sin \delta_c \right) \cdot f' \quad (13)$$

where δ_r and δ_b—the angles of incidence of the optical radiation of the blue and red spectral components on the diffraction grating, respectively. Then the value of $H_{0/0}$, equivalent to the value of H in Formula (6), can be written as

$$H_{0/0} = (A - B) - (y_{AC} + y_{DB}) \quad (14)$$

Considering expressions (12) and (13):

$$H_{0/0} = H_{1/-1} - f' \cdot \left(\frac{1}{d} \cdot (\lambda_c - \lambda_K) + \sin \delta_c + \sin \delta_K \right) \quad (15)$$

Denote $\lambda_b - \lambda_r = \Delta\lambda$ and the value $\frac{1}{d} = N$ the number of grooves of the diffraction grating in a unit of length. Then expression (15) can be written as:

$$H_{0/0} = H_{1/-1} - f' \cdot (N \cdot \Delta\lambda + \sin \delta_c + \sin \delta_K) \quad (16)$$

Formula (16) is a mathematical description of the conversion of a linear difference between 1/−1 images of the order of a diffraction pattern and simple images. By selecting the period of the diffraction grating so that the images do not overlap, it is possible to eliminate the CMOS sensor cross-links effect.

3 Conclusion

Using the dispersion method to determine the vertical temperature gradient is an effective way to correct the measurement results in an optical-electronic system for monitoring the spatial position of objects.

The paper proposes a method to eliminate the effect of cross-linking of the sensor on the error in determining the vertical temperature gradient. The mathematical description of the diffraction method for the implementation of the dispersion method is given.

With a lack of energy in the required maxima of the diffraction pattern, it is possible to adjust the brightness of the optical radiation source of various spectral components independently of each other.

Further studies are aimed at an experimental evaluation of the diffraction method for implementing the dispersion method.

References

1. Maletsky, L.P., Sun, J., Morton, N.A.: Accuracy of an optical active-marker system to track the relative motion of rigid bodies. J. Biomech. **40**(3), 682–685 (2007)
2. Wiles, A.D., Thompson, D.G., Frantz, D.D.: Accuracy assessment and interpretation for optical tracking systems. In: Medical Imaging 2004: Visualization, Image-Guided Procedures, and Display, vol. 5367, pp. 421–432. International Society for Optics and Photonics (2004)
3. Schöffel, P.J., et al.: Accuracy of a commercial optical 3D surface imaging system for realignment of patients for radiotherapy of the thorax. Phys. Med. Biol. **52**(13), 3949 (2007)
4. Sun, T., Xing, F., You, Z.: Optical system error analysis and calibration method of high-accuracy star trackers. Sensors **13**(4), 4598–4623 (2013)
5. Nekrylov, I.S. et al.: Choosing parameters of active reference mark optical-electronic systems spatial position control. 20th International Symposium on Precision Engineering Measurements and Instrumentation, vol. 11053, p. 110534H. International Society for Optics and Photonics (2019)
6. Hill, R.J., Clifford, S.F.: Modified spectrum of atmospheric temperature fluctuations and its application to optical propagation. JOSA **68**(7), 892–899 (1978)
7. Beland, R.R.: Propagation through atmospheric optical turbulence. Atmos. Propag. Radiat. **2**, 157–232 (1993)
8. Roddier FV (1981) The effects of atmospheric turbulence in optical astronomy. In: Progress in optics, vol. 19, pp. 281–376. Elsevier
9. Fried, D.L.: Optical resolution through a randomly inhomogeneous medium for very long and very short exposures. JOSA **56**(10), 1372–1379 (1966)
10. Andrews, L.C., Phillips, R.L.: Laser beam propagation through random media, vol. 152. SPIE press, Bellingham, WA (2005)
11. Schmidt, J.D.: Numerical simulation of optical wave propagation with examples in MATLAB. SPIE, Bellingham, Washington, USA (2010)
12. Li, C., et al.: Minimization of region-scalable fitting energy for image segmentation. IEEE Trans. Image Proces. **17**(10), 1940–1949 (2008)
13. Richardson, W.H.: Bayesian-based iterative method of image restoration. JoSA **62**(1), 55–59 (1972)
14. Bosco, A., Mancuso, M.: Noise filter for Bayer pattern image data : пат. 7369165 США (2008)
15. Lukac, R., Plataniotis, K.N., Hatzinakos, D.: Color image zooming on the Bayer pattern. IEEE Trans. Circ. Syst. Video Tech. **15**(11), 1475–1492 (2005)
16. Nekrylov I. S. et al.: The research of the cross-links effect influence in the color matrix photodetector on an error of the air tract vertical temperature gradient determination. In: Optical Measurement Systems for Industrial Inspection X, vol. 10329, p. 103294L. – International Society for Optics and Photonics (2017)
17. Ma, X., Li, M., He, J.J.: CMOS-compatible integrated spectrometer based on echelle diffraction grating and MSM photodetector array. IEEE Photonics J. **5**(2), 6600807 (2013)
18. Stork, D.G., Gill, P.R.: Lensless ultra-miniature CMOS computational imagers and sensors. Proc. Sensorcomm. 186–190 (2013)

Optoelectronic SEMS for Preventing Object Destruction

Anton A. Nogin and Igor A. Konyakhin

Abstract Control and monitoring of supporting structures are important part of ensuring security in the work area of robots. Late detection of deformations leads to the destruction of the object and possible material damage. *The purpose* of the work is the development SEMS for remote control of supporting structures of infrastructure objects to prevent destruction in the area of robot operation. *The result* is an optoelectronic device capable to control angular and linear movements of the object being monitored and warning of critical displacements. The proposed solution is integrated into the overall monitoring system. The possibility of creating a single-channel four-coordinate device capable of real-time monitoring of the position of an object and warning of critical deformations and displacements has been proved. The sensitivity and accuracy of the device is comparable with commercially available two-coordinate devices. The proposed option is small in size, more compact and can be installed at a distance of up to 5 m. Mathematical models have been developed to evaluate the effect of various measurement error on the result. Specially designed software allows to compensate for errors within the device and expands the metrological range of measurements with solving the problem of inoperability in the case of overlapping image in the field of analysis through the use of Hough Transformation. A passive reflector that does not require power supply is installed at the monitoring point. The practical significance lies in the possibility of using SEMS with the optical-electronic part for solving problems of monitoring and ensuring the safety of both robots and humans.

Keywords Optoelectronic devices · Non-contact monitoring · Angular and linear displacement · Autocollimator · SEMS for monitoring

---

A. A. Nogin (✉) · I. A. Konyakhin
ITMO University, St. Petersburg, Russia
e-mail: anogin@itmo.ru

I. A. Konyakhin
e-mail: igor@grv.ifmo.ru

© Springer Nature Switzerland AG 2020

A. E. Gorodetskiy and I. L. Tarasova (eds.), *Smart Electromechanical Systems*, Studies in Systems, Decision and Control 261, https://doi.org/10.1007/978-3-030-32710-1_15

1 Introduction

The supporting structure is an important element of the safety of any building and structure. However, time, the external environment, man-made factors constantly influence the supporting structures gradually reducing their reliability. Design errors can also cause structural wear. Any changes in the structure are associated with its deformation and a change in its position in space. Registration of the displacement of the supporting structure allows detecting defects in a timely manner, as well as preventing or minimizing damage and loss of life [1–3].

As a rule, the building has several supporting structures, which requires a comprehensive assessment of reliability. This task can be solved by optical-electronic SEMS, and combining them into a single network appears a comprehensive solution to the problem of monitoring.

2 General Principles and the Possibility of Devices

To solve the problem of controlling the position of the supporting structure it is proposed to use an optical-electronic autoreflection system. This system is distributed and consists of two parts. The first part—the control element is located on the monitoring object. The control element is a pyromidal prism with angles of 90° at the apex. This element is passive, does not require power at the control point. It is rigidly fixed on the controlled object and its displacement uniquely corresponds to the displacement of the supporting structure. The second part is an autoreflection unit, which is one of the variants of the autocollimation scheme.

2.1 Control Element

The developed control element is called the pyromidal reflector. In its essence and mathematical description, it is a combination of two two-sided mirrors, where the axes of the mirrors are perpendicular to each other. This combination allows you to simultaneously control 4 coordinates, namely: the movement along the axes OX and OZ, as well as turns around these axes.

When moving such a control element along the OX′ or OZ′ axes, the image is shifted by two times. In this case, the distance between the two marks does not change. When the control element is rotated around the OX′ or OZ′ axes, i.e. angles Θ1, Θ2 also occurs moving labels, but according to a more complex law. The matrix method is used to calculate these angles.

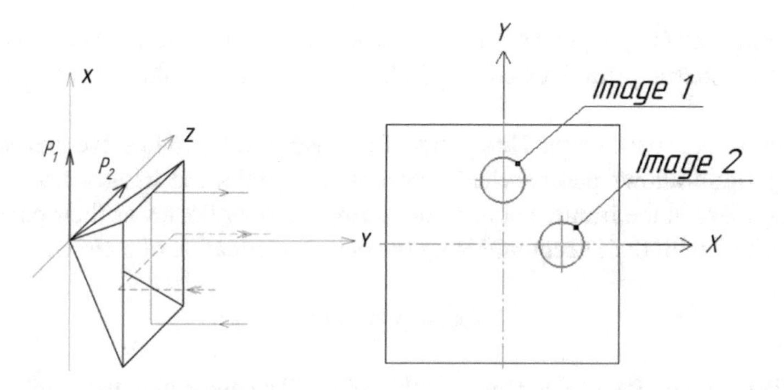

Fig. 1 Control element and the image formed by it

2.2 Autoreflection Systems and Single Field Analysis

Autoreflection system is a classic optical system consisting of a lens and a receiver of optical radiation—CMOS matrix. At the top of the last optical surface of the lens is a radiating diode, the radiation of which is reflected from the control element. It should be noted that the system operates in a converging beam of rays, which allows controlling linear displacements [4, 5].

In order to minimize the size of the system, only one receiving-emitting channel is used, and therefore the concept of a single field of analysis arises. In this field there are simultaneously two images (Fig. 1), which has its own characteristics.

When working in a single field of analysis, there may be situations of intersection of images that currently cause inoperable system. In connection with this, it is necessary to develop a special algorithm capable of simultaneously measuring the coordinates of both images and functioning in a situation of intersection of images.

3 Algorithm for Processing Measurements in a Single Analysis Field

The problem of overriding turned out to be very important not only because it led to the device's inoperability, but also because it limited the sensitivity of the device. In addition, because of this problem, a situation may arise that does not detect the initial stage of deformation of the object.

Calculation of mark movement in the analysis plane, when the control element is turned, is the main stage of the operation of the angle measuring system. To do this, it is necessary to recognize the mark, that is, to distinguish it from the background and other objects in the image, and also to measure the coordinates of its center. To solve the problem, a priori known information about the geometric form of the mark is used. In this system, the mark has a circular shape, which makes it stable to noise and invariant to rotation, scale, etc. A circle can be described by an equation with three parameters (x, y, r), where r is the radius of the circle. Thus, the primary task is to detect geometric primitives corresponding to a given combination of parameters [6, 7].

An algorithm based on the Hough transform was developed to solve this problem. This algorithm allows parametrically describing the desired images and not only detecting them in the frame, but also measuring the coordinates as their parameters are described with their account. Using the classical equation of a circle,

$$(x^2 + y^2) = r^2 \tag{1}$$

a sub-pixel calculation of coordinates takes place. The algorithm also applies a gradient estimate for the voting procedure for the most probable center of the circle. The application of all these methods, taking into account the various filterings, allows one to measure the angular and linear displacements of the object with high accuracy, including in the situation of image intersection. Algorithm also maintains its operation under flare and parasitic glare caused by the mirror nature of the control element.

4 Experimental Studies

Experimental studies were conducted in the laboratory. The autoreflection unit consisted of a lens with a focal length of 400 mm and an infrared emitting diode. The working distance was 3 m. The control element was installed on moving platforms providing accuracy of 0.1 mm displacements along the axes and 1 angular minute for turning around the axes.

Figure 2 shows the frames obtained using the projected autoreflection unit. As can be seen from the presented images, the algorithm successfully solves the problem of detecting and measuring the center of coordinates both in the case of the normal functioning of the system and in the case of intersection of images.

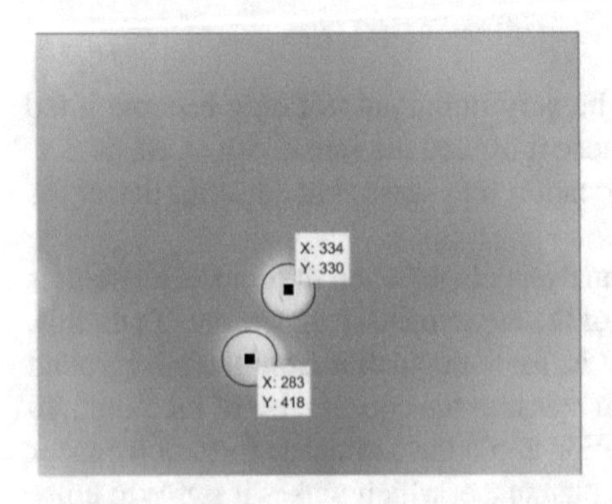

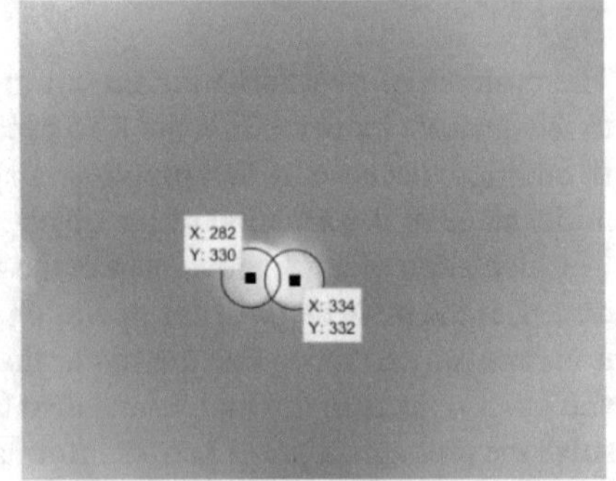

Fig. 2 Examples of the program window

As can be seen from the presented frames, the algorithm can recognize and measure coordinates. For the operator, coordinates rounded to the integer are displayed on the screen, but the capabilities of the algorithm make it possible to calculate up to 3 decimal places, that is, calculations can be called subpixel. To improve the accuracy, it is also planned to rotate the receiving matrix and several software techniques.

With the help of the designed system, it was possible to measure linear displacements in the range from 0 to 20 mm. As can be seen from Fig. 3, the dependence of linear measurements is linear. Increasing the working distance with sufficient energy will increase the measurement accuracy because the transfer coefficient for linear measurements is two.

Also, measurements of angular displacements were performed in a wide range from 0 to 28 angular minutes. The measurement results are presented in Fig. 4.

As can be seen from the graphs, the system successfully measures the angular and linear coordinates with a sufficiently high accuracy.

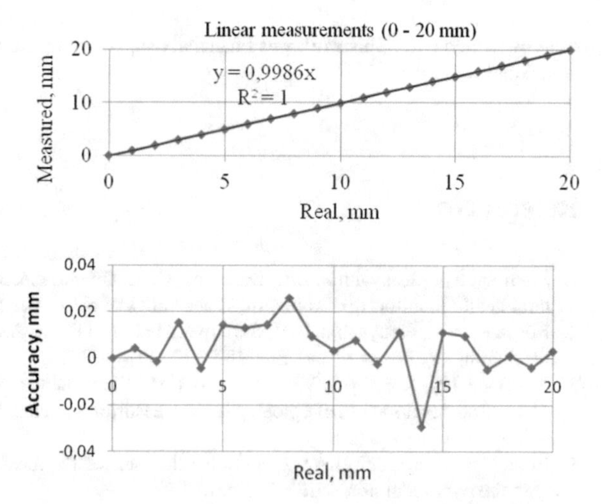

Fig. 3 The results of measurements of linear displacements

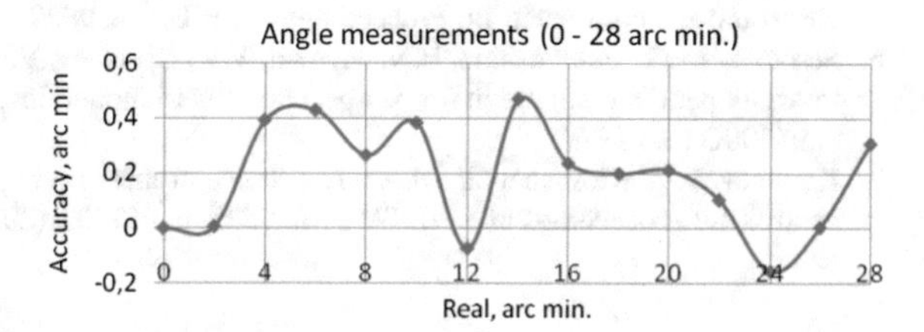

Fig. 4 The results of measurements of angular displacements

5 Conclusion

The achieved results confirm the theoretical possibility of using single-channel angle-measuring autocollimator for measuring several angles at the same time, and also for working with new types of control elements that improve the metrological characteristics. The functioning of the autoreflection scheme and the developed mirror element has been proven for measuring linear displacement. A device capable of simultaneously controlling two linear and two angular coordinates has been developed.

The algorithm solves the problems of measuring the coordinates, when overlapping expanded the metrological possibilities of using such sensors. The developed software solves the measurement problem and recognizes images in the event of their overlap. The accuracy of measuring the coordinates of this algorithm meets the requirements for the accuracy of the entire device, allowing you to register displacements measured in angular seconds.

Acknowledgements This work was financially supported by Government of the Russian Federation, Grant (08-08).

References

1. Mikheev, S.V., Konyakhin, I.A., Barsukov, O.A.: Optical-electronic system for real-time structural health monitoring of roofs. In: Proceedings of SPIE, vol. 9896, p. 98961C (2016)
2. Korotaev, V.V., Pantyushin, A.V., Serikova, M.G., Anisimov, A.G.: Deflection measuring system for floating dry docks. Ocean Eng. IET **117**, 39–44 (2016)
3. Nekrylov, I.S., Korotayev, V.V., Denisov, V.M., Kleshchenok, M.A.: Modern approaches to the design and development of optoelectronic measuring systems. In: Proceedings of SPIE—2016, vol. 9889, p. 988920 (2016)
4. Hoang, P., Konyakhin, I.A.: Autocollimation sensor for measuring the angular deformations with the pyramidal prismatic reflector (2017)
5. Konyakhin, I.A., Moiseeva, A.A., Moiseev, E.A.: Configurations of the reflector for optical-electronic autocollimator. In: Proceedings of SPIE, vol. 9889, p. 98891S (2016)
6. Serikova, M.G., Pantyushina, E.N., Zyuzin, V.V., Korotaev, V.V., Rodrigues, J.J.P.C.: Accurate invariant pattern recognition for perspective camera model. In: Proceedings of SPIE, vol. 9530, p. 95300O (2015)
7. Korotaev, V.V., Kleshchenok, M.A.: The choice of marks for systems with noncontact position control. In: Proceedings of SPIE IET, vol. 9525, p. 95253T (2015)

Airspace Monitoring in the Event of a Critical Situation with the Help of an Active Optical-Electronic Device as Part of a Robotic Complex

Leonid V. Smirnov and Victoria A. Ryzhova

Abstract *Problem statement*: The interaction of the Intelligent Electromechanical System (SEMS) with lidars is highly relevant at this time. The symbiosis of these areas will allow to properly monitoring potentially dangerous territories and zones of industrial facilities. The expediency of using robotized systems is conditioned by the ability to transmit real-time information about the state of the probing space to the appropriate authorities for taking prompt and adequate measures. *Purpose of research*: The introduction of lidar technology in the robotic meteorological complex. For the successful implementation and identification of the weather phenomenon in the meteorological robotic complex, it is necessary to specify the main components of lidar sensing. *Results*: Paramount will be given the parameters of the atmosphere, namely, the ultraviolet region of the optical range. After that, substances indicators characterizing this phenomenon will be given. Then you need to set the threshold power of the backscatter signal on the photo detector, which will allow identifying substances with a concentration of up to ppb. *Practical significance*: The using of this technology will further improve the efficiency of detection and identification of weather phenomena in particular and in general. Information from the complex, transmitted to special services, will allow for timely response to weather events and minimizing their consequences.

Keywords UV · Hurricane · Atmosphere · SEMS · Backscatter power

1 Introduction

The interaction of the Intelligent Electromechanical System (SEMS) with lidars is very relevant nowadays. The symbiosis of these systems will allow to properly monitor potentially dangerous areas and zones of industrial facilities. This article will consider the application of similar system for remote sensing of an emerging hurricane, located nearby complex of engineering objects. The expediency of using robotized systems is due to the ability to transmit real-time information to the appropriate

L. V. Smirnov (✉) · V. A. Ryzhova
ITMO University, Saint Petersburg, Russia
e-mail: e*as13@ro.ru

© Springer Nature Switzerland AG 2020

A. E. Gorodetskiy and I. L. Tarasova (eds.), *Smart Electromechanical Systems*, Studies in Systems, Decision and Control 261, https://doi.org/10.1007/978-3-030-32710-1_16

authorities on the state of the probing space in order to take prompt and adequate decisions.

In addition, the introduction of lidar technology in the robotic complex is innovative. To successfully integrate and identify a weather phenomena in a meteorological robotic complex, it is necessary to specify the main components of lidar sensing.

First of all you need to set the parameters of the atmosphere. After that, you need to specify substances-indicators which are characterizing this phenomenon. Then you need to set the threshold power of the backscatter signal on the photodetector, which will allow identifying substances with a concentration of up to ppb.

2 Spectral Range

For the successful integration of the system in the field of sensing firstly it is necessary to pay attention to the key factors in this matter. Such key factors include the composition of the atmosphere and the distribution of the atmosphere in the optical range of the spectrum. The composition of the atmosphere directly affects the sensing method, since some of the key methods are based on the choice of the atmospheric reference component. The atmospheric transparency windows allow you to work in the field of sensing with the least signal loss, this is necessary to achieve the maximum response from the object under study in the selected spectral range.

Table 1 shows the main components of the Earth's atmosphere; it can be seen from the table that the amount of nitrogen in the atmosphere is the highest, so using this component as a reference for some sensing methods is the best option.

$$M(\lambda) = (2^*h^*c\^{}2)/((\lambda\^{}5)^*(EXP(h^*c/(\lambda^*k^*T) - 1)) \quad (1)$$

where, $M(\lambda)$ is the luminosity at the dyne of the wave, h is the Planck constant, c is the speed of light, k is the Boltzmann constant, T is the temperature on the Kelvin scale, λ is the wavelength.

In order to isolate the intensity of solar radiation, we should go back to the Planck luminosity Formula (1). According to this formula, we obtain the spectral characteristic of the absorption of an absolutely black body, shown in Fig. 1. It is well known, that at a temperature of 6500 K the spectral component as close as possible to the radiation emitted by the sun [1].

Table 1 The components of the Earth's atmosphere

Substance	Percentage component (%)
Nitrogen	78.1
Oxygen	20.96
Argon	0.94
Other components	0.001

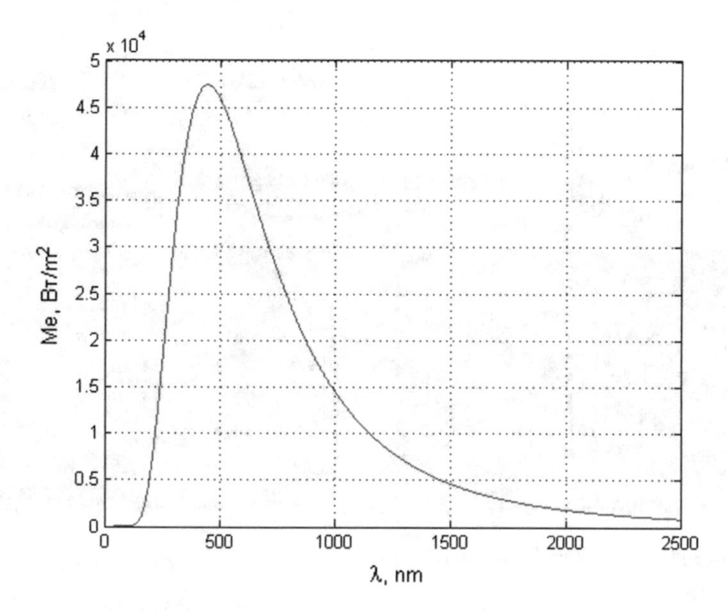

Fig. 1 The intensity of radiation of solar illumination [1]

The study of the spectral characteristics of solar radiation suggests that the best option for sensing is the ultraviolet optical range. In addition, solar illumination in this area is minimal, which leads to an increase in the reflected signal from the object of study.

Thus, the first criterion for successful integration of the system is using of the ultraviolet optical range (UV) from 200 to 380 nm.

3 Substance Indicators

Hurricane, as a weather phenomenon, is a swirling air flow, mixed with sand, smoke and other particles, except that in the center of the hurricane there are areas of low pressure. Based on the information given in source [2], it follows that the vast majority of hurricanes originate on the sea surface.

Figures 2 and 3 clearly demonstrate not only the most significant stages of the formation of a hurricane, but also its structure. From these illustrations and the information above, it can be concluded that a hurricane carries a large amount of sea salt. Natural sea salt contains about 90–95% NaCl (sodium chloride) and up to 2% of other minerals: magnesium salts, calcium salts, potassium salts, manganese salts, phosphorus salts, iodine salts and other substances [4].

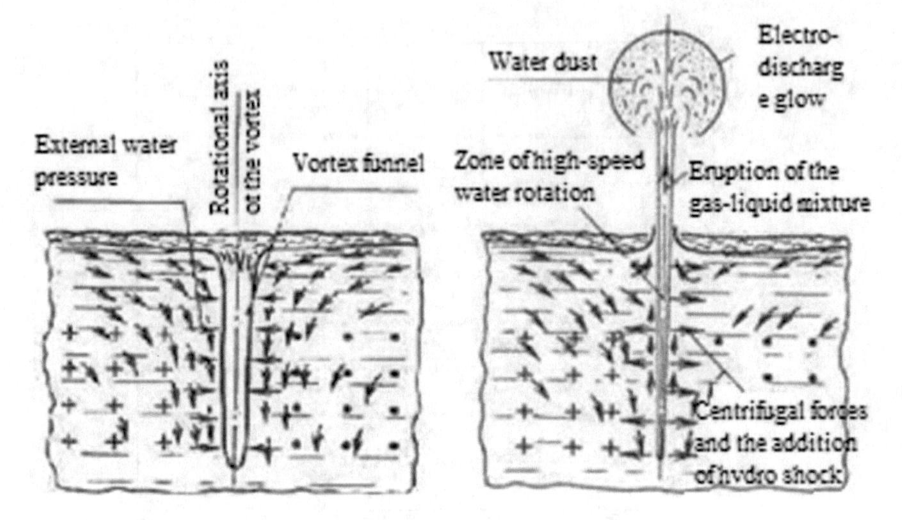

Fig. 2 Key generation stage hurricane [3]

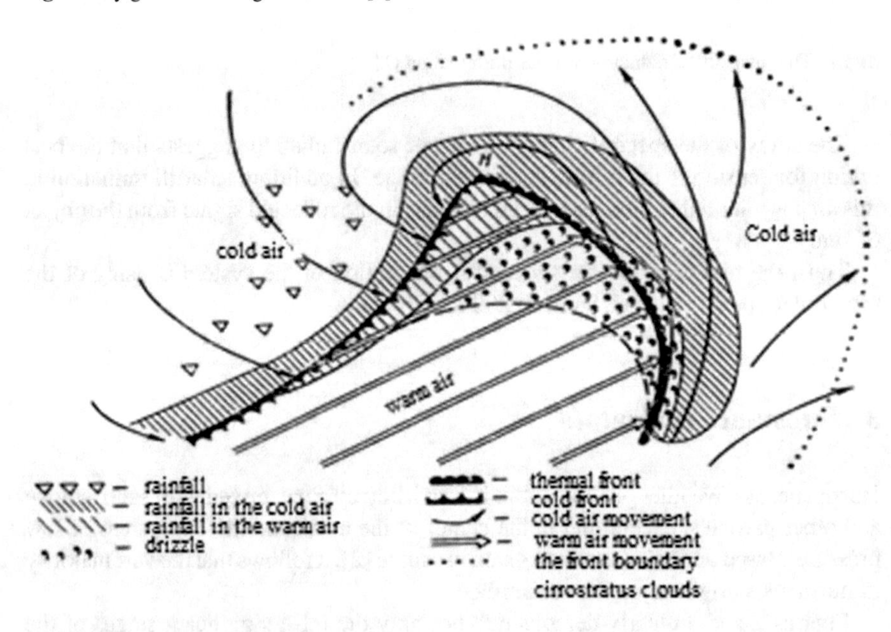

Fig. 3 The components of a hurricane [3]

Table 2 Ionic composition of sea salt [4]

Substance	Content (mg/l)	Salinity (%)
Chlorides (Cl^-)	19,385	55.03
Sodium (Na^+)	10,752	30.59
Magnesium (Mg^{2+})	1,295	3.68
Calcium (Ca^{2+})	416	1.18
Potassium (K^+)	390	1.11

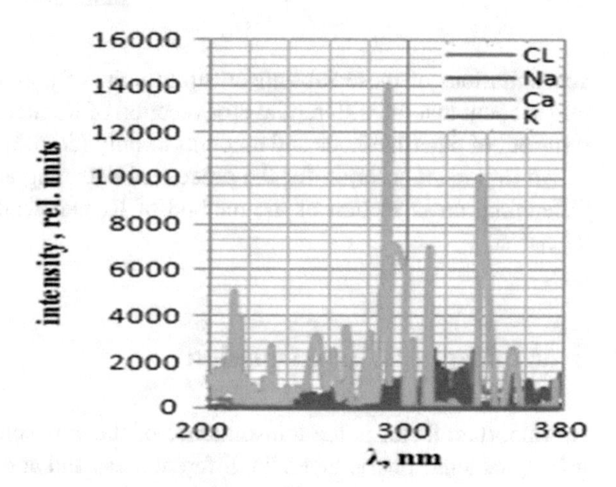

Fig. 4 The intensity of the substances of the indicators in the UV range [5]

Table 2 shows the most significant components of sea salt, yet sodium and chlorides have the greatest influence, besides increasing concentrations of other substances, such as calcium, magnesium and potassium, may allow the detection of sea salt in the air.

Figure 4 shows the radiation intensity, in relative units, for indicator substances in the ultraviolet region of the spectrum. As a consequence, all these substances should be chosen as indicators, namely chlorine, sodium, magnesium, calcium and potassium. Such a wide range of indicators will make it possible to reliably judge an increase in the amount of sea salt in the atmosphere.

4 Features of Remote Sensing

When using a laser remote sensing sensor, it becomes possible to carry out a detailed analysis of the objects and environmental components under study according to their spectral characteristics [6].

Among the methods of remote laser sensing most attention is paid to the method of Raman scattering, Fig. 5. Using the method of Raman scattering, it is possible not only to detect a wide range of indicator substances, but also to detect ultra-low

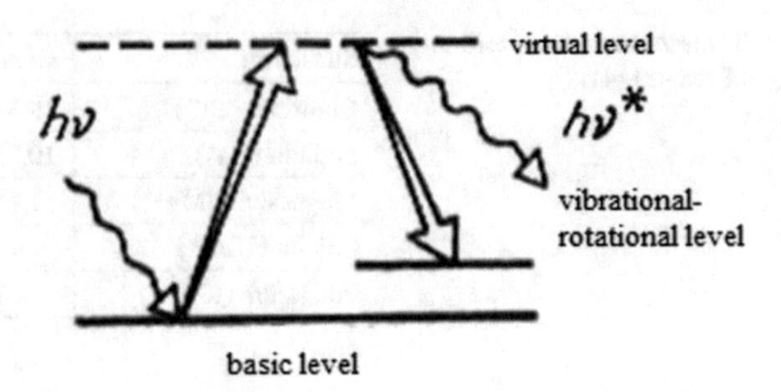

Fig. 5 The principle of the method of Raman scattering [6]

concentrations of these substances, up to units of ppb. It is also necessary to take into account that the differential cross section of Raman scattering is smaller than it turns out of other methods and amounts to only (dσ/dΩ)10-30-10-27 $см^2$/sr [6].

An important criterion for the process of detecting an incipient hurricane is the differential cross section of the method of Raman scattering, which is 10-30-10-27 $см^2$/sr.

5 Atmospheric Transmission

An important factor is the transmittance of the atmosphere. In view of the heterogeneity of atmospheric layers in different areas and at different altitudes above sea level, this parameter should be determined in advance. No less important component is the humidity and water content on the studied routes. For the calculation, the Elder-Strong method is used in the temperature range from −15 to +50 °C.

Consider the case when radiation occurs at a wavelength of 1064 nm. The selected wavelength is in the infrared region of the spectrum, which corresponds to the first harmonic of a YAG-Nd solid-state laser.

This solid-state laser was chosen as the source of optical radiation for solving the tasks posed for remote sensing and detection of indicator substances in open space.

The calculation of the capacity of the atmosphere by the method of Elder—Strong [6–8] is calculated by Formula (2).

$$\tau_n = t_0 - K_1 \cdot \lg(\omega \cdot 100_0) \quad (2)$$

where t_0 and K_1 is a coefficient depending on the wavelength, ω is the water content in meters.

In practice, the concept of water content is used to calculate the absorption of radiation in the atmosphere, i.e. the amount of water deposited in a single section layer of length L:

$$\omega = \frac{L \cdot f \cdot a_\varphi}{\gamma \cdot 10^6}, \quad (3)$$

Table 3 The values of humidity, water content and transmittance at different temperatures

t (°C)	a (g\m^3)	ω (mm)	τ (%)
−15	1.6	6.4	92.99
−10	2.35	9.4	90.24
−5	3.41	13.64	87.57
0	4.86	19.44	85.03
5	6.32	25.28	83.15
10	9.41	37.64	80.30
20	17.32	69.28	75.92
30	30.38	121.52	71.90
40	51.1	204.4	68.17
50	82.8	331.2	64.71

where L is the distance in meters; f is relative humidity, in relative units; a_φ is absolute humidity; γ is specific gravity of water, g/m^3.

With height, the concentration of water vapor rapidly decreases and can be calculated:

$$a_\varphi = a_0 \cdot 10^{-H/K}, \tag{4}$$

where a_0 is the humidity at sea level; H—height, in meters; K—coefficient (K = 5000, if the height in meters).

Calculation of the atmospheric transmittance on a horizontal highway with a length of 5000 m in the research temperature range was from −15 to + 50 ° C. The results of the calculation make it possible to assess the change in the atmospheric transmittance over the entire temperature range. The obtained data of the atmospheric transmission coefficient, the water content and the humidity of the environment are given in Table 3. Figures 6, 7 and 8 clearly demonstrates the dependence of the

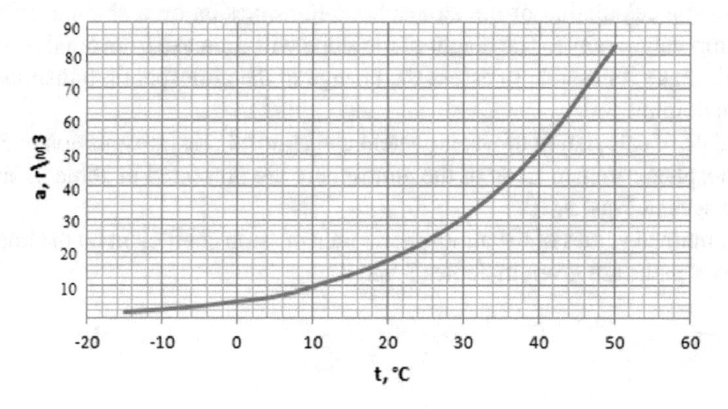

Fig. 6 Dependence of air humidity on the temperature on the horizontal route

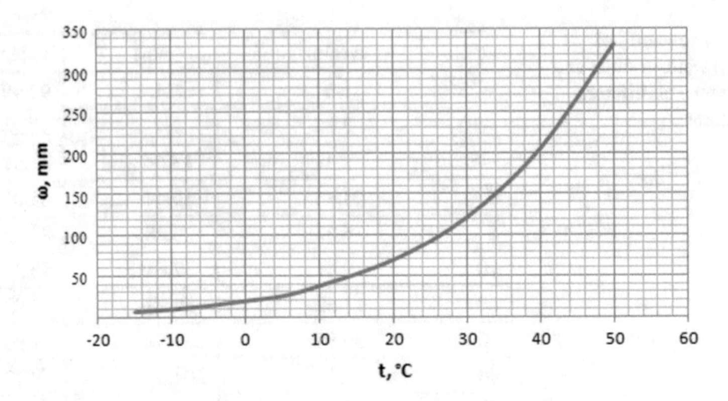

Fig. 7 Dependence of water content on temperature on a horizontal road

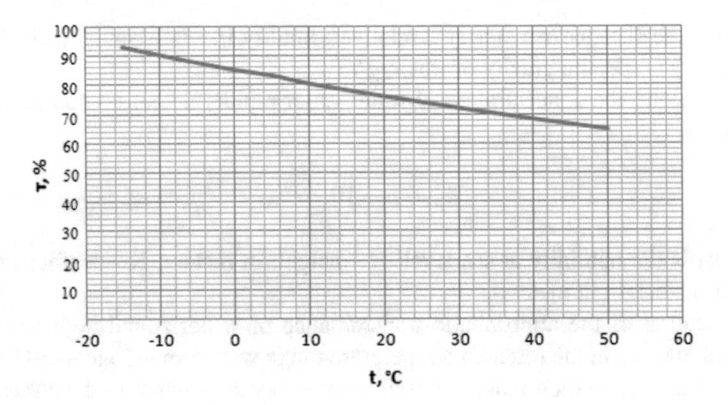

Fig. 8 Dependence of the atmospheric transmittance on the temperature on the horizontal route

atmospheric transmittance over the entire temperature range.

Perform a calculation of the atmospheric transmission on a sloping path in the same temperature range. The height of the slope will decrease from 5000 to 1000 m. This will make it possible to assess the change in the atmospheric transmittance at different heights.

The data of calculation of water content and atmospheric transmission coefficient and atmospheric transmission of the atmosphere are presented in Table 4, and their dependences in Figs. 9, 10.

The obtained values of the atmospheric transmission coefficient on the horizontal and sloped paths are given in Tables 3 and 4.

Table 4 Water values and transmittance at various temperatures

	OX = 5000 m; OY = 5000 m		OX = 5000 m; OY = 4000 m		OX = 5000 m; OY = 3000 m	
t (°C)	ω (mm)	τ (%)	ω (mm)	τ (%)	ω (mm)	τ (%)
−15	3.06	98.28	3.08	98.21	14.20	87.28
−10	4.49	95.52	4.53	95.46	20.85	84.53
−5	6.52	92.85	6.58	92.79	30.26	81.86
0	9.30	90.31	9.38	90.25	43.13	79.32
5	12.09	88.43	12.20	88.37	56.09	77.44
10	18.00	85.58	18.16	85.52	83.52	74.59
20	33.15	81.21	33.43	81.15	153.73	70.21
30	58.13	77.18	58.64	77.12	269.65	66.19
40	97.79	73.45	98.65	73.39	453.56	62.46
50	158.45	70.00	159.84	69.93	734.94	59.01

	OX = 5000 m; OY = 2000 m		OX = 5000 m; OY = 1000 m	
t (°C)	ω (mm)	τ (%)	ω (mm)	τ (%)
−15	5.51	94.071	24.341	83.42
−10	8.08	91.32	35.751	80.67
−5	11.73	88.65	51.881	78.00
0	16.72	86.11	73.941	75.46
5	21.75	84.23	96.151	73.58
10	32.37	81.38	143.161	70.72
20	59.59	77.01	263.511	66.35
30	104.53	72.98	462.20	62.33
40	175.83	69.25	777.44	58.60
50	284.90	65.79	1259.74	55.15

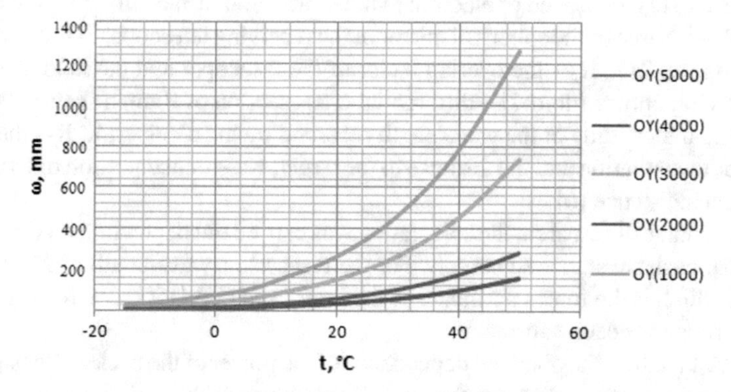

Fig. 9 Dependence of water content on temperature on an inclined route

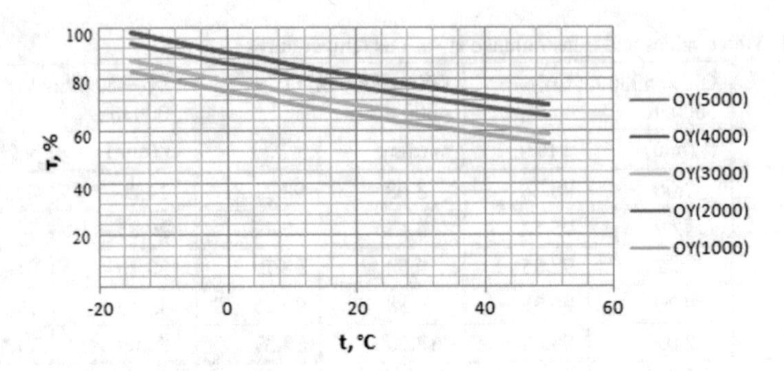

Fig. 10 Dependences of atmospheric transmittance on temperature on an inclined path

6 The Threshold Power

The result of this work was the calculation of the basic lidar equation for the power of the backscatter signal (5). The calculation was carried out for three different substances, namely for chlorine, sodium chloride and calcium chloride.

$$P(\lambda, R) = PLK1\Delta RTA0T(\lambda, R)(d\sigma/d\Omega)NaR2 \quad (5)$$

$$K1 = \xi(\lambda)\xi p(\lambda) \quad (6)$$

$$\Delta R = cTd/2 \quad (7)$$

P(λ, R)—power at the wavelength recorded by the receiver; PL—laser power at wavelength; K1—lidar constant (6); ξ(λ)—spectral transmission coefficient of the optical system; ξp(λ)—spectral transmittance of optical radiation receiver; ΔR—distance step (7); c—speed of electromagnetic radiation in vacuum; τd—pulse duration; A0—the area of the spherical mirror of the receiving telescope; T—atmospheric transmittance; T(λ, R)—the overlap factor of the laser spot and the field of view of the receiving optics; (dσ/dΩ)—differential cross section of Raman scattering of the molecules under study at the wavelength recorded by the receiver; λ, R—the wavelength of the radiation and the distance to the target; Na—concentration of molecules of the test substance [6].

In the course of research, the calculation was carried out depending on the sensing range, for each substance separately. For the research, a wavelength of 266 nm was chosen, which is the fourth harmonic of the laser used (Y3Al5O12 − Nd3+) and the sensing range for each substance.

Figure 11 shows the obtained dependence of the power of the backscatter signal on the photodetector for indicator substances depending on the sensing range. From this dependence it ensues an obvious fact that while decreasing the distance to the sensing

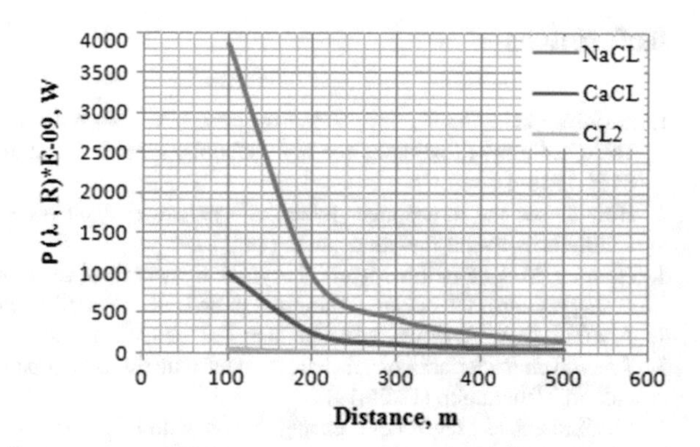

Fig. 11 Backscatter signal strength versus distance [6]

Table 5 The power recorded by the receiver at a wavelength of 266 nm

Distance (m)	P(λ, R), WNaCL	P(λ, R), WCaCL	P(λ, R), WCL2
100	3.86784E−06	9.97579E−07	3.18795E−08
200	9.66961E−07	2.49395E−07	7.96986E−09
300	4.2976E−07	1.10842E−07	3.54216E−09
400	2.4174E−07	6.23487E−08	1.99247E−09
500	1.54714E−07	3.99032E−08	1.27518E−09

object, the response of the power of the backscatter signal on the photodetector is increasing.

Table 5 shows the low threshold values of the indicators of each of the substances necessary to automate the process, depending on the sensing range.

7 Conclusions

To integrate successfully the remote sensing system with the robotic complex, it is necessary to take into account a number of important parameters of the remote sensing system. The first important criterion is the reduction of the optical range to the region from 200 to 380 nm. Then set up the system to detect indicator substances such as chlorine, sodium, magnesium, calcium and potassium. An important criterion for the process of detecting an incipient hurricane is the differential cross section of the method of Raman scattering, which is 10-30-10-27 см^2/sr. The atmospheric transmittance in different temperature limits is an equally important component of further work, the data are given in Tables 3 and 4.

The threshold power of the backscatter signal on the photodetector will allow to properly monitor the received signal; the data are given in Table 5.

References

1. Ishanin, G.G., Chelibanov, V.P., Korotaev, V.V. (eds.): St. Petersburg [and others]: Lan: yl., Table. 24 cm—(Textbooks for high schools, special literature), p. 303 (2014). ISBN 978-5-8114-1048-4
2. How to see the atmosphere [electronic resource] open access. https://www.nasa.gov/image-feature/how-to-see-the-atmosphere (19.05.19)
3. Herman, M.A.: Satellite Meteorology: Fundamentals of Cosmos. Methods of Research in Meteorology. Gidrometeoizdat, Leningrad, p. 367: ill., maps; 22 centimeters (1975)
4. Alyokin, O.A., Lyakhin, Y.I.: Chemistry of the Ocean. Gidrometeoizdat, Leningrad (1984)
5. Spectral characteristics of substances [electronic resource] open access. http://www.spectralcalc.com/info/about.php (12.06.19)
6. Mezheris, R.: Laser remote sensing (Trans. with English). Mir, Moscow, p. 550, Ill (1987)
7. Ishanin, G.G., Pankov, E.D., Andreev, A.L., Polschikov, G.V.: Sources and receivers of radiation: a manual for students of optical specialties of universities. SPb.: Polytechnic, p. 240, Il (1991)
8. Kozintsev, V.I., et al.: Fundamentals of quantitative laser analysis. Izd-vo MGTU im. NE Bauman, Moscow, T. 464 (2006)

Optical-Electronic System for Measuring Spatial Coordinates of an Object by Reference Marks

Hoang Anh Phuong, Alexey A. Gorbachev, Igor A. Konyakhin and Tong Minh Hoa

Abstract *Problem statement*: Optical-electronic systems can be used to control the elements of large objects such as radio telescope mirrors, floating docks, bridges etc. Often, large objects have smart electromechanical systems (SEMS) to adjust the position of their elements in space. The processing unit uses reference marks to calculate the spatial position of the object's elements and gives commands to the adjustment mechanisms to set the elements at a given position. An external force may be exerted on the measuring system and the system may rotate about an arbitrary axis. Therefore, the spatial coordinates will be calculated with an error. *Purpose of research*: Development of an algorithm to compensate for the random rotation of the measuring system. *Results*: Algorithm for finding the direction of the rotation axis and the rotation angle of the measuring system is proposed. The results of calculating the spatial coordinates on the theoretical and experimental models were compared. The developed algorithm in combination with the optical scheme allows to create measuring systems with selective invariance. *Practical significance*: The developed algorithm allows to increase the accuracy of measuring the spatial coordinates of the object. The proposed optical-electronic system can be used together with such systems as SEMS and automatic control systems.

Keywords SEMS · Optical-electronic system · Measuring system · Reference mark · Rotation parameters estimation

H. A. Phuong (✉) · A. A. Gorbachev (✉) · I. A. Konyakhin · T. Minh Hoa
ITMO University, 49 Kronverksky Pr, St. Petersburg 197101, Russia
e-mail: hoanganhphuongitmo@gmail.com

A. A. Gorbachev
e-mail: gorbachev@niuitmo.ru

I. A. Konyakhin
e-mail: igor@grv.ifmo.ru

T. Minh Hoa
e-mail: hoa.chiton@mail.ru

© Springer Nature Switzerland AG 2020

A. E. Gorodetskiy and I. L. Tarasova (eds.), *Smart Electromechanical Systems*, Studies in Systems, Decision and Control 261, https://doi.org/10.1007/978-3-030-32710-1_17

1 Introduction

Floating dry docks are large scale structures designed for building and maintaining ships. Dimensions of floating dry docks (length, width, and depth) depend on the size of ships to be docked. As sizes of modern ships tend to grow, the length of the modern docks can exceed 200 m, while the height and width can reach 50 and 100 m [1]. As it is a huge structure, a dry floating dock experiences heavy loads both from its own weight and from a docked ship. This leads to uneven deformations of the dock structure. Reliable measurement of a deflection magnitude refers to one of the most important tasks in floating dry dock operation. Current and accurate information about the deflection can prevent structural damage and failure by activating alarms and on-board deflection compensation systems [2].

When the measurement of deflection is performed in outside for the inspection of real structures, the influence of external conditions (because of temperature deformation) can lead to rotate the measurement unit. In this case, the deflection of the object cannot be measured because the effect of the measurement unit rotation is included in the measured deflection.

For control the measurement unit rotation it is suggested to use smart electromechanical systems, like High-Speed Parallel-kinematic Micropostioner with Controller or Hexapods [3]. According to the design the measurement unit is positioned on a mobile platform of the hexapod. The base platform of hexapod is placed in the middle of a wing deck of the dock (Fig. 1).

The collimation marks 1, 2 (Fig. 2) generates the collimation optical beam and directs it on beam-deflection system 3. The reflected beam passes through an lens 4 and it is received by the photoreceiving matrix 5, which is placed in the focal plane of lens 4. The video-frame from photoreceiving matrix 5 is calculated by the digital microprocessor 6 to obtain the reference mark coordinates and the rotation parameters of the measurement unit. After that, the digital microprocessor 6 gives a signal to

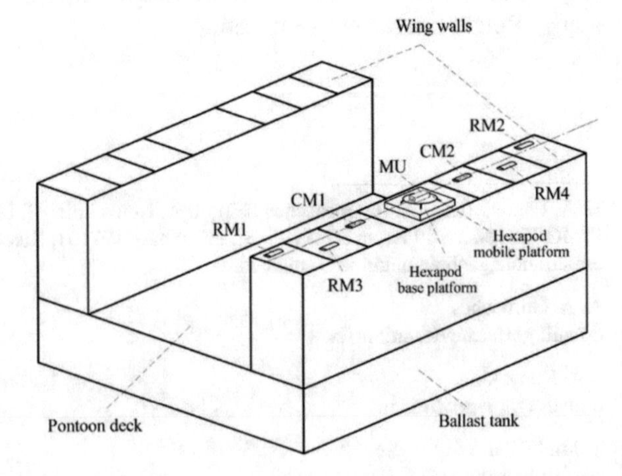

Fig. 1 The measuring systems of floating dock, RM—reference mark, CM—collimation mark, MU—measurement unit

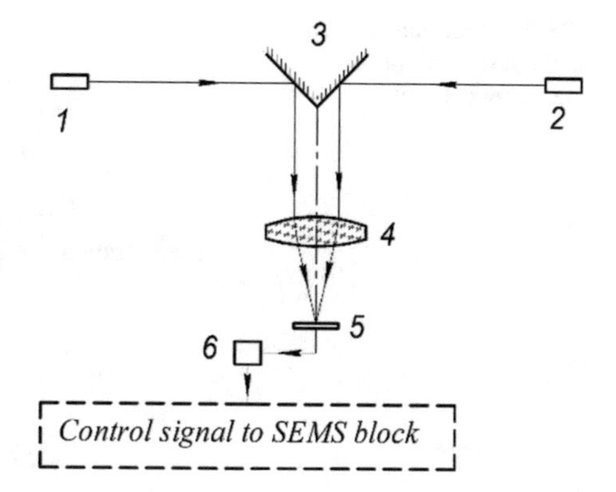

Fig. 2 Structure of the deflection measuring system for floating dry docks: 1, 2—source of optic radiation (reference marks and collimation marks), 3—beam-deflection system, 4—lens, 5—photo receiving matrix, 6—microprocessor system

SEMS system to define the movement to correct the rotation of the measurement unit.

2 The Structure of Electro-Optical Deflection Measuring System

The deflection measurement task for floating dry docks can be resolved by an implementation of a measuring system, which is capable of measuring a spatial position of predefined points along the wing deck of the dock. In this paper, an electro-optical system for transverse deflection measuring system for floating dry docks is considered (Fig. 1) [4, 5]. The system consists of a set of reference marks based on light emitting diodes (LED) and a measurement unit containing a camera and a beam-deflection system for two measurement channels for monitoring two opposite directions. All elements of the measurement unit are rigidly fixed to each other. Also, on the top deck of wing wall it is possible to install collimation marks CM1 and CM2 to measure the angular position of the measurement unit (Fig. 1).

A deflection of the floating dock causes a linear shift of the reference marks. As the marks are projected onto the camera sensor, the images with marks are captured and analyzed to give the shift of the marks. The developed software automatically detects the marks images and measures their location at the camera's sensors. The required accuracy of transverse displacement measurement of ±5 mm in the range of ±150 mm at distance of 100 m [2].

The coordinate $(X_i,\ Y_i,\ Z_i)$ of the reference marks (Fig. 3) in the coordinate system *XYZ* can be obtained by the following formulas:

$$\begin{pmatrix} X_i \\ Y_i \\ Z_i \end{pmatrix} = \mathbf{M}_i^{-1} \cdot \mathbf{T}_{x_c y_c z_c \to XYZ} \cdot \begin{pmatrix} -(Z_i + z_0)/f' & 0 & 0 \\ 0 & -(Z_i + z_0)/f' & 0 \\ 0 & 0 & 1 \end{pmatrix} \cdot \begin{pmatrix} x_i' \\ y_i' \\ 1 \end{pmatrix} \quad (1)$$

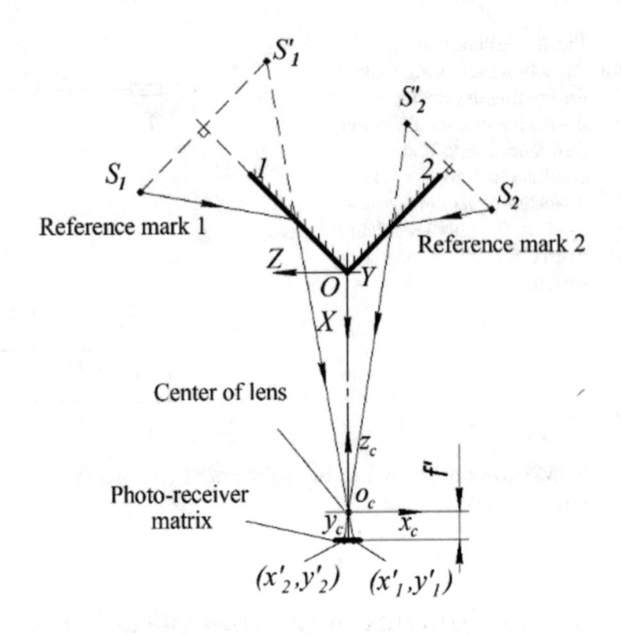

Fig. 3 Measuring system without rotation of the measurement unit

where $i = 1, 2$—the measuring channel number, $\mathbf{M}_i$—the reflection matrix of the ith flat mirror; $\mathbf{T}_{x_c y_c z_c \to XYZ}$—transformation matrix of coordinate system *XYZ* to camera coordinate system $x_c y_c z_c$; z_0—the coordinates of the origin of the coordinate system *XYZ* in camera coordinate system $x_c y_c z_c$ on z_c-axis; Z_i—the distance from the reference mark to measurement unit in the ith measuring channel; $\left(x_i', y_i'\right)$—the image coordinate of reference mark on the photo receiving matrix in the ith measuring channel.

According to the found coordinates of the reference marks, we can determine and construct the profile of the dock deflection [2].

As we said before, the temperature deformation can lead to rotate the measurement unit, this will change the parameter values $\mathbf{M}$, $\mathbf{T}_{x_c y_c z_c \to XYZ}$, z_c in expression (1) and as a result, it causes an error in determining the coordinates of the reference marks. Thus, it is necessary to eliminate the error of the measuring system caused by the rotation of the measurement unit.

3 The Algorithm for Rotation Parameters Estimation of the Measurement Unit

Consider the rotation of the measurement unit in the coordinate system *XYZ* around an arbitrary axis $\mathbf{L}$ by an angle φ. The direction of the rotation axis is determined in the world coordinate system *XYZ* by the zenith η and azimuthal χ angles (Fig. 4) with the unit directional vector $\mathbf{L} = \left(l_x, l_y, l_z\right)$, where $l_x = \sin\eta \cdot \sin\chi, l_y = \cos\eta$,

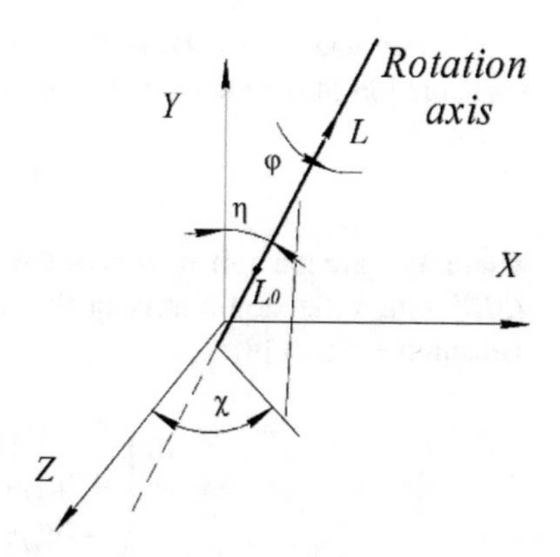

Fig. 4 Rotation parameters: zenith angle η, azimuthal angle χ and rotation angle φ

$l_z = \sin\,\eta \cdot \cos\,\chi$. The rotation axis **L** passes through the point $L_0(x_{L_0},\ y_{L_0},\ z_{L_0})$. The rotation parameter η, χ, φ can be estimate using collimation marks (Fig. 4).

In the coordinate system *XYZ* the unit normal vector of the reflecting surface 1 and 2 before the rotation of the measurement unit: $\mathbf{N}_1 = (\sin(\pi/4),\ 0,\ \cos(\pi/4))$ and $\mathbf{N}_2 = (\sin(\pi/4),\ 0,\ -\cos(\pi/4))$ (Fig. 5).

The unit normal vectors $\mathbf{N}'_1$, $\mathbf{N}'_2$ of the reflecting surface 1 and 2 after the rotation of the measurement unit can be determined by the following formula:

$$\mathbf{N}'_i = \mathbf{R} \cdot \mathbf{N}_i \tag{2}$$

where *i*—the measuring channel number, $i = 1, 2$, **R**—rotation matrix about axis **L** [6, 7]:

$$\mathbf{R} = \begin{bmatrix} l_x^2 \cdot (1-\cos\varphi) + \cos\varphi & l_x \cdot l_y \cdot (1-\cos\varphi) - l_z \cdot \sin\varphi & l_x \cdot l_z \cdot (1-\cos\varphi) + l_y \cdot \sin\varphi \\ l_x \cdot l_z \cdot (1-\cos\varphi) + l_z \cdot \sin\varphi & l_y^2 \cdot (1-\cos\varphi) + \cos\varphi & l_y \cdot l_z \cdot (1-\cos\varphi) - l_x \cdot \sin\varphi \\ l_x \cdot l_z \cdot (1-\cos\varphi) - l_y \cdot \sin\varphi & l_y \cdot l_z \cdot (1-\cos\varphi) + l_x \cdot \sin\varphi & l_z^2 \cdot (1-\cos\varphi) + \cos\varphi \end{bmatrix} \tag{3}$$

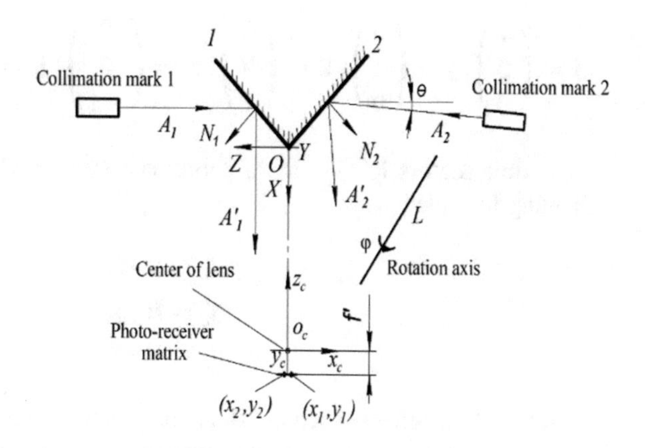

Fig. 5 Using collimation marks for rotation parameters estimation of the measurement unit

The unit vectors $\mathbf{B}_1$, $\mathbf{B}_2$ of the reflected rays from the reflecting surface 1 and 2 after rotation of the measurement unit are determined using the following formula:

$$\mathbf{B}_i = \mathbf{M}'_i \cdot \mathbf{A}_i \tag{4}$$

where $\mathbf{A}_i$—are the unit vectors of the incident beams on the reflecting surface 1 and 2, $\mathbf{M}'_i$—the reflection matrix of the reflecting surface 1 and 2 after rotation of the measurement unit [8]:

$$\mathbf{M}'_i = \begin{pmatrix} 1 - 2n'^2_{xi} & -2n'_{xi}n'_{yi} & -2n'_{xi}n'_{zi} \\ -2n'_{xi}n'_{yi} & 1 - 2n'^2_{yi} & -2n'_{yi}n'_{zi} \\ -2n'_{xi}n'_{zi} & -2n'_{yi}n'_{zi} & 1 - 2n'^2_{zi} \end{pmatrix}$$

where $(n'_{xi}, n'_{yi}, n'_{zi})$—are three components of the unit normal vectors $\mathbf{N}'_i$ in the coordinate system XYZ.

It should be pointed out that, in the process of rotation around any axis, the image plane rotates along with the mirror system, and therefore should be transformed to the camera coordinate system $x'_c y'_c z'_c$ after rotation of the measurement unit:

$${}^c\mathbf{B}_i = \mathbf{T}_{XYZ \to x'_c y'_c z'_c} \cdot \mathbf{B}_i \tag{5}$$

where $\mathbf{T}_{XYZ \to x'_c y'_c z'_c}$—the transformation matrix of coordinate system XYZ to $x'_c y'_c z'_c$

$$\mathbf{T}_{XYZ \to x'_c y'_c z'_c} = \begin{pmatrix} \mathbf{i}'_c \cdot \mathbf{i} & \mathbf{i}'_c \cdot \mathbf{j} & \mathbf{i}'_c \cdot \mathbf{k} \\ \mathbf{j}'_c \cdot \mathbf{i} & \mathbf{j}'_c \cdot \mathbf{j} & \mathbf{j}'_c \cdot \mathbf{k} \\ \mathbf{k}'_c \cdot \mathbf{i} & \mathbf{k}'_c \cdot \mathbf{j} & \mathbf{k}'_c \cdot \mathbf{k} \end{pmatrix} \tag{6}$$

where $\mathbf{i},\ \mathbf{j},\ \mathbf{k};\ \mathbf{i}_c,\ \mathbf{j}_c,\ \mathbf{k}_c;\ \mathbf{i}'_c,\ \mathbf{j}'_c,\ \mathbf{k}'_c$—the unit vectors of coordinate systems XYZ, $x_c y_c z_c$, $x'_c y'_c z'_c$, respectively; $(\mathbf{i}'_c \cdot \mathbf{i})$ denotes scalar product of unit vectors $\mathbf{i}'_c$ and $\mathbf{i}$;

$$\mathbf{i} = \begin{pmatrix} 1 \\ 0 \\ 0 \end{pmatrix}; \mathbf{j} = \begin{pmatrix} 0 \\ 1 \\ 0 \end{pmatrix}; \mathbf{k} = \begin{pmatrix} 0 \\ 0 \\ 1 \end{pmatrix}; \mathbf{i}_c = \begin{pmatrix} 0 \\ 0 \\ -1 \end{pmatrix}; \mathbf{j}_c = \begin{pmatrix} 0 \\ -1 \\ 0 \end{pmatrix}; \mathbf{k}_c = \begin{pmatrix} -1 \\ 0 \\ 0 \end{pmatrix}$$

The unit vectors $\mathbf{i}'_c,\ \mathbf{j}'_c,\ \mathbf{k}'_c$ in coordinate system XYZ can be obtained by the following formula:

$$\begin{aligned} \mathbf{i}'_c &= \mathbf{R} \cdot \mathbf{i}_c \\ \mathbf{j}'_c &= \mathbf{R} \cdot \mathbf{j}_c \\ \mathbf{k}'_c &= \mathbf{R} \cdot \mathbf{k}_c \end{aligned}$$

Images of the reflected beams ${}^c\mathbf{B}_i$ are projected on the photoreceiving matrix. The image coordinates in the coordinate system $x'_c y'_c z'_c$ depending on unknown rotation parameters can be determined by the following formula:

$$\begin{cases} x_i(\eta, \chi, \varphi) = f' \cdot \mathrm{tg}\left[\arcsin\sqrt{x_{Bi}^2 + y_{Bi}^2}\right] \cdot \frac{x_{Bi}}{\sqrt{x_{Bi}^2 + y_{Bi}^2}} \\ y_i(\eta, \chi, \varphi) = f' \cdot \mathrm{tg}\left[\arcsin\sqrt{x_{Bi}^2 + y_{Bi}^2}\right] \cdot \frac{y_{Bi}}{\sqrt{x_{Bi}^2 + y_{Bi}^2}} \end{cases} \tag{7}$$

where (x_{Bi}, y_{Bi})—two components of the unit vectors ${}^{c}\mathbf{B}_i$ along the x-axis and y-axis in the coordinate system $x'_c y'_c z'_c$, f'—focal length of the lens. Thus, the rotation parameters η, χ, φ of the measurement unit can be determined as a result of solving system of nonlinear equations (7) with known image coordinates x_i, y_i on the photoreceiving matrix.

4 The Algorithm for Reference Mark Coordinate Calculation with Unknown Rotation Axis Position of the Measurement Unit

The image coordinate $(x'_i; y'_i)$ of the reference marks on the photoreceiving matrix in coordinate system $x'_c y'_c z'_c$ can be calculated using following formula:

$$\begin{pmatrix} x'_i \\ y'_i \\ 1 \end{pmatrix} = \begin{pmatrix} -f'/z_{S'_i} & 0 & 0\,0 \\ 0 & -f'/z_{S'_i} & 0\,0 \\ 0 & 0 & 0\,1 \end{pmatrix} \cdot \mathbf{T}_{XYZ \to x'_c y'_c z'_c} \cdot \begin{pmatrix} \mathbf{M}'_i & -2 \cdot d_i \cdot \mathbf{N}_i \\ 0 & 1 \end{pmatrix} \cdot \begin{pmatrix} \mathbf{S}_i \\ 1 \end{pmatrix} \tag{8}$$

where $i = 1, 2$; $S_i = \left(X_{S_i}, Y_{S_i}, Z_{S_i}\right)^T$—coordinate of ith reference mark in coordinate system XYZ; T—symbol for matrix transpose; $\mathbf{N}_i$—the unit normal vectors of the ith reflecting surface after the rotation of the measurement unit; d_i—the perpendicular distance from the origin coordinate system XYZ to ith reflecting surface after rotation of the measurement unit; $\mathbf{M}_i$—the reflection matrix of the ith reflecting surface after rotation of the measurement unit; $\mathbf{T}_{XYZ \to x'_c y'_c z'_c}$—the transformation matrix of coordinate system XYZ to $x'_c y'_c z'_c$; f'—focal length of the lens; $z_{S'_i}$—the mirror image coordinate of ith reference mark in coordinate system $x'_c y'_c z'_c$ along z'_c-axis. Formulas for determining d_i, $\mathbf{T}_{XYZ \to x'_c y'_c z'_c}$, $z_{S'_i}$ after the rotation of the measurement unit are given in paper [9].

In the expression (8) there are the following variables: coordinate $\left(X_{S_1}, Y_{S_1}, Z_{S_1}\right)$, $\left(X_{S_2}, Y_{S_2}, Z_{S_2}\right)$ of the reference mark 1 and reference mark 2 (Fig. 6); rotation parameter η, χ, φ of the measurement unit, coordinate $\left(x_{L_0}, y_{L_0}, z_{L_0}\right)$ of point L_0, through which the rotation axis of the measurement unit passes. Since reference marks only shift in the plane parallel to the OXY plane, so we assume that the

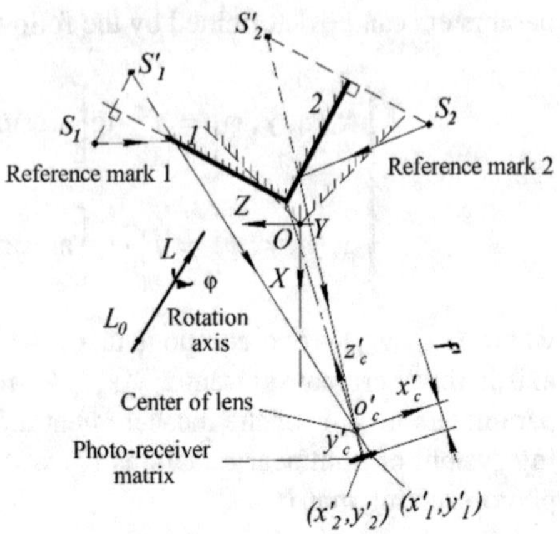

Fig. 6 Measuring system with rotation of the measurement unit around L-axis

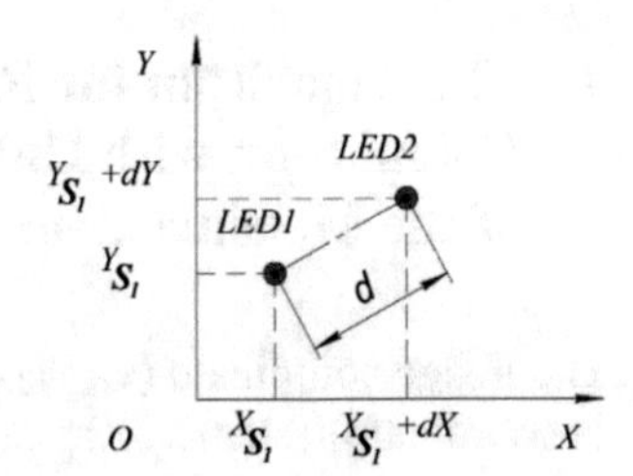

Fig. 7 Structure of the reference mark, where $d = \sqrt{dX^2 + dY^2}$—the distance between LED is known

values of the distance Z_{S_1}, Z_{S_2} are known, and the rotation parameters η, χ, φ of the measurement unit are defined above using collimation marks. Therefore, we need to find the 7 following unknown variables: X_{S_1}, Y_{S_1}, X_{S_2}, Y_{S_2}, x_{L_0}, y_{L_0}, z_{L_0}. It can be seen that from each expression (8) we have two equations. To find the 7 unknown variables above, we will need at least 7 independent equations. To obtain this, the structure of each reference mark containing two LED with known distance between LED is used (Fig. 7). Thus, the unknown variables X_{S_1}, Y_{S_1}, X_{S_2}, Y_{S_2}, x_{L_0}, y_{L_0}, z_{L_0} can be determined as a result of solving system of nonlinear equations (8) for 4 LED from 2 measuring channels.

5 Experiment and Results

Some experimental studies were conducted in the laboratory in order to validate the method described in previous section. The experimental setup is shown in Fig. 8. The measuring system is mounted on an optical bench 5 and contains two reference marks 1, LED power supplies 2, measurement unit 3 and an information processing unit 4 (personal computer).

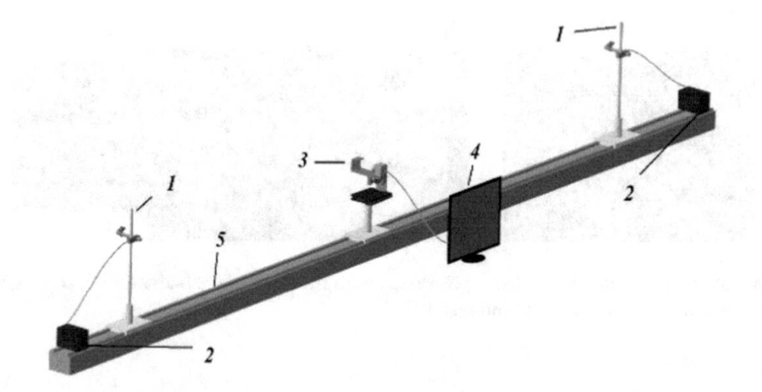

Fig. 8 Experimental setup for measuring the reference marks coordinate

The measurement unit is mounted on a rotation stage 3 providing accuracy of 1 angular minute for turning around the axes. The measurement unit contains a CMOS camera 1($f' = 12$ mm), a beam-deflection system 2 (Fig. 9a). The reference mark contains two LED with distance d = 16.5 mm between LED (Fig. 9b).

Figure 10 shows the images obtained in this experiment. The developed software automatically detects the marks images and measures their location at the camera sensor. The following algorithm is used for the detection of the mark image. First, an image with the switched on mark is captured (Fig. 10a). Then an image with the switched of mark is captured. As a result, a frame-to-frame difference which reveals the mark location is calculated. As can be seen from the cropped region of interest, the algorithm can recognize and calculate the image coordinates of the reference marks.

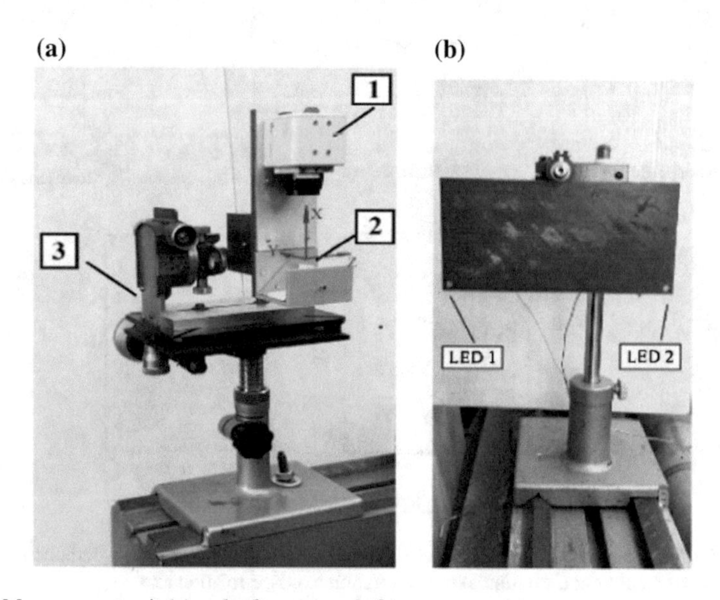

Fig. 9 Measurement unit (**a**) and reference mark (**b**)

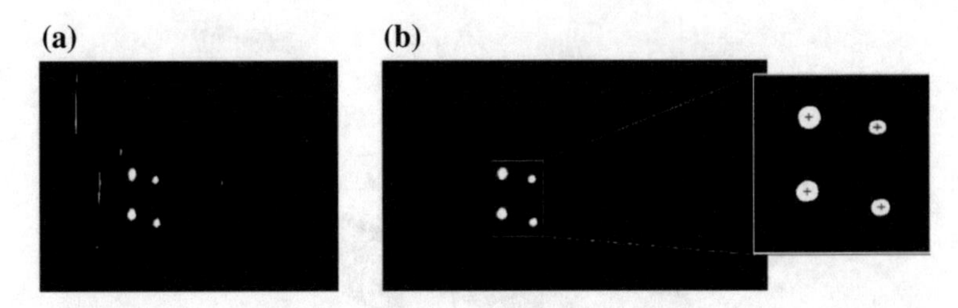

Fig. 10 Images of reference marks: **a** raw image with the mark and a background; **b** images after processing with cropped region of interest

The images of the reference marks without the rotation of the measurement unit are acquired to obtain the actual reference marks coordinate $\mathbf{S}_1^{act} = (-225.1, -46.7, 3445)^T$. Next, the measurement unit is rotated in turn around following axes: rotation axis 1 with $\mathbf{L} = (1, 0, 0)^T$ and $\mathbf{L_0} = (0, -97.8, 6.4)^T$; rotation axis 2 with $\mathbf{L} = (0, 1, 0)^T$ and $\mathbf{L_0} = (-72.1, 0, 61.1)^T$; rotation axis 3 with $\mathbf{L} = (0, 0, 1)^T$ and $\mathbf{L_0} = (0, -86.5, 0)^T$. Then the reference marks coordinate after the rotation measurement unit is calculated.

Differences between the reference mark coordinate before (actual) and after the rotation of the measurement unit is calculated and plotted in Fig. 11. As can be seen

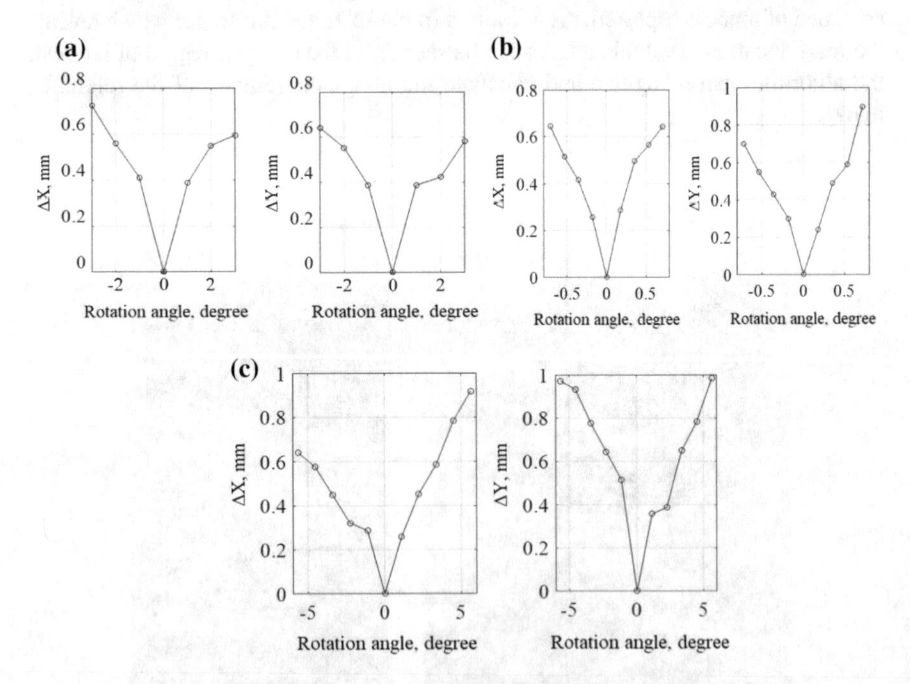

Fig. 11 Differences (ΔX, ΔY) between the reference mark coordinate with and without the rotation of the measurement unit; **a** rotation axis 1, **b** rotation axis 2, **c** rotation axis 3

from the graphs, the system successfully measures the reference mark coordinate with a sufficiently high accuracy (less than 1 mm) with the rotation of the measurement unit.

6 Conclusion

An optical-electronic system for measuring spatial coordinates of an object by reference marks is presented in this paper. A method for determining rotation parameters of the measurement unit was developed. The using SEMS system allows to correct the rotation of the measurement unit and reduce the measuring error of reference mark coordinate.

References

1. Gaythwaite, J.W.: Design of marine facilities for the berthing, mooring, and repair of vessels, p. 531 (2004)
2. Korotaev, V.V., Pantiushin, A.V., Serikova, M.G., Anisimov, A.G.: Deflection measuring system for floating dry docks. Ocean Eng. **117**, 39–44 (2016). https://doi.org/10.1016/j.oceaneng.2016.03.012
3. Gorodetskiy, A.E. (ed.): Smart electromechanical systems. In: Studies in Systems, Decision and Control, vol. 49. Springer International Publishing, Switzerland, p. 277 (2016). http://dx.doi/org/10.1007/978-3-319-27547-5_4
4. Gorbachev, A.A., Hoang, A.P.: Invariant electro-optical system for deflection measurement of floating docks. In: Proceedings of SPIE, vol. 10329, p. 103294F (2017). https://doi.org/10.1117/12.2269118
5. Hoang, A.P., Gorbachev, A.A.: An electro-optical system for transverse displacement measurement with rotation parameters estimation of the measurement unit. In: Proceedings of SPIE, vol. 11059, p. 1105918 (2019)
6. Kovács, E.: Rotation about an arbitrary axis and reflection through an arbitrary plane. In: Annales Mathematicae et Informaticae, vol. 40, pp. 175–186 (2012)
7. Korn, G.A., Korn, T.M.: Mathematical Handbook for Scientists and Engineers: Definitions, Theorems, and Formulas for Reference and Review. Dover Publications, New York, 1152 (2000)
8. Vanderwerf, D.F.: Applied Prismatic and Reflective Optics. SPIE, Bellingham, Washington, 303 (2010)
9. Hoang, A.P., Gorbachev, A.A., Konyakhin, I.A.: Image displacement analysis for electro-optical system for deflection measurement of floating docks. In: Proceedings of SPIE, vol. 11053, p. 110534A (2019). https://doi.org/10.1117/12.2512563

Simulation of Situation Control

Software Complex Testing Models for the Evaluation of Group Intelligence Robots

Andrey E. Gorodetskiy and Irina L. Tarasova

Abstract *Problem statement*: Questions of group interaction of intelligent electromechanical systems (SEMS) in complex robotic systems (CRS) play an important role in the analysis of their capabilities in performing various tasks. The paper proposes a solution to the problem of group intelligence assessment based on the results of test computer simulation of CRS. This problem occurs if necessary matching the best candidates to the RTK group from existing set of modules SEMS, for example, when you want to perform merge/split parts of RTK, the introduction of the group new robot or removal of existing or producing other transformation groups SEMS associated with the implementation of technological tasks. *Purpose of research*: Development of software complex for testing group intelligence assessment models in CRS based on SEMS in various environments. *Results*: A generalized structure of the software and hardware complex for testing models of intelligent robot groups and expert subsystems providing an assessment of group intelligence based on the results of test tests have been developed. The principles of teaching robots and methods of model correction based on the results of test tests are proposed. *Practical importance*: The generalized structure of the software and hardware complex for the evaluation of group intelligence CRS provides high functionality of models of groups of robots, taking into account the ideology of SEMS, which allows to evaluate group intelligence by computer modeling, taking into account the characteristics of the group members and the environment of their functioning.

Keywords Robotic systems · Group of robots · SEMS · Central nervous system · Fuzzy mathematical modeling · Group intelligence · Test computer modeling · Structure of the software complex · Expert subsystems · Models of robots and the environment

---

A. E. Gorodetskiy · I. L. Tarasova (✉)
Institute of Problems of Mechanical Engineering, Russian Academy of Sciences, St. Petersburg, Russia
e-mail: g17265@yandex.ru

A. E. Gorodetskiy
e-mail: g27764@yandex.ru

© Springer Nature Switzerland AG 2020

A. E. Gorodetskiy and I. L. Tarasova (eds.), *Smart Electromechanical Systems*, Studies in Systems, Decision and Control 261, https://doi.org/10.1007/978-3-030-32710-1_18

1 Introduction

Currently, the problem of optimizing the interaction of robots in the group, taking into account their intelligence when working together is becoming increasingly important. In complex robotic systems (CRS), built on the basis of intelligent Electromechanical systems (SEMS) [1] it is necessary to analyze not only the behavior of a single robot, which has due to the presence of the central nervous system (CNS) [2] appropriate behavior, but also the dynamics of the interaction of SEMS in RS. In assessing the performance of such CRS have to solve a wide range of problems. These problems include:

- evaluation of the ability of the SEMS group consisting of CRS for proper decision-making under uncertainty [3],
- determination of the optimal number of group members,
- evaluation of compatibility of group members taking into account static and dynamic features of each robot,
- taking into account the peculiarities of decision-making in the CNS of individual robots, etc.

Procedures that are commonly used to evaluate the efficiency and performance of a single robot can be based on testing and assessing the quality of decisions made, in accordance with a particular scale of its ability to reason [4]. When assessing the quality of work when solving a joint task by a group of robots, testing is complicated taking into account the so-called "psychological features", just as the work of a human operator in the group is evaluated [5].

A well-known vector approach to the evaluation of intelligence [6], based on the calculation of the assessment system of artificial intelligence with the use of test results. Such vector estimates usually contain static probabilistic components that evaluate the ability to solve fuzzy applied problems, and dynamic probabilistic components that evaluate the system's ability to self-study [7]. It is obvious that this approach can be used for an objective assessment of the suitability of certain robots to work in a particular RS. In this case, the comparison and selection of the best candidates from the set of tested robots can be based on the numerical evaluation of the results of dynamic testing of group work. In this case, the test results are recorded in different time intervals, which are divided into the entire test run. This is due to the specific differences between group and individual work activities. Therefore, the recommended formulas for calculating the components of the vector estimation of the intelligence of an individual robot cannot be directly applied to the assessment of the suitability of this robot for group interaction. In particular, as shown in [8, 9], there are factors such as group reaction time, the time required to make correct or important joint decisions, and other static and dynamic characteristics.

By analogy with human-machine systems (HMS) to assess the intelligence of CNS robots in CRS it is necessary to carry out its computer simulation with identification and evaluation of decisions in complex dynamic systems under conditions of uncertainty [10]. However, in this case, the structure of the mathematical model should include methods for evaluating the parameters characterizing the group intelligence of CRS. In addition, when assessing the quality of work when solving a joint task by a group of robots, testing is complicated taking into account the so-called "psychological characteristics", just as the work of a human operator in the group is evaluated. Therefore, it is important to create new systems and algorithms for testing groups of robots to assess the group intelligence team of working together robots based on SEMS, taking into account their "psychological" features.

In this paper we propose an approach based on fuzzy mathematical and computer modeling of the behavior of a robots group in the CRS and the dynamics of the environment of choice [11]. The use of the results of test modeling of a group of robots in the CRS in the system introduced in [3] estimates allows to take into account the characteristics of the CNS of each robot group and the psychology of the group in solving the joint problem. First, these estimates have static and dynamic components that take into account both the ability to make decisions under uncertainty and the ability to learn in the process. Secondly, they have components that characterize the psychological stability of groups. In some cases, vector estimates can be replaced by average estimates, which make it easy to rank from best to worst groups.

2 A Typical Structure of a Software System Testing of Models of Groups of Robots

Software testing complex (SCT), which provides testing of models of different groups of robots in order to assess their group intelligence, has structure of the tree type (Fig. 1). The SCT contains a supervisor (1) that supports dialogue with the operator (2) and a human-machine interface (3). The supervisor is designed to select operating modes and control software modules. These include the following modules:

- assembly and adjustment of models of control object (AAMCO) (4),
- assembly and adjustment of models of the environment (AAME) (5),

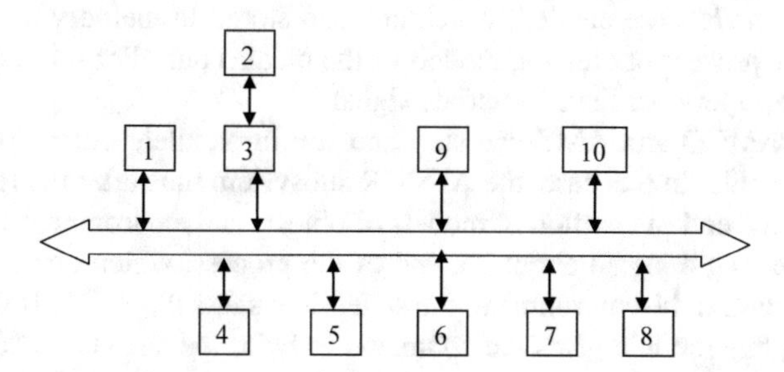

Fig. 1 Structure of the software package. 1—Supervisor, 2—the operator, 3—a human-machine interface, 4—assembly and adjustment of models of control object, 5—assembly and adjustment of models of the environment, 6—assembly of test models groups of robots, 7—testing groups of robots, 8—model training and correction, 9—text tasks, 10—storage of results of work of all blocks

- assembly of test models groups of robots (ATMGR) (6),
- testing groups of robots (TGR) (7),
- model training and correction (MTC) (8),
- text tasks (9),
- storage of results of work of all blocks (10).

Each module is connected to the supervisor via a controller. Blocks 4, 5, 6, 7, 8 are expert systems that contain a database (DB), knowledge base (KB) and inference engine (IE) for the model elements of the corresponding modules. In addition, module 4 contains the models of control objects, module 5—model environment, module 6—test model groups of robots, module 7—the random number generator, program testing, computer assessments of intelligence and module 8—corrective amendments of the training program. Module 9 is a database of test tasks'. Module 10—storage of the results of all blocks.

The system of testing robot models works as follows.

Before starting work, the operator 2 via the man-machine interface 3 controlled by 1 supervisor fills in the module 9 test tasks'. In each expert system of modules 4, 5, 6, 7, 8, the DB and the KB are filled with relevant information. In addition, the operator can adjust the DB and KB, as well as conduct training of expert systems on the results of test tests.

The supervisor then starts the AAMCO module. The controller of this module gives the command to create an object model. IE substitutes from the database the available data in the selected structure of the object model in the KB rules and creates a model of the object, which is stored in the memory module 10. Similarly, other object models are created, which are also stored in module 10. At the end of the process of creating object models, the module controller sends a corresponding signal to the supervisor.

After receiving a signal about the end of the process, which creates models of control objects, the supervisor starts the AAME module. The controller of this module gives the command to create an environment model. IE substitutes from the database the available data on the elements of the environment in the KB rules and creates a model of the environment, keeping it in the memory block 10. Similarly, other environment models are created, which are also stored in memory module 10. At the end of the process of creating models of the media controller subsystem AAME reports to the supervisor the appropriate signal.

Blocks AAMCO and AME can start and run in parallel, which speeds up the system as a whole. In this case, the ATMGR subsystem runs after the receipt of the signal about the end of creation of models of objects and environments.

After receiving a signal about the end of the process, which creates models of control objects and the environment, the supervisor starts the ATMGR module. The controller of this module gives the command to build the test task model from the control object models and environment models that are stored in the memory module 10. IE substitutes from the database the available data on the assembly of the tested model in the rules of the KB and creates a model of the test task, which is also stored in the memory module 10. At the end of the assembly process of the tested model of the robot group, the controller of the module 6 transmits a corresponding signal to the supervisor 1.

After receiving a signal about the end of the process, which carries out the assembly of the tested model of the group of robots, the supervisor starts the TGR module. The controller of this unit gives the command for testing, launching a random number generator and test programs. In the block of test tests IE substitutes from the database data on the situation in the test problem in the rules of the KB and it forms a sequence of situations in the test problem on the basis of a random sample from the random number generator. The module accumulates statistics of test results and calculates estimates of group intelligence. At the end of the process of calculating the estimates of group intelligence controller module TGR transmits the corresponding signal to the supervisor.

After receiving a signal about the end of the testing process, the supervisor transmits the test results via the human-machine interface to the operator. After analyzing the test results, the operator makes a decision on the completion of the work, or on the need to adjust the testing.

If necessary, the supervisor starts the MTC system by issuing the appropriate command to the MTC module controller. The controller of this module gives the command for correction. IE, substituting data from the database about possible adjustments in the rules the KB determines the sequence of correction and type of correction test patterns and transmits the corresponding information in AAMCO module, which introduces the corrective amendments in these models.

After adjustment at the request of the operator through the human-machine interface, the supervisor can instruct the controller of the module to start training, which analyzes the results of the correction, comparing the intelligence estimates obtained before and after the correction, and enter the successful values of the corrections in the database with the simultaneous correction of the relevant rules of the KB of the MTC module.

3 Assembly and Adjustment of Models of Control Object

The module AAMCO under the control of the controller provides for the assembly and adjustment of models of three types of intelligent robots: robot-forklift for warehouse work, robot-car at the intersection and robot-assembler mirror antenna system. The database of this module stores the structures and elements of the models of these robots. The rules are stored in KB:

- selection of the structure of different types of robots;
- selection of models of component parts of robots for the selected structure;
- formation of Central nervous systems of different robot models;
- connection of robot component parts models according to the selected structure.

The structures of robot models contain a list of their elements and connections between them, indicating all inputs and outputs for each type of robot. Elements of any robot model represent data on the state of the model: initial, required, current, permissible, permissible control actions and models of component parts of robots.

Initial state data is information about the initial coordinates and characteristics of the robot, as well as its components. For example, hands and hands with fingers for a robot forklift. In addition, additional information may be available on the initial state of the environmental sensors and the model status feedback sensors.

The data on the required state of the model contains a list of step-by-step transitions from the initial states to the required end states and transition constraints.

Data about the current state of the model contain information from the environment sensors and feedback state of the model.

Data on the permissible state of the model contains the permissible coordinates and characteristics of the robot, as well as its components.

In addition to these data on the state of the robot model and its structure, contains the values of permissible control actions and information about the components of the robot models. Information about the components of robot models contains parameters, static and dynamic characteristics of control systems models: electric drives, actuators and feedback sensors.

In KB there are rules for selecting and assembling the structure of the robot model, which are rules of the "if-then" type. The operator sets the robot type via the human-machine interface and the supervisor. IE substitutes the data type of the robot, selected from the database, the rules of BL, and the selection of its structure in general or the assembly of component parts.

The formation of the Central nervous system of robot models is carried out by means of control actions in the models of control systems, based on the analysis of the initial, current, permissible and required state, as well as the components of the robot and permissible control actions.

Assembly of robot models from their components is carried out in accordance with the rules of connection of models of robot components (control systems, electric drives, actuators, feedback sensors) selected structure, which is described by the specified parameters and characteristics.

Similarly, on the instructions of the operator through the human-machine interface and supervisor IE creates models of other types of robots, placing them in the repository of the results of all blocks (Fig. 1 module 10).

4 Assembly and Adjustment of Models of the Environment

The AAME module under the control of the controller provides for the assembly and configuration of models of three types of environments in which the work of groups of intelligent robots. These include warehouses, road junctions and near-earth orbits of satellites. Accordingly, the database of this module has three sections, where the structures of environments with elements of models of these environments are stored. KB also has three sections, where the stored selection rules in the model structure of the environment, rules for the selection of the models of elements for the chosen structure and rules of the association of the models of elements in the shared model of the environment functioning group of intelligent robots.

Environment structures contain for each type of environment a list of media model elements and the relationships between them, indicating all inputs and outputs.

The following software modules are the elements of the storage environment model:

- dimensions of premises,
- schemes of cargo location, their coordinates and dimensions,
- scheme of passages between the goods their coordinates and dimensions,
- schemes of departure and entry from the warehouse, their coordinates and dimensions,
- schemes of cargo delivery from the warehouse, their coordinates and dimensions,
- a set of parameters of the required states of warehouse models with a list of step-by-step transitions from initial to final states,
- list of restrictions during transitions,
- a set of parameters of the current state of the warehouse models with information from environmental sensors,
- the set of valid states of the models of warehouses with valid coordinates and characteristics of the warehouses as a whole and their component parts.

Elements of the model environment "crossroads" are:

- types of intersections,
- types of streets adjacent to the intersection,
- availability of traffic lights,
- types of traffic lights,
- schemes of the initial location of robots-cars,
- the dimensions of the robot car,
- current coordinates, direction and speed of robot cars,
- the list of restrictions on the movement of robots-cars,
- a set of permissible coordinates and speeds of robots-cars, etc.

The following software modules are the elements of the "near-earth orbits of satellites" environment model:

- types, parameters and characteristics of orbits,
- types, the parameters of the disturbing factors,
- static and dynamic characteristics of disturbing factors,
- number and type of satellites,
- current, valid and desired satellite coordinates
- the speed of satellites and installed on them elements of the mirror system,
- types, parameters and characteristics of the mirror system,
- permissible control actions, etc.

Rules for selecting the structure of the model of the functioning environment contain rules of the "if-then" type for the choice of the scheme for the given operator through the human-machine interface and the supervisor of the type of environment, rules for selecting models of environment elements for the selected scheme and rules for combining the selected models of environment elements in the General model of the environment of the group of intelligent robots.

IE is a software module that, under the control of the controller, performs a logical conclusion from the database pre-filled with current facts according to the rules of the BR in accordance with the laws of formal logic. The IE first selects from the BS the rules for selecting the structure of the environment model for the type of robot group specified by the operator and the type of environment for its functioning. Then, from the selected rules, selects the rules that correspond to the current facts of the database, distributes the selected rules by priorities and selects the structure of the environment model, which corresponds to the rule with the highest priority. At the next stage, the IE selects rules for choosing models of environment elements from the database, selects those that correspond to the selected environment structure and current database facts, distributes the selected rules by priorities and selects those models of environment elements that correspond to the rules with the highest priority. Then IE chooses from the KB rules of combining the selected models of elements of the environment into a General model of the environment for the functioning of a group of intelligent robots for the selected group of robots and the environment of its functioning, placing it in a module where all the results are stored. Similarly, on the instructions of the operator through the human-machine interface and supervisor creates models of other types of environments functioning groups of robots, placing them in a module where all the results are stored.

5 The Assembly of Test Models Groups of Robots

In module ATMGR under the controller provides for the assembly of the models of group three types of intelligent robots: robot forklift, the robot vehicle and robot collector mirror antenna system for the proper functioning in three environments: warehouses, intersections of roads and low-earth orbit satellites. Accordingly, the database has three sections, which store the layout of the members of different groups of robots in the possible environments of their operation, their coordinates, robot model schemes, as well as static and dynamic characteristics. In addition, the database stores the initial, permissible, current and final coordinates of robots, permissible control actions and goals of functioning in different environments, as well as options for group interaction algorithms for different target tasks.

The KB contains the following "if-then" rules:

- selection of the structure of the robot group models for the purpose of functioning,
- selection of algorithms for interaction of robot models in the group under the selected structure,
- formation of a group model of the same type and different types of models of robots and their environments.

IE is a software module that performs the output from the pre-filled current facts from the database to the rules of the KB in accordance with the laws of formal logic. IE first selects from the KB selection rules for the model structure of groups of robots taking into account the purpose of operation. Then, from the selected rules, the ones that correspond to the current facts of the database (the layout of the members of the group of robots in the model of the environment of their functioning, their coordinates, parameters of robot models, as well as their static and dynamic characteristics) are selected. IE distributes the selected rules by priority and chooses the structure of the model groups of robots, which corresponds to the rule with the highest priority. At the next stage, the IE selects the rules for choosing algorithms for interaction of robot models for the selected structure and purpose of operation, taking into account the current facts (parameters of robot models, their static and dynamic characteristics, initial, permissible, current and final coordinates of robot models, permissible control actions). Then IE distributes the selected rules according to priorities and selects those algorithms for interaction of robot models in the environment, which correspond to the rules with the highest priority. Then, the IE selects from the BR the rules for the formation of a group model from the same and different types of robot models and their environments, taking into account the location of groups in the surrounding space and robot movements in the specified directions. The operator prioritizes the selected rules and selects the rules of assembling the model of the group of robots in the environment, which correspond to the rules with the highest priority. After that, the group model and the environment of its functioning are assembled. This model is stored in the result store.

6 Testing Groups of Robots

The TGR module provides for testing of models of groups of robots-forklifts, robots-cars and robots assemblers, functioning in the appropriate models of environments. For example, warehouses, intersections of roads and near-earth orbits of satellites. The tested models of robot groups are formed with the help of the module "assemblies of tested models of robot groups" and stored in module 10 (see Fig. 1). The test results are used to calculate estimates of their group intelligence.

The elements of the DB module are: test task, possible limitations and obstacles, their parameters and characteristics, algorithms of test tests, formulas for calculating estimates of group intelligence with adjustable coefficients, test time, the number of attempts to perform the test task, types of failures and accidents and other collisions, formulas for calculating their number.

To calculate the estimates of group intelligence, you can use the following factors:

– coefficient of intellectual abilities of the jth group of robots [3]:

$$Q_j = \frac{1}{FN}\sum_{i=1}^{F}\sum_{n=1}^{N} q_{inj} \quad (1)$$

– creative coefficient of the jth group of robots:

$$V_j = \frac{1}{FN}\sum_{i=1}^{F}\sum_{n=1}^{N}\frac{1}{\Delta T}_{inj} \quad (2)$$

where ΔT_{inj}—time of the *n*th attempt to solve the *i*th problem by the *j*th group of robots, $i = 1, 2, 3, \ldots F$.
– the coefficient of motivational involvement of M_j of the *j*th group of robots is also calculated by Formula (2), but all test problems should be approximately of the same complexity.

By combining the entered coefficients, it is possible to calculate the intelligence coefficient of the *j*th group of robots in RTC:

$$IQ_j = k_q Q_j + k_v V_j + k_M M_j, \quad (3)$$

where k_q, k_v and k_M—adjustable weight coefficients set by the operator conducting the simulation test, depending on the purpose of the test RTC.

The KB is the following set of rules:

– selection of test tasks,
– selection of possible constraints and obstacles with their characteristics and parameters,
– selection of test algorithms,
– choice of formulas to calculate estimates of group intelligence, and custom indexes.

IE is a software module that performs a logical inference from the KB rules in accordance with the current facts from the database and the laws of formal logic. At the beginning of the IE selects from the KB rules that are responsible for the selection of test tasks for the test objectives set by the operator and stored in module 10 (see Fig. 1) group model and operating environment. Then, from the selected rules are selected rules that correspond to the current facts (types of obstacles and restrictions, test time, the number of attempts to perform the test task, the types of failures and accidents and other collisions), distributes the selected rules on the priorities and selects the test task, which corresponds to the rules with the highest priority. At the next stage, the IE selects the rules of selection of possible constraints and obstacles with their characteristics and parameters for the selected test task and stored in the TGR module (see Fig. 1) the group model and the operating environment. From the selected rules, the IE selects the rules that correspond to the current facts (possible constraints and obstacles, their parameters and characteristics) from the database, distributes the selected rules according to priorities and selects those constraints and obstacles with their parameters and characteristics that correspond to the rules with the highest priority. Next, the IE selects from the BS rules for the selection of algorithms for test tests, based on the selected test problem and possible constraints and obstacles. From the selected rules, the IE selects the rules that correspond to the current facts (formulas for calculating the estimates of group intelligence) from the database, distributes the selected rules according to priorities, selects the algorithm of test tests to which the rules with the highest priority correspond and transmits it to the test module. At the same time, the identifier of the selected algorithm is transmitted to the module for calculating intelligence estimates. Then, the IE selects from the KB the rules for the selection of formulas for calculating the estimates of group intelligence and adjustable coefficients for the selected algorithm of test tests. From the selected rules, the IE selects the rules that correspond to the current facts (types of failures and accidents and other collisions, formulas for calculating their number). It then distributes the selected rule according to the priorities and selects the formulas to calculate estimates of group intelligence and custom coefficients that correspond to the rules with the highest priority and transmits them to the module of calculation of estimates of the intelligence of TGR.

The module of test tests for the chosen algorithm of test tests by means of the random number generator substitutes obstacles and restrictions with their parameters and characteristics (probability of occurrence, coordinates, etc.) into the algorithm of test tests. In addition, the test module accumulates statistics of collisions and other collisions. Then passes it to the module for calculating intelligence estimates. In the TGR module under the control of the controller on a set of test statistics are calculated estimates of group intelligence on the selected IE formulas and stored in a special memory section of module 10, which also stores the identifier of the test algorithm used in this case, and in the memory of the training module.

7 Model Training and Correction

In module MTC provides for the calculation of the amendments according to the results of testing and correction models produced in modules of AAMCO, AAME and ATMGR. DB contains the acceptable and desired estimates of group intelligence test for different tasks and sets of possible amendments of the models. The KB contains rules such as "if-then" selection of corrections for the analysis of the values of the calculated estimates.

A inferencing engine (IE) is a software module that performs logical inference from a pre-filled with current facts from a DB according to the rules from the KB according to the laws of formal logic. In the beginning, IE selects all the correction rules from the KB, corresponding to the identifier of the used test algorithm stored in module 9, on the controller command launched on the basis of the operator's request through the human-machine interface and the supervisor. From the selected rules, the IE selects the rules that correspond to the current facts (valid and desired estimates of group intelligence corresponding to the identifier of the test algorithm used). Then, it prioritizes the selected rules, selects the selection rules that correspond to the rules with the highest priority, and passes them to the correction module. The selected corrections can be made to the tested models through the correction module at the operator's request. In addition, at the request of the operator, you can run the training module, which analyzes the results of the correction, comparing the intelligence estimates obtained before and after the correction, and the successful values of the corrections are recorded in the database of the training and correction module with the simultaneous correction of the relevant rules in the KB.

8 Conclusion

The generalized structure of the software package for testing models of groups of intelligent robots, including expert systems for creating dynamic models of interacting robots and environments. It allows you to evaluate the group intelligence of CRS by computer modeling, taking into account the characteristics of the group members and the environment of their functioning. The introduction of estimates of the results of the test simulation of a group of robots in the CRS in solving a joint problem can take into account the characteristics of each robot group and the psychology of the group. The supply of the complex with the training and correction module allows the operator to make the correction of robot models based on the results of testing and thereby ensure the improvement of the quality of CRS work.

It seems that a more subtle modeling of CRS characteristics and the study of the nature of variability in the time of the proposed group assessments based on the results of testing of candidates will help to identify a number of new important group psychological features for work in CRS, such as indecision, nervousness, patience, scrupulousness, etc. [8]. It is also important to note that the assessment of the group intelligence of robots should be improved in terms of assessing learning, professionalism, psychological stability and compatibility with the human observer (operator) or the head-controller of the technological process. Such an assessment can in many cases be key to making the right decisions under uncertainty.

Acknowledgements This work was financially supported by Russian Foundation for Basic Research, Grants 16-29-04424, 18-01-00076 and 19-08-00079.

References

1. Gorodetskiy, A.E.: Smart Electromechanical Systems, 277 p. Springer International Publishing (2016). https://doi.org/10.1007/978-3-319-27547-5
2. Gorodetskiy, A.E., Kurbanov, V.G.: Smart Electromechanical Systems: The Central Nervous System. Studies in Systems, Decision and Control, vol. 95, 270 p. Springer International Publishing Switzerland (2017). https://doi.org/10.1007/978-3-319-53327-8
3. Gorodetskiy, A.E., Tarasova, I.L.: Estimates of the group intelligence of robots in robotic systems. In: Gorodetskiy, A.E., Tarasova, I.L. (eds.) Smart Electromechanical Systems: Group Interaction. Studies in Systems, Decision and Control, vol. 174, 337 p. Springer International Publishing, 2019 (2018). https://doi.org/10.1007/978-3-319-99759-9-13

4. Gorodetskiy, A.E., Tarasova, I.L.: Intelligent software for automated testing of sensor systems [Intellektual'ny'e programmny'e sredstva dlya avtomatizirovanny'x ispy'tanij sensorny'x sistem]. In: Gorodetskiy, A.E., Kurbanov, V.G. (eds.) Physical Metrology: Theoretical and Applied Aspects, pp. 68–74. KN (1996)
5. Gorodetskiy, A.E., Al-Kasasbeh, R.T.: Vector assessments of group activities of operators [Vektorny'e ocenki gruppovoj deyatel'nosti operatorov]. In: Modern Problems of Social and Economic Development and Information Technologies, pp. 63–79. Baku (2004)
6. Al-Kasasbeh, R.T.: Statistical-similar model of organization work for small group information system operators. In: Proceeding of International Carpathian Conference ICCC, pp. 217–224 (1988)
7. Popechetelev, E.P.: Training for Studying Group, pp. 65–67. Leningrad Technical University News, Leningrad (1988)
8. Antonets, V.A., Anishkina, N.M.: Measurements and Perception in Man/Machine Systems, 12 p. Preprint of IAP RAS, N 518, Nizhny Novgorod (1999)
9. Yoshida, K., Yokobayashi, M.: Development of AI-Based Simulation System for Man-Machine System Behavior in Accidental Situations of Nuclear Power Plant. Japan
10. Gorodetsky, A.E., Tarasova, I.L.: Fuzzy Mathematical Modeling of Poorly Formalized Processes and Systems [Nechetkoe matematicheskoe modelirovanie ploho formalizuemyh processov i sistem], 336 p. SPb Publishing House of Polytechnical Institute (2010). (in Russian)
11. Gorodetsky, A.: Fundamentals of the Theory of Intelligent Control Systems [Osnovy teorii intellektual'nyh sistem upravlenija], 313 p. LAP LAMBERT Academic Publishing GmbH@Co. KG Publ. (2011)

Test Model of a Warehouse Loader Robot for Situational Control Analysis System

Andrey Yu. Kuchmin

Abstract *Introduction*: In the modern world, more and more tasks are performed by robotic systems with artificial intelligence. Of great interest is the creation of control systems for distributed groups of robots when operating under conditions of non-stationarity of the external environment and interaction of robots, as well as insufficient information about the situations that arise. One of the most promising directions for the synthesis of such systems is the use of situational control methods. Modern intellectual systems calculate control actions taking into account forecasting, using adequate models of control objects. The development of such models should meet the following criteria. Such a model should adequately describe the behavior of the control object, should be compact and economical in terms of computation, since such models are repeatedly used to predict in one iteration of the control calculation cycle. Testing and analysis of situational control systems are important tasks that are often solved by simulation methods using control object models. Therefore, the development and analysis of situational control systems of groups of robots and the creation of adequate models of the dynamics and kinematics of these robots are actual. *Purpose*: To develop a simplified kinematic and dynamic model of a warehouse loader robot for use in the analysis system of situational control of a group of robots. Methods: Creating a library of modules based on object-oriented modeling methods. The library allows you to simulate the kinematics and dynamics of various configurations of robots based on intelligent electromechanical modules with a parallel kinematic scheme (SEMS). *Results*: Kinematic and dynamic models of a warehouse loader robot created from modules with parallel kinematic scheme (SEMS) are proposed. The robot is equipped with two universal adaptive grips that simulate human hands. These models can be used to calculate the trajectories of the robots and calculate the movements of universal adaptive grippers. These models can be used as virtual robots for testing situational control systems of a group of robots. *Practical significance*: The research results are supposed to be used in the development of test equipment of situational control systems for a group of robots.

A. Yu. Kuchmin (✉)
Institute of Problems of Mechanical Engineering, Russian Academy of Sciences, St. Petersburg, Russia
e-mail: radiotelescope@yandex.ru

© Springer Nature Switzerland AG 2020

A. E. Gorodetskiy and I. L. Tarasova (eds.), *Smart Electromechanical Systems*, Studies in Systems, Decision and Control 261, https://doi.org/10.1007/978-3-030-32710-1_19

Keywords SEMS · Warehouse loader robot · Situational control · Model · Dynamics · Kinematics

1 Introduction

For many industrial enterprises, trade enterprises, an important issue is the handling and storage of a huge variety of goods, such as: tools, components for production, spare parts, etc. The need for continuous monitoring, accounting and security of data storage of components leads to the creation of automated warehouse control systems [1, 2]. In non-automated warehouses, accounting and cargo handling are difficult to optimize and usually take a long time. A large amount of human resources is spent on this process, and the area of the warehouse is inefficiently used. The use of robotic systems [3, 4] allows for rational allocation of warehouse resources and reducing the staff to several operators of the robotic complex. This significantly increases the quality of accounting and the speed of cargo handling.

There are various structures of such robotic complexes, but an integral part of them are robots loaders of various designs, for example, forklifter, anthropomorphic, robot manipulators, etc. [1–4]. This study aims to create a universal tech solution that allows you to create mobile robotic systems based on SEMS modules [5]. The article deals with the tasks of developing models of an anthropomorphic loader robot for a warehouse constructed of such modules.

Universal SEMS modules are electromechanical systems of parallel architecture similar to n-pods. Advantages, applications and description of these mechanisms are reflected in the publications [5–12]. The main difference from the classical view is the use of self-organizing modules with a distributed control system, built on the principles of optimal control, self-organization and agent-based approach. The proposed approach allows for a set of modules created using the same technology to develop designs for solving certain production problems.

The several mobile robots are supposed to be used in the warehouse, the question arises of synthesizing group situational control. It is also necessary to develop algorithms for controlling the robot loader when performing various operations. Most modern control systems use in their structure predictive models of control objects. These models should be used to test the laws governing the control of robots at the production stage. Therefore, special attention should be paid to the creation of models of such robots as control objects.

2 Description of the Object of Study

A modern large warehouse is a complex technical structure, which consists of many different subsystems, for example, a complex of buildings, a set of processed cargoes, information support systems, etc. and elements of a certain structure that are combined to perform specific functions of transforming material flows. Robots are used to facilitate the implementation of laborious and monotonous work on unloading and sorting of various goods in warehouses.

The robot we are considering is anthropomorphic. The robot consists of the following elements:

- a mobile platform on wheels on which the robot places the goods during transportation;
- the body of the robot, the design of which allows it to make: a turn around the vertical axis at a limited angle, tilts in different directions, changing the height of the robot;
- two arms are attached to the body, at the ends of which there are universal grippers [13];
- each universal gripper consists of a palm and three fingers;
- each finger has three phalanges;
- the robot has a moving head, on which a system of cameras and range-measuring equipment is installed.

The design of the hands allows you to change their length. The universal gripper palm can change its size symmetrically, the phalanges of the fingers can change its length and thickness to ensure a better grip quality. The design of the robot is modular, it is composed of standard modules (base units) with a universal controller. The base unit is an electromechanical device parallel kinematic scheme, resembling hexapods. The base unit consists of two platforms, each of which can change its size symmetrically through a system of actuators. The platforms are interconnected with six actuators. Each actuator is attached to the platform using hinges mounted on its ends. Each actuator consists of a screw jack based on a ball-screw pair, a gearbox and a DC motor. Each actuator is equipped with a linear displacement sensor, an angular velocity sensor of the electric motor and a force sensor.

From the point of view of the control task, the warehouse loader truck is a dynamic object that sequentially performs a series of operations (receiving instructions from the operator, processing current information about his position, recognizing obstacles, loads, developing the optimal path for moving loads, etc.) in a constantly changing external environment, information about which he receives through sensors, for example, range-measuring equipment and video cameras. In various modes of operation, the warehouse loader truck is described by various mathematical models. When moving around the warehouse, it is advisable to describe the dynamics of the warehouse loader as a material point moving on a plane. Since the object functions in a space of a complex configuration (the presence of other robots by a loader, the presence of obstacles, people, etc.), the dimensions and geometry of the robot and the elements of its surrounding environment should be taken into account. When selecting, capturing and moving goods, detailed models of kinematics and dynamics of similar robots as a complex electromechanical object by a parallel kinematic scheme are required. The modular arrangement of the robot allows you to simplify this task by creating models of the base unit. It is necessary to reach a compromise between the complexity of the model and its applicability in the on-board computer of the robot. These models are supposed to be used as predictive for developing control instructions for situational control. It is also supposed to use these models in the system of synthesis and testing of group control of groups of robots.

3 Kinematic and Dynamic Models of the Base Unit

Detailed models of kinematics and dynamics of similar electromechanical structures with fixed bases were considered in [14, 15]. A distinctive feature of this base unit is the ability to symmetrically change the size of the platforms using actuators, which complicates the model and does not allow using the ready-made formulas described in [14, 15]. Let us use the ideas described in [14] and modify them for our case.

We introduce the basic definitions and formulas of the kinematic model according to [14]. So the upper platform will be called the adaptive platform (AP), the lower one will be called the base platform (BP).

Introduce into consideration the basic coordinate system (BCS) $E_0 = (\mathbf{o}_0, [\mathbf{e}_0])$, where $\mathbf{o}_0$—the origin of coordinates of BCS; $[\mathbf{e}_0]$—the three unitary vectors of BCS. For angles, vectors and rotation matrices, the subscript is the number of the coordinate system (CS), the superscript is the CS number, relative to which the angular and linear positions are determined, the second superscript indicates the CS number, in which the coordinates of vectors are calculated. Rotation matrices $\mathbf{c}^i_j$ have the form $\mathbf{c}^i_j\left(\varphi^i_j\right) = \mathbf{c}_1(\beta^i_j)\cdot\mathbf{c}_2(\theta^i_j)\cdot\mathbf{c}_3(\alpha^i_j)$, where $\varphi^i_j = \left[\begin{array}{ccc}\beta^i_j & \theta^i_j & \alpha^i_j\end{array}\right]^{\mathrm{T}}$, β^i_j, θ^i_j and α^i_j—the angles of elementary rotations about the x, y and z axes respectively; the matrices of elementary rotations have the form:

$$\mathbf{c}_1(\beta^i_j) = \begin{bmatrix} 1 & 0 & 0 \\ 0 & \cos(\beta^i_j) & -\sin(\beta^i_j) \\ 0 & \sin(\beta^i_j) & \cos(\beta^i_j) \end{bmatrix};\ \mathbf{c}_2(\theta^i_j) = \begin{bmatrix} \cos(\theta^i_j) & 0 & \sin(\theta^i_j) \\ 0 & 1 & 0 \\ -\sin(\theta^i_j) & 0 & \cos(\theta^i_j) \end{bmatrix};$$

$$\mathbf{c}_3(\alpha^{jc}_j) = \begin{bmatrix} \cos(\alpha^i_j) & -\sin(\alpha^i_j) & 0 \\ \sin(\alpha^i_j) & \cos(\alpha^i_j) & 0 \\ 0 & 0 & 1 \end{bmatrix}.$$

Here the indices i, j are the designations of the corresponding coordinate systems.

The base platform moves in three linear and three angular coordinates relative to the BCS, can also symmetrically change the linear size. The adaptive platform is moved by actuators in three linear and three angular coordinates relative to the base platform and can change its linear size symmetrically.

We introduce a body-oriented coordinate system of base platform (BPCS) $E_{1b} = (\mathbf{o}_{1b}, [\mathbf{e}_{1b}])$, where $\mathbf{o}_{1b}$—the origin of coordinates of BPCS, which relative to the BCS is determined by the coordinate column $\mathbf{r}^{0,0}_{1b} = \left[x^{0,0}_{1b}, y^{0,0}_{1b}, z^{0,0}_{1b}\right]^{\mathrm{T}}$; $[\mathbf{e}_{1b}]$—the three unitary vectors of BPCS, the orientation of the BPCS relative to BCS is determined by the angles of the elementary rotations $\boldsymbol{\varphi}_{1b}^{0}$. The number "1" denotes the base unit number, for example, the model equations are obtained for the module with the number "1".

To set the initial position of the AP, we introduce the coordinate system of the initial position of the AP (CS AP0) $E_{1c} = (\mathbf{o}_{1c}, [\mathbf{e}_{1c}])$, where $\mathbf{o}_{1c}$—the origin of coordinates of CS AP0, which relative to the BPCS is determined by the coordinate column $\mathbf{r}_{1c}^{1b,1b} = \left[x_{1c}^{1b,1b}, y_{1c}^{1b,1b}, z_{1c}^{1b,1b}\right]^{\mathrm{T}}$; $[\mathbf{e}_{1c}]$—the three unitary vectors of CS AP0; the orientation of the CS AP0 relative to BPCS is determined by the angles of the elementary rotations $\boldsymbol{\varphi}_{1c}^{1b}$.

To set the position of the AP, we introduce the body-oriented coordinate system of adaptive platform (APCS) $E_1 = (\mathbf{o}_1, [\mathbf{e}_1])$, where $\mathbf{o}_1$—the origin of coordinates of APCS, which relative to the CS AP0 is determined by the coordinate column $\mathbf{r}_1^{1c,1c}$; $[\mathbf{e}_1]$—the three unitary vectors of APCS; the orientation of the APCS relative to CS AP0 is determined by the angles of the elementary rotations $\boldsymbol{\varphi}_1^{1c}$.

The position of the APCS relative to the BCS is described by a vector whose coordinate column in the BCS can be calculated by the formula:

$$\mathbf{r}_1^{0,0} = \mathbf{r}_{1b}^{0,0} + \mathbf{c}_{1b}^{0}\left[\mathbf{r}_{1c}^{1b,1b} + \mathbf{c}_{1c}^{1b}\mathbf{r}_1^{1c,1c}\right], \tag{1}$$

and the angular position is described by a rotation matrix:

$$\mathbf{c}_1^0 = \mathbf{c}_{1b}^0\mathbf{c}_{1c}^{1b}\mathbf{c}_1^{1c}. \tag{2}$$

It should be noted, that the origin of the coordinates of the BPCS, CS AP0, APCS combined with the centers of inertia of the respective platforms. The points at which the origin of the coordinates are located do not have physical reference points on the platforms due to the design of the platforms and are determined by calculation.

The coordinates of the hinge attachment on the BP are defined in the BPCS: $\mathbf{r}_{1bj1}^{1b,1b}, \mathbf{r}_{1bj2}^{1b,1b}, \ldots, \mathbf{r}_{1bj6}^{1b,1b}$, where the subscript denotes the hinge number on the BP. The last digit in the subscript indicates the sequence number of the hinge on the BP. On the BP there are 6 hinges in a circle with a radius R_{1b}. The BP radius can vary, so the coordinates of the hinges are a function of the BP radius: $\mathbf{r}_{1bj1}^{1b,1b}(R_{1b}), \mathbf{r}_{1bj2}^{1b,1b}(R_{1b}), \ldots, \mathbf{r}_{1bj6}^{1b,1b}(R_{1b})$.

Similarly, we introduce the coordinates of the mounting of the hinges on the adaptive platform in the APCS: $\mathbf{r}_{1j1}^{1,1}, \mathbf{r}_{1j2}^{1,1}, \ldots, \mathbf{r}_{1j6}^{1,1}$, where the subscript indicates the number of the hinge on the AP. The last digit in the subscript indicates the sequence number of the hinge on the AP. On the AP there are 6 hinges on a circle with a radius R_1. The radius of the AP can change, so the coordinates of the hinges are a function of the radius of the AP: $\mathbf{r}_{1j1}^{1,1}(R_1), \mathbf{r}_{1j2}^{1,1}(R_1), \ldots, \mathbf{r}_{1j6}^{1,1}(R_1)$.

The current lengths of the actuators can be defined as the distances between the respective hinges of the base and the AP by the formula: $l_{1ji}^{1bji} = \left|\mathbf{r}_{1ji}^{1bji,1b}\right| = \left|\mathbf{r}_{1ji}^{1b,1b}(R_1) - \mathbf{r}_{1bji}^{1b,1b}(R_{1b})\right|$, $i = 1\cdots6$, where $\mathbf{r}_{1ji}^{1b,1b}$—the coordinates of the point of attachment of the hinges on the AP in the BPCS, which are calculated as follows: $\mathbf{r}_{1ji}^{1b,1b} = \mathbf{r}_{1c}^{1b,1b} + \mathbf{c}_{1c}^{1b}\left[\mathbf{r}_{1}^{1c,1c} + \mathbf{c}_{1}^{1c}\mathbf{r}_{1ji}^{1,1}(R_1)\right]$, $i = 1\cdots 6$. As a result, the expression for the current length will take the form:

$$l_{1ji}^{1bji} = \left|\mathbf{r}_{1c}^{1b,1b} + \mathbf{c}_{1c}^{1b}\left[\mathbf{r}_{1}^{1c,1c} + \mathbf{c}_{1}^{1c}\mathbf{r}_{1ji}^{1,1}(R_1)\right] - \mathbf{r}_{1bji}^{1b,1b}(R_{1b})\right|$$
$$= l_{0i} + \Delta l_{ai}(\psi_i), \quad i = 1\cdots 6, \tag{3}$$

where l_{0i}—the initial lengths of the actuators, $l_{0i} = \left|\mathbf{r}_{1c}^{1b,1b} + \mathbf{c}_{1c}^{1b}\mathbf{r}_{1ji}^{1,1}(R_1(0)) - \mathbf{r}_{1bji}^{1b,1b}(R_{1b}(0))\right|$; Δl_{ai}—current extensions of actuator rods; $\Delta l_{ai} = \tau_i + \frac{\psi_i}{I_i}$, τ_i—actuator deformation, ψ_i—rotation angle of the actuator motor rotor, I_i—gear ratio, $R_1(0)$, $R_{1b}(0)$—the initial values of the radii R_1, R_{1b} respectively.

Expressions (1)–(3) determine the geometry of the base unit as a function of the controlled parameters R_1, R_{1b}, ψ_i. Depending on the task, these expressions can be used both to find the required program values of the parameters R_1, R_{1b}, ψ_i, and the position of the AP of the base unit from known changes in these parameters.

To calculate the program values of the parameters R_1, R_{1b}, ψ_i from known values $\mathbf{r}_1^{0,0}$, $\mathbf{r}_{1b}^{0,0}$, $\mathbf{c}_{1b}^{0}$, $\mathbf{c}_{1c}^{1b}$ it is necessary to determine $\mathbf{r}_1^{1c,1c}$ and $\mathbf{c}_1^{1c}$ due to the formulas:

$$\mathbf{r}_1^{1c,1c} = \mathbf{c}_{1c}^{1b,\mathrm{T}}\left[\mathbf{c}_{1b}^{0,\mathrm{T}}\left[\mathbf{r}_1^{0,0} - \mathbf{r}_{1b}^{0,0}\right] - \mathbf{r}_{1c}^{1b,1b}\right],$$
$$\mathbf{c}_1^{1c} = \mathbf{c}_{1c}^{1b,\mathrm{T}}\mathbf{c}_{1b}^{0,\mathrm{T}}\mathbf{c}_1^{0}, \tag{4}$$

then it is necessary to solve numerically the system of equations (3). It should be noted that in (3) the number of equations is 6, and the required parameters are 8. Thus, we have a lot of possible solutions and must develop a strategy for control of the base unit, for example, by the criterion of maximum speed. In order to simplify calculations, you can initially work out the laws of changing of R_1, R_{1b}, based on the task solved by the robot.

The solution of the direct problem of determining the position of the AP in BCS by parameters R_1, R_{1b}, ψ_i leads to ambiguity due to the structure of the system of equations (3). The voluntary combinations of these parameters can lead to inconsistency of the equation in (3). One of the methods to overcome the ambiguity is linearization of (3) with respect to the programmatic trajectory of the basic unit, then $\mathbf{r}_1^{1c,1c}$ and $\boldsymbol{\varphi}_1^{1c}$ can be calculated as:

$$\begin{bmatrix}\mathbf{r}_1^{1c,1c} \\ \boldsymbol{\varphi}_1^{1c}\end{bmatrix} = \mathbf{A}\begin{bmatrix} R_{1b} \\ R_1 \\ \boldsymbol{\psi}\end{bmatrix}, \tag{5}$$

where $\boldsymbol{\psi}$—vector composed of ψ_i, $\mathbf{A}$—matrix composed of the parameters of equations (3).

The relationship between the speeds of the base unit will find by differentiating $\mathbf{r}_1^{0,0}$ and $\mathbf{c}_1^0$. Taking into account the properties of skew-symmetric matrices, we obtain the expressions for the linear $\mathbf{v}_1^{0,0}$ and angular $\boldsymbol{\omega}_1^{0,0}$ velocities of the AP in the BCS:

$$\mathbf{v}_1^{0,0} = \mathbf{v}_{1b}^{0,0} + \mathbf{c}_{1b}^0 \left\langle \mathbf{r}_{1c}^{1b,1b} + \mathbf{c}_{1c}^{1b} \mathbf{r}_1^{1c,1c} \right\rangle^{\mathrm{T}} \boldsymbol{\omega}_{1b}^{0,1b} + \mathbf{c}_{1b}^0 \mathbf{c}_{1c}^{1b} \mathbf{v}_1^{1c,1c},$$
$$\boldsymbol{\omega}_1^{0,0} = \mathbf{c}_{1b}^0 \boldsymbol{\omega}_{1b}^{0,1b} + \mathbf{c}_1^0 \boldsymbol{\omega}_1^{1c,1}, \tag{6}$$

where $\mathbf{v}_{1b}^{0,0}$—the linear velocities of BP in BCS; $\boldsymbol{\omega}_{1b}^{0,1b}$—the angular velocities of BP in BCS; <…>—skew-symmetric matrix of the form $\left\langle [x, y, z]^{\mathrm{T}} \right\rangle = \begin{bmatrix} 0 & -z & y \\ z & 0 & -x \\ -y & x & 0 \end{bmatrix}$; $\mathbf{v}_1^{1c,1c}$—the linear velocities of AP relevant to CS AP0 in APCS; $\boldsymbol{\omega}_1^{1c,1}$—the angular velocities of AP relevant to CS AP0 in APCS.

The angular velocities $\boldsymbol{\omega}_1^{0,0}$, $\boldsymbol{\omega}_{1b}^{0,1b}$ and $\boldsymbol{\omega}_1^{1c,1}$ can be determined through the elementary rotation velocities $\dot{\boldsymbol{\varphi}}_1^0$, $\dot{\boldsymbol{\varphi}}_{1b}^0$ and $\dot{\boldsymbol{\varphi}}_1^{1c}$:

$$\boldsymbol{\omega}_{1b}^{0,1b} = \boldsymbol{\varepsilon}_{1b}^0 \dot{\boldsymbol{\varphi}}_{1b}^0; \ \boldsymbol{\omega}_1^{1c,1} = \boldsymbol{\varepsilon}_1^{1c} \dot{\boldsymbol{\varphi}}_1^{1c}; \ \boldsymbol{\omega}_1^{0,0} = \mathbf{c}_1^0 \boldsymbol{\varepsilon}_1^0 \dot{\boldsymbol{\varphi}}_1^0, \tag{7}$$

where $\boldsymbol{\varepsilon}_j^i$—the Euler matrices of type $\boldsymbol{\varepsilon}_j^i = \left[\mathbf{c}_3^{\mathrm{T}}\left(\alpha_j^i\right) \mathbf{c}_2^{\mathrm{T}}\left(\theta_j^i\right) \mathbf{e}_x | \mathbf{c}_3^{\mathrm{T}}\left(\alpha_j^i\right) \mathbf{e}_y | \mathbf{e}_z \right]$ from the corresponding angles, $\mathbf{e}_x = [1\ 0\ 0]^{\mathrm{T}}$, $\mathbf{e}_y = [0\ 1\ 0]^{\mathrm{T}}$, $\mathbf{e}_z = [0\ 0\ 1]^{\mathrm{T}}$.

We obtain the rate of change of the lengths of the actuators $\dot{l}_{1ji}^{1bji}$ by differentiating (3):

$$\dot{l}_{1ji}^{1bji} = \frac{\mathbf{v}_{1ji}^{1bji,1b,\mathrm{T}} \mathbf{r}_{1ji}^{1bji,1b}}{l_{1ji}^{1bji}} = v_{ai} + \dot{\tau}_i, \quad i = 1 \cdots 6, \tag{8}$$

where v_{ai}—the velocity of actuator, which in the case of a lead screw can be determined by the formula $v_{ai} = \frac{\Omega_i}{I_i}$, Ω_i—the motor angular velocity; $\dot{\tau}_i$—the actuator strain rate. The velocity of the relative translational motion of the hinges of the ith actuator in the BPCS find the formula:

$$\mathbf{v}_{1ji}^{1bji,1b} = \mathbf{c}_{1c}^{1b} \mathbf{v}_1^{1c,1c} + \mathbf{c}_{1c}^{1b} \mathbf{c}_1^{1c} \dot{\mathbf{r}}_{1ji}^{1,1}(\dot{R}_1) + \mathbf{c}_{1c}^{1b} \mathbf{c}_1^{1c} \left\langle \mathbf{r}_{1ji}^{1,1}(R_1) \right\rangle^{\mathrm{T}} \boldsymbol{\varepsilon}_1^{1c} \dot{\boldsymbol{\varphi}}_1^{1c} - \dot{\mathbf{r}}_{1bji}^{1b,1b}(\dot{R}_{1b}), \tag{9}$$

where $\dot{\mathbf{r}}_{1ji}^{1,1}(\dot{R}_1)$ and $\dot{\mathbf{r}}_{1bji}^{1b,1b}(\dot{R}_{1b})$—the velocities of change of position of the respective hinges of AP and PB; $\dot{R}_1$ and $\dot{R}_{1b}$—the rate of change of the radii of the AP and BP, respectively.

The simultaneous solution of equations (3) and (8) in small deviations from the programmed AP trajectory makes it possible to determine $\mathbf{r}_1^{1c,1c}$, φ_1^{1c}, $\dot{\varphi}_1^{1c}$, $\mathbf{v}_1^{1c,1c}$ for known values of controlled parameters R_1, R_{1b}, ψ_i, $\dot{R}_1$, $\dot{R}_{1b}$ and Ω_i.

To calculate the program values of parameters R_1, R_{1b}, ψ_i, $\dot{R}_1$, $\dot{R}_{1b}$ and Ω_i, it is necessary to simultaneously solve Eqs. (3) and (8). It should be noted that the number of equations is 12, and the desired parameters are 16. In order to simplify calculations, you can initially work out the laws of change of R_1, R_{1b}, and $\dot{R}_1$, $\dot{R}_{1b}$, based on the task solved by the robot.

The controlled parameters of the base unit can be described with the help of functional dependencies on the control actions of the universal controller.

Equations (1)–(9) are a simplified kinematic model of the base unit of the warehouse loader robot.

The models of the dynamics of the basic unit SEMS with platforms of constant size are discussed in detail in [14, 15]. It is necessary to modify the equations from [14, 15] for our case:

$$\mathbf{\Theta}\dot{\mathbf{V}}_1^{0,1} + \dot{\Theta}\mathbf{V}_1^{0,1} + \mathbf{\Phi}\mathbf{\Theta}\mathbf{V}_1^{0,1} = \Xi,$$
$$\mathbf{V}_1^{0,1} = \begin{bmatrix} \mathbf{v}_1^{0,1} \\ \boldsymbol{\omega}_1^{0,1} \end{bmatrix}, \ \Xi = \begin{bmatrix} \mathbf{F}_1 \\ \mathbf{M}_1 \end{bmatrix}, \ \mathbf{\Phi} = \begin{bmatrix} \langle \boldsymbol{\omega}_1^{0,1} \rangle & 0 \\ \langle \mathbf{v}_1^{0,1} \rangle & \langle \boldsymbol{\omega}_1^{0,1} \rangle \end{bmatrix}, \ \mathbf{\Theta} = \begin{bmatrix} \mathbf{m}_1 & 0 \\ 0 & \mathbf{I}_1 \end{bmatrix}, \ \dot{\Theta} = \begin{bmatrix} 0 & 0 \\ 0 & \dot{\mathbf{I}}_1 \end{bmatrix}, \quad (10)$$
$$\mathbf{m}_1 = diag\left(m_1 \ m_1 \ m_1 \right),$$

where $\mathbf{\Theta}$—the inertia matrix AP; $\mathbf{v}_1^{0,1}$—linear velocities of AP relative to BCS in projections on the axis APCS $\mathbf{v}_1^{0,1} = \mathbf{c}_1^{0,\mathrm{T}}\mathbf{v}_{1b}^{0,0} + \mathbf{c}_1^{0,\mathrm{T}}\mathbf{c}_{1b}^{0}\left\langle \mathbf{r}_{1c}^{1b,1b} + \mathbf{c}_{1c}^{1b}\mathbf{r}_1^{1c,1c} \right\rangle^{\mathrm{T}} \boldsymbol{\omega}_{1b}^{0,1b} + \mathbf{c}_1^{1c,\mathrm{T}}\mathbf{v}_1^{1c,1c}$; $\boldsymbol{\omega}_1^{0,1}$—the angular velocities of AP relative to BCS in projections on the axis APCS $\boldsymbol{\omega}_1^{0,1} = \mathbf{c}_1^{0,\mathrm{T}}\mathbf{c}_{1b}^{0}\boldsymbol{\omega}_{1b}^{0,1b} + \boldsymbol{\omega}_1^{1c,1}$; $\mathbf{V}_1^{0,1}$—the AP velocity vector; $\mathbf{F}_1$—the sum vector of forces acting on the AP in the APCS; $\mathbf{M}_1$—the sum vector of torques acting on the AP in the APCS; Ξ—the sum vector of forces and torques acting on the AP in the APCS; m_1—the mass of AP; $\mathbf{I}_1$—the moments of inertia of AP; $\dot{\mathbf{I}}_1$—the matrix of the rates of change of the moments of inertia, which are caused by the change in the dimensions of the platform. The matrix $\dot{\mathbf{I}}_1$ is a function of parameters R_1 and $\dot{R}_1$. Forces and torques have expressions similar to the case when the dimensions of the platforms of the base unit do not change [14, 15].

Given the kinematic scheme of the base unit, the velocity vector can be defined as follows:

$$\mathbf{V}_1^{0,1} = \mathbf{M}_1^{1c}\dot{\mathbf{q}}_1^{1c} + \mathbf{L}_1^{1c,\mathrm{T}}\mathbf{L}_{1c}^{1b,\mathrm{T}}\mathbf{M}_{1b}^{0}\dot{\mathbf{q}}_{1b}^{0},$$
$$\mathbf{M}_1^{1c} = \begin{bmatrix} \mathbf{c}_1^{1c,\mathrm{T}} & 0 \\ 0 & \boldsymbol{\varepsilon}_1^{1c} \end{bmatrix}, \ \mathbf{L}_1^{1c} = \begin{bmatrix} \mathbf{c}_1^{1c} & 0 \\ \langle \mathbf{r}_1^{1c,1c} \rangle \mathbf{c}_1^{1c} & \mathbf{c}_1^{1c} \end{bmatrix},$$
$$\mathbf{L}_{1c}^{1b} = \begin{bmatrix} \mathbf{c}_{1c}^{1b} & 0 \\ \langle \mathbf{r}_{1c}^{1b,1b} \rangle \mathbf{c}_{1c}^{1b} & \mathbf{c}_{1c}^{1b} \end{bmatrix}, \ \mathbf{M}_{1b}^{0} = \begin{bmatrix} \mathbf{c}_{1b}^{0,\mathrm{T}} & 0 \\ 0 & \boldsymbol{\varepsilon}_{1b}^{0} \end{bmatrix}, \ \mathbf{q}_1^{1c} = \begin{bmatrix} \mathbf{r}_1^{1c,1c} \\ \varphi_1^{1c} \end{bmatrix}, \ \mathbf{q}_{1b}^{0} = \begin{bmatrix} \mathbf{r}_{1b}^{0,0} \\ \varphi_{1b}^{0} \end{bmatrix}, \quad (11)$$

where $\mathbf{M}_1^{1c}$ and $\mathbf{M}_{1b}^{0}$—the matrices of relationships of velocities of platforms and generalized velocities; $\mathbf{L}_1^{1c}$ and $\mathbf{L}_{1c}^{1b}$—the coordinate system transformation matrices; $\mathbf{q}_1^{1c}$—the generalized coordinates of AP; $\mathbf{q}_{1b}^{0}$—the generalized coordinates of BP.

4 Warehouse Loader Robot Model

The warehouse loader robot, which we are considering, has a modular design, an element of which is the basic unit of SEMS, whose platforms can change their dimensions symmetrically. The design and models of which are described in detail in the previous section. The structure of the robot model is shown in Fig. 1. Since in Fig. 1 all the vectors are specified in the BCS, for simplicity, the superscripts of the vectors are omitted.

Each block is described by a kinematic model (1)–(9) and a model of dynamics (10)–(11). The blocks are fixed in such a way that the upper platform of the previous block is the lower platform of the subsequent one. In the case of hands and the first phalanxes of the fingers, the lower platforms of their blocks have a fixed size. The robot head unit has a fixed size of the upper platform. For modeling, unified program blocks are used that implement the models described above.

Each unit has the following input signals:

- the vector of control actions $\mathbf{u}_i$, which are control signals supplied to the actuator motors;
- the vector of linear coordinates, which determine the position of the BP of the simulated block in the BCS;
- the vector of angular coordinates, which determine the position of the BP of the simulated block in the BCS;
- the vector of linear velocities of the BP of the simulated block in BCS;
- the vector of angular velocities of the elementary rotations of the BP of the simulated block in the BCS;
- the radius of the BP;
- the rate of change of the radius of the BP;
- the vector of reactions and external loadings.

The output signals of the block are:

- the vector of linear coordinates, which determine the position of the AP of the simulated block in the BCS;
- the vector of angular coordinates, which determine the position of the AP of the simulated block in the BCS;
- the vector of linear velocities of the AP of the simulated block in BCS;
- the vector of angular velocities of the elementary rotations of the AP of the simulated block in the BCS.
- the velocities of lengthening rod actuators;
- the lengths of actuator rods;
- the radius of the AP;
- the rate of change of the radius of the AP;
- the vector of reactions.

As parameters in the block are specified:

- the initial position of the elements of the warehouse loader robot;

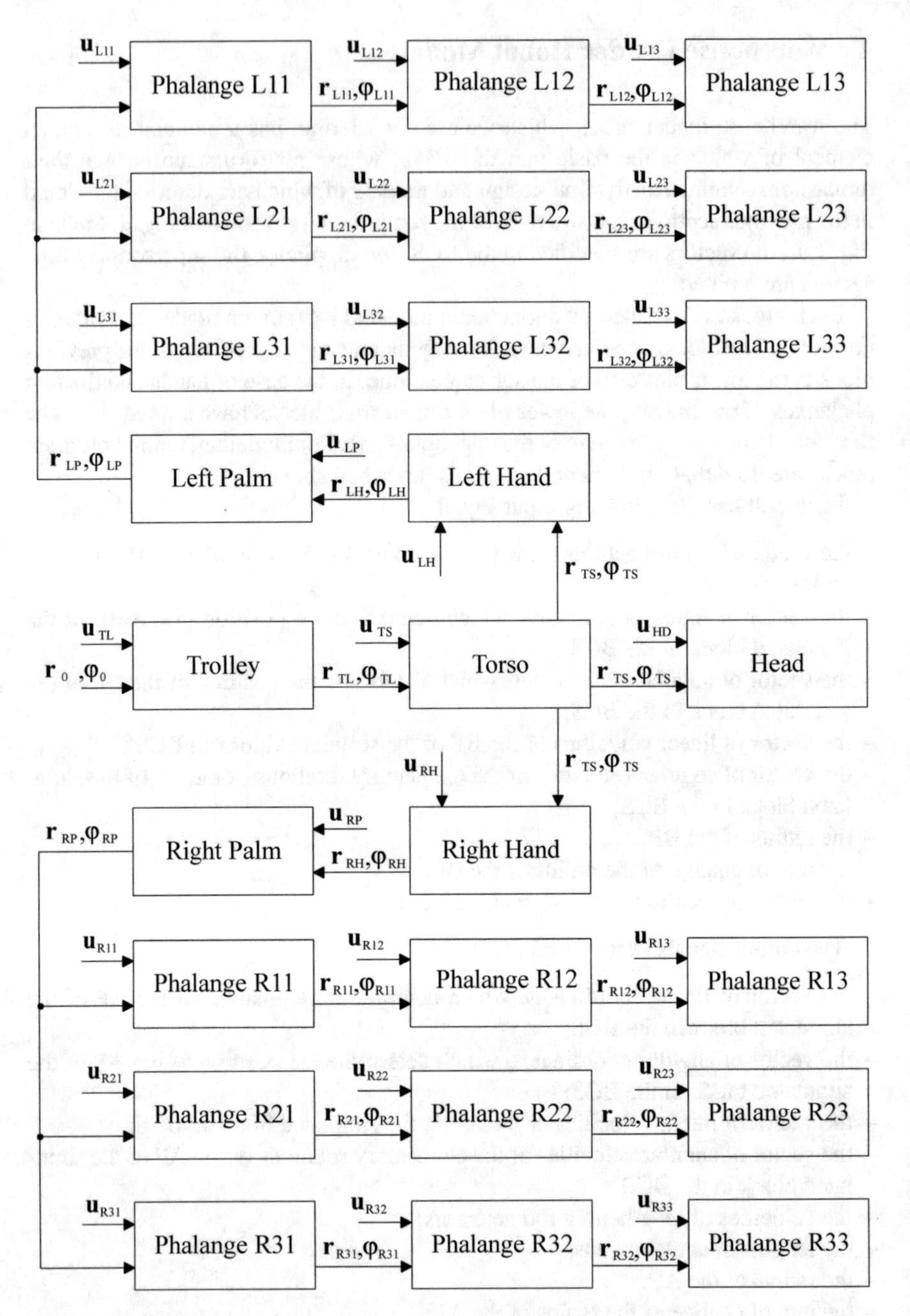

Fig. 1 The structure of the model of the warehouse loader robot

- the masses and moments of inertia of the block platforms at the initial moment of time;
- the initial values of the radii of the platforms;
- the initial position of the platform hinges;
- the initial length of the rod actuators;
- time-constants and gain coefficients for actuators electric drives.

The following numbering of blocks is adopted according to Fig. 1: a trolley (TL); torso (TS); head (HD); left hand (LH); left palm (LP); right hand (RH); right palm (RP); left phalanges (L11–L33); right phalanges (R11–R33).

5 Construction of a Simplified Dynamic Model of a Warehouse Loader Robot and Methods for Identifying Its Parameters

The study of SEMS modules allows us to conclude that they are electromechanical systems with slowly varying parameters. And it is possible to perform the linearization of the original non-linear equations by analogy with the basic blocks with unchanged geometry of the platforms. In practice, such a model can be obtained using identification methods [16–21]. The identification task includes the determination of such system parameters as moments of inertia, stiffness, damping and friction coefficients. In the general case, the object model is described by the equation:

$$\mathbf{\Theta}(\mathbf{q})\ddot{\mathbf{q}} + \left[\mathbf{B}(\mathbf{q}, \dot{\mathbf{q}}) + \mathbf{D}(\mathbf{q})\right]\dot{\mathbf{q}} + \mathbf{C}(\mathbf{q})\mathbf{q} = \mathbf{u} + \mathbf{f},$$

where $\mathbf{q}, \ \dot{\mathbf{q}}, \ \ddot{\mathbf{q}}$—the generalized coordinates, generalized velocities and generalized accelerations of the mechanical system respectively; $\mathbf{\Theta}$—the inertia matrix; $\mathbf{B}$—the friction coefficient matrix; $\mathbf{D}$—the damping matrix; $\mathbf{C}$—the stiffness matrix; $\mathbf{u}$—the control, $\mathbf{f}$—the external loadings.

The low dynamics of changes in the system parameters allows the system to be linearized at stationary points, and a lot of linear models are obtained for different regions of the phase space:

$$\mathbf{\Theta}_i\Delta\ddot{\mathbf{q}} + [\mathbf{B}_i + \mathbf{D}_i]\Delta\dot{\mathbf{q}} + \mathbf{C}_i\Delta\mathbf{q} + \mathbf{O}_i = \Delta\mathbf{u} + \Delta\mathbf{f},$$

where $\mathbf{\Theta}_i$, $\mathbf{B}_i$, $\mathbf{D}_i$, $\mathbf{C}_i$—the matrix of inertia, friction coefficients, damping, stiffness of the linearized system for the *i*th stationary point respectively; $\mathbf{O}_i$—constant component; $\Delta\mathbf{q}, \ \Delta\dot{\mathbf{q}}, \ \Delta\ddot{\mathbf{q}}, \ \Delta\mathbf{u}, \ \Delta\mathbf{f}$—small increments of generalized coordinates, velocities, acceleration, control and external actions, respectively.

The presence of unmeasurable external influences complicates the identification process, so they try to choose such operating modes of the system when its influence is minimal. In this case, the identification method is described as follows:

1. The measured coordinate of the state vector x and the control signal are selected u.
2. The transfer function of the system is based on the selected variables:

$$W(s) = \frac{x(s)}{u(s)} = \mathbf{H}_1\left[\mathbf{\Theta}_i s^2 + [\mathbf{B}_i + \mathbf{D}_i]s + \mathbf{C}_i\right]^{-1}\mathbf{H}_2,$$

   where $\mathbf{H}_1$—the measured output matrix, $\mathbf{H}_2$—the control matrix, s—the complex variable.
3. If the main source of information is frequency characteristics, then the system model will be sought in the form:

$$W(j\omega) = \mathbf{H}_1\left[\mathbf{C}_i - \mathbf{\Theta}_i\omega^2 + [\mathbf{B}_i + \mathbf{D}_i]\omega j\right]^{-1}\mathbf{H}_2,$$

   where j—the imaginary unit, ω—the frequency. Then the identification task is formulated as:

$$J = \min_{\mathbf{C}_i,\mathbf{\Theta}_i,\mathbf{B}_i,\mathbf{D}_i}\left\{\sum_{\omega=\omega_L}^{\omega_U}\left[\left(|W(j\omega)| - |W^*(j\omega)|\right)^2 + \lambda\left(\arg W(j\omega) - \arg W^*(j\omega)\right)^2\right]\right\},$$

   with restrictions on the measurement ranges of model parameters, where $W^*(j\omega)$—the experimentally obtained characteristic of object, λ—the weight coefficient, ω_L, ω_U—the upper and lower limits of the frequency range.

If time signals are used as a source of information, then it is necessary to build a model in the form of a discrete filter. Then a time period of quantization is selected based on the dynamics of the object. With the help of the z-transformation, a transition is made from the transfer function of a continuous system to a discrete one. The filter parameters are evaluated using deterministic, least-squares identifiers. Then a system of equations is compiled, where each parameter of the difference filter is expressed by functional dependence through the parameters of the original system. This system of equations has redundancy and has many solutions, so it is necessary to take into account the ranges of changes in the parameters of the mechanical system.

Simultaneous estimation of moments of inertia and stiffness leads to ambiguity. So in practice, the values of the moments of inertia are set, which can be calculated using finite element models of structures, and the ranges of changes in the stiffness coefficients of the system are also determined by the models.

The considered technique is focused on estimating equivalent stiffness, however, when modeling mechanical systems, it is convenient to use in the description of systems partial stiffnesses that weakly depend on changes in generalized coordinates [22]. In linear systems, the relationship between the matrix of equivalent stiffness and partial stiffness is established by the expression:

$$\mathbf{\Gamma}^{\mathrm{T}}\mathbf{C}_P\mathbf{\Gamma} = \mathbf{C}_i,$$

where $\mathbf{\Gamma}$—the topological matrix of the system, $\mathbf{C}_P$—partial stiffness matrix. The solution of this equation is performed taking into account the ranges of partial stiffness variation by linear programming methods. In the general case, a topological matrix is a function of generalized coordinates, where a change in the line of action of an elastic coupling in space is taken into account.

6 Conclusion

The article deals with the development of kinematic and dynamic models of a warehouse loader robot created from modules with a parallel kinematic scheme (SEMS) with a changing geometry of the platforms. A method for constructing a simplified model of a fork lift robot by linearizing the equations of the dynamics of the base unit at a stationary point is proposed. The method of identification of parameters of such a simplified model is considered. The use of such simplified models will allow them to be used in the on-board computer of the robot for predicting and optimizing trajectories and controlling adaptive grippers.

Acknowledgements This work was financially supported by Russian Foundation for Basic Research, Grant 19-08-00079.

References

1. Bonkenburg, T.: Robotics in Logistics. A DPDHL Perspective on Implications and Use Cases for the Logistics Industry, Mar 2016. https://www.dhl.com/content/dam/downloads/g0/about_us/logistics_insights/dhl_trendreport_robotics.pdf. July 04, 2019
2. McCrea, B.: Mobility & Robotics in the Warehouse. Modern Materials Handling, 96 p., Mar 2016
3. Ackerman Warehousing Forum, vol. 31, no. 5, April 2016. https://www.locusrobotics.com/wp-content/uploads/Vol-31-Number-5-April2016.pdf. July 04, 2019
4. Scassellati, B., Tsui, K.M.: Co-Robots: Humans and Robots Operating as Partners. Handbook of Science and Technology Convergence. Springer International Publishing Switzerland (2015). https://doi.org/10.1007/978-3-319-04033-2_27-1. https://link.springer.com/content/pdf/10.1007%2F978-3-319-04033-2_27-1.pdf. July 04, 2019
5. Gorodetskiy, A.E. (ed.): Smart Electromechanical Systems, 277 p. Springer International Publishing Switzerland (2016)
6. Gorodetskiy, A.E., Kurbanov, V.G. (eds.): Smart Electromechanical Systems: The Central Nervous System, 266 p. Springer International Publishing AG (2017)
7. Volkomorov, S.V., Kaganov, Yu.T., Karpenko, A.P.: Modelling and Optimization of Some Parallel Mechanisms, 32 p. New Technologies, Moscow (2010). (in Russian)
8. Glazunov, V.A., Koliskor, A.Sh., Kraynev, A.F.: Spatial Mechanisms of Parallel Structure, 96 p. Science, Moscow (1991). (in Russian)
9. Zenkevich, S.L., Yushchenko, A.S.: Bases of Control of Handling Robots, 480 p. MSTU, Moscow (2004). (in Russian)
10. Merlet, J.P.: Parallel Robots (Solid Mechanics and Its Applications). Springer, Berlin (2004)

11. Heylo, S.V., Glazunov, V.A., Palochkin, S.V.: Handling Mechanisms of Parallel Structure. The Dynamic Analysis and Management, 86 p. MGUDT, Moscow (2014). (in Russian)
12. Agapov, V.A., Gorodetsky, A.E., Kuchmin, A.Yu., Selivanova, E.N.: Medical microrobot. Patent for the Invention of RUS 2469752 5/20/2011. (in Russian)
13. Gorodetskij, A.E., Kurbanov, V.G., Tarasova, I.L.: Adaptive gripping device. Patent for the Invention of RUS 2624278 C1, July 12, 2016. (in Russian). https://www1.fips.ru/wps/PA_FipsPub/res/Doc/IZPM/RUNWC1/000/000/002/624/278/%D0%98%D0%97-02624278-00001/DOCUMENT.PDF. July 04, 2019
14. Kuchmin, A.Yu., Dubarenko, V.V.: Linearized model of the mechanism with parallel structure. In: Gorodetskiy, A.E., Kurbanov, V.G. (eds.) Smart Electromechanical Systems: The Central Nervous System, 266 p. Springer International Publishing AG (2017). https://doi.org/10.1007/978-3-319-53327-8_13
15. Artemenko, Yu.N., Agapov, V.A., Dubarenko, V.V., Kuchmin, A.Yu.: Co-operative control of subdish actuators of radio telescope. Informatsionno-upravliaiushchie sistemy **4**, 2–9 (2012). (in Russian)
16. Isermann, R., Münchhof, M.: Identification of Dynamic Systems. An Introduction with Applications, 705 p. Springer (2011)
17. Mzyk, G.: Combined Parametric-Nonparametric Identification of Block-Oriented Systems, 238 p. Springer (2014)
18. Boutalis, Y., Theodoridis, D., Kottas, T., Christodoulou, M.A.: System Identification and Adaptive Control. Theory and Applications of the Neurofuzzy and Fuzzy Cognitive Network Models, 313 p. Springer (2014)
19. Grop, D.: Methods of Identification Systems, 302 p. Springer-Verlag (1979)
20. Karabutov, N.N.: Strukturnaia identifikatsiia sistem: Analiz dinamicheskikh struktur [Structural Identification of Systems: Analysis of Dynamic Structures], 160 p. MGIU, Moscow (2008). (in Russian)
21. Kuchmin, A.Yu.: Identification of dynamics of modules SEMS. In: Gorodetskiy, A.E., Tarasova, I.L. (eds.) Smart Electromechanical Systems. Group Interaction, pp. 193–202. Springer Nature Switzerland AG (2019). https://doi.org/10.1007/978-3-319-99759-9_16
22. Kuchmin, A.Yu., Dubarenko, V.V.: Definition of a rigidity of a hexapod. In: Gorodetskiy, A.E. (ed.) Smart Electromechanical Systems, 277 p. Springer International Publishing Switzerland (2016)

Decision Making an Autonomous Robot Based on Matrix Solution of Systems of Logical Equations that Describe the Environment of Choice for Situational Control

Andrey E. Gorodetskiy and Irina L. Tarasova

Abstract *Problem statement*: Progress in the development of modern industrial society is associated with the intellectualization of robotic systems, allowing to make informed decisions in group interaction in a dynamic environment of choice. *Purpose of research*: Providing autonomous robots with the ability to understand the language of sensations to enable decision-making regarding appropriate behavior in a group of autonomous robots under conditions of uncertainty. *Results*: A method of processing sensory information in the Central nervous system of the robot in order to obtain pragmatic information about the environment of choice is proposed. The methods of decision-making based on pragmatic information using matrix solutions of systems of logical equations in the description of optimization problems in the form of binary relations are described. *Practical significance*: The possibility of conscious decision-making by an Autonomous robot based on the analysis of the coordinator's goal of functioning, as well as the behavior and intentions of neighboring robots, the selection of semantic data on the environment pragmatic data related to the purpose of functioning, and the choice of genetic algorithms that most optimally lead to the goal of functioning is shown.

Keywords Situational control · Groups of autonomous robots · Environment of choice · Central nervous system of the robot · Quantization · Fuzzification · Images and images · Sensory · Syntactic · Semantic and pragmatic data · Deterministic · Stochastic and not fully defined constraints · Synthesis of the algorithm for finding the optimal solution. Progress in this area is associated with the development of intellectualization of robots · Allowing to make informed decisions in group interaction in a dynamic environment of choice

---

A. E. Gorodetskiy (✉) · I. L. Tarasova
Institute of Problems of Mechanical Engineering, Russian Academy of Sciences, St. Petersburg, Russia
e-mail: g27764@yandex.ru

I. L. Tarasova
e-mail: g17265@yandex.ru

© Springer Nature Switzerland AG 2020

A. E. Gorodetskiy and I. L. Tarasova (eds.), *Smart Electromechanical Systems*, Studies in Systems, Decision and Control 261, https://doi.org/10.1007/978-3-030-32710-1_20

1 Introduction

One of the main features of the development of modern industrial society is the mass use of robots and other robotic machines, both in production processes and in everyday life.

The functioning of autonomous intelligent robots is based on information from sensor systems regarding the environment and the state of the robot itself. Without this information, their automatic control systems (ACS) will not be able to determine the goals of operation and achieve these goals [1]. However, in order for these robots to be able, without human intervention, to formulate tasks and successfully perform them, they must have the ability to understand the language of sensations, i.e. to have feelings such as "friend–foe", "dangerous–safe", "beloved–unloved", "pleasant–unpleasant", etc., formed as a result of solving systems of logical equations [2].

While having the ability to the formation of a database of images based on the language of sensations in the central nervous system of the robot (CNSR) [3], you receive the possibility of independent decision-making regarding appropriate behavior [4]. At the same time, the formation of the behavior of an autonomous robot is based on pragmatic information obtained by successive transformation of the measuring information from the robot sensors into syntactic, then semantic and finally pragmatic [5]. As a result, the top-level system, i.e. the system of situational control, operating semantic [6] and pragmatic [7] information, form the behavior of the robot in the group, i.e. determine the sequence of its actions, which in an ever-changing environment are necessary to achieve this goal.

2 Structure of CNSR

The central nervous system (CNS) robot is based on the analogy with the CNS of a human who has sensory organs that perceive information about the environment and their own state. Then this information is converted into nerve impulses and enters the brain for analysis and development of a response. In this scheme similar to the following: RECEPTOR → NERVOUS CIRCUIT → region → IMPULSE → RESPONSE [8]. Signals from the outside process areas of the cerebral cortex, which are the core of the human central nervous system. Each sense organ corresponds to certain areas of the cerebral cortex. In them there is a comprehension of information. Another important property of human perception of the world is the presence of links between the perceived information from different senses. However, despite numerous studies in this area, there are still no sufficiently substantiated mathematical models of human CNS functioning, which prevents the creation of full-fledged CNSR [3]. Currently, the solution to the problem of creating the CNSR is reduced, first of all, to the study and development of chains such as the following scheme: (sensors, sensors and other measuring systems of the robot) → (information channel of reception of

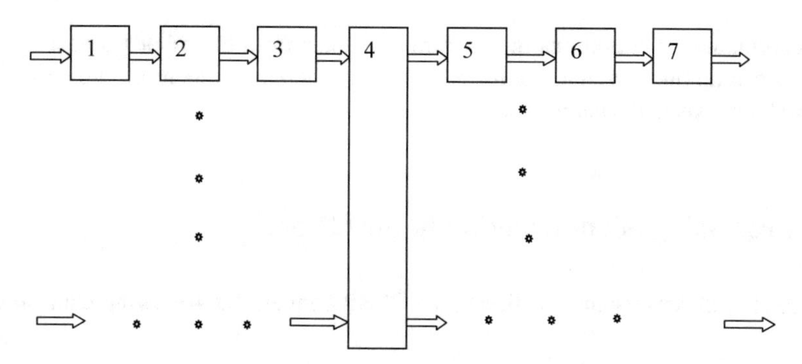

Fig. 1 Scheme CNSR. 1—Measuring system (sensors), 2—channel of transmission of measuring signals, 3—block of pre-processing of measuring signals, 4—block of fuzzification, recognition and decision-making, 5—channel of transmission of control signals, 6—block of formation of control actions, 7—working bodies of the robot

signals and their primary processing) → (combining signals, recognition, classification, decision-making) → (information channel of transmission of control signals, their transformation and formation of control action) → (movement, stretching and other actions of the working bodies of the robot) (see Fig. 1). In this case, the task of creating the CNSR is divided into two interrelated subtasks. The first is the creation of more advanced sensors and technical sensing systems. The second is the creation of software tools that provide robots with the ability to understand the language of sensations and to form behavioral processes based on the analysis of sensations, i.e. to provide robots with the opportunity not only to reflexive reasoning, but also to conscious ones based on the formation of pragmatic data.

The tasks of creating technical sensation systems (block 1) for robots have been sufficiently studied and there are numerous technical products for this purpose, for example [9–11]. Little studied and almost poorly used are measuring systems that simulate smell and taste [12]. Methods and means of transmission and conversion of signals in order to isolate information parameters from the signal-to-noise mixture have been studied well enough and there is a significant number of different technical means of transmission and filtration (blocks 2, 3, 5) of signals [13]. Also, the technical means of forming control actions (block 6) are well studied and developed [14].

Methods of data (signals) fuzzification, logical inference and logical-interval, logical-probabilistic and logical-linguistic methods of recognition and decision-making are known [15]. However, the solution of the problem of creating mathematical and software tools that allow robots to provide the possibility of reflexive and conscious reasoning significantly approximating the sign systems of the latter to those used in their daily practice, to be at the very initial stage, limited to modeling behavioral processes based on the analysis of sensations.

Thus, the available technical solutions at this stage allow us to begin the creation of simplified prototypes of the central nervous system. One of the most promising options for the mathematical implementation of block 4 can be the logical and mathematical implementation of the formation of behavioral processes based on the transformation of sensations in the form of signals from the robot's sensor system into semantic and pragmatic information about the environment of choice and its logical analysis in decision-making.

3 Processing of Information in the CNSR

Processing of sensory information in the CNSR contains the following main stages.

3.1 Quantization of the Surrounding Space in the Field of View of the Sensor System of the Robot with the Assignment of the Resulting Pixels Names in the Form of a Pixel Number

To do this, place the center of coordinates of the Euclidean space E^3 at the center of gravity of the robot. Then determine the boundaries of the sensitivity zones of the sensor system: $-X, +X, -Y, +Y, -Z, +Z$. The result is a three-dimensional space $C \subset E^3$. This space is divided into quanta on the X-axis with a step h_x, Y-axis with a step h_y and Z-axis with a step h_z. Quantum in the interval $-X$ to X, assign the number $i = -I = X/h_x, \ldots - 1, 0, 1, \ldots, X/h_x = I$. Quantum in the interval $-Y$ to Y, assign the number $j = -J = Y/h_y, \ldots - 1, 0, 1, \ldots, Y/h_y = J$. Quanta in the interval $-Z$ to Z are given the number $k = -K = Z/h_z, \ldots - 1, 0, 1, \ldots, Z/h_z = K$. As a result, all space C will be divided into a set $P = 2I * 2J * 2K$ pixels $p_{ijk} \in P$. Each p_{ijk} pixel will correspond to the information measured by the CNS sensory system about the senses of sight, hearing, smell, taste, touch, etc.

3.2 Fuzzification of Sensory Information Work on Each Pixel of the Surrounding Space and the Formation in the Memory of the CNSR Display of the Surrounding Space in the Form of Pixels with Their Coordinates and Fuzzified Data

To do this, the sensors can be combined into groups that form the following robot senses like a person: vision in the form of a set V; hearing in the form of a set R; smell in the form of a set S; taste in the form of a set G; touch in the form of a set T. In each of the introduced sets it is possible to allocate the subsets forming them characterizing properties of the observed or studied object:

$$V_i \subset V, R_i \subset R, S_i \subset S, G_i \subset G, T_i \subset T. \quad (1)$$

The set of such subsets depends on the set of sensors that form the senses of a particular robot. For example, to view you can enter the following sub-sets: $V1$ is the brightness of the image; $V2$ is a color image; $V3$—the flashing frequency; $V4$—the rate of change of brightness; $V5$—speed color changes, etc. Hearing can be introduced following subsets: $R1$ volume, $R2$ is the tone; $R3$—interval; $R4$—rate approximation; $R5$—removal rate; $R6$—direction. The following subsets can be introduced for smell: $S1$—type of smell; $S2$—intensity of smell; $S3$—rate of increase or decrease of smell; $S4$—rate of change of type of smell; $S5$—interval of smell, etc. For taste, the following subsets can be introduced: $G1$—type of taste; $G2$—the power of taste; $G3$—the rate of change in taste, etc. For touch, the following subsets can be introduced: $T1$—surface evenness; $T2$—surface dryness; $T3$—surface temperature, etc.

Of the signals from the sensors of the robot for each pixel $\mathrm{p_{ijk}}$ by fuzzification of the extracted logical data (v_{ijk}; r_{ijk}; s_{ijk}; g_{ijk}; t_{ijk}). For example, for $V1$, we obtain the following logical variables by fuzzification: v_{11k} is "very weak brightness", v_{12k} is "weak brightness", v_{13k} is "normal brightness", v_{14k} is "strong brightness" and v_{15k} is "very strong brightness". This means that for every logical variable is associated to its inherent attribute. In the simplest case, this attribute is the interval. In more complex cases, the attributes can be probabilities ($P\{v_{ijk} = 1\}$) or membership functions $\mu(v_{ijk})$. Consequently, data from robot sensors are stored in memory in the form of logical-interval, logical-probabilistic, or logical-linguistic variables [16].

3.3 Formation of Images in the Display of the Surrounding Space for Each Sense Organ of the Robot

The formation of images is made to display the surrounding space for each sense organ separately. In particular for vision will be the display of the surrounding space $C_V \subset C$, for hearing $C_R \subset C$, for smelling $C_S \subset C$, for taste $C_G \subset C$ and for touch $C_T \subset C$. In each of these maps, you can combine adjacent pixels with equal values of logical variables and similar values of their attributes. Then in the spaces of the senses $\mathrm{C_V}$, $\mathrm{C_R}$, $\mathrm{C_S}$, $\mathrm{C_G}$ and $\mathrm{C_T}$ we get sets of images Im_V, Im_R, Im_S, Im_G, Im_T with certain contours. Since the attributes of logical variables can be intervals, probabilities, membership functions, etc., for each type of attribute it is necessary to enter a measure of proximity δ_Δ, δ_P, δ_μ. После операции объединения пикселей в наборы изображений Im_V, Im_R, Im_S, Im_G, Im_T в каждом пространстве органов чувств C_V, C_R, C_S, C_G и C_T можно изобразить контуры изображений и каждому контуру присвоить имя (Fig. 1). As a result, there will be five cards K_V, K_R, K_S, K_G and K_N with sets of image contours Im_V, Im_R, Im_S, Im_G, Im_T.

3.4 Formation of Imagers by Combining Images from Different Senses and Assigning Names to Images in the Form of English Words

Usually, the formation of imagers from images consists in operations on images such as the intersection or union of ordered sets and the assignment of the result to a particular reference imager stored in the database. If no suitable imager is found in the database, this imager combination is given the name of the new imager that is placed in the database for temporary storage. With repeated repetition of this new imager during the operation of the robot, this imager becomes a reference and it is assigned a permanent new name.

To assign a combination of imagers to a particular standard, you must enter a measure of the proximity of ordered sets. Among the most well-known proximity measures, the following binary functional relations can be distinguished [16]:

– evaluation of the maximum deviation of the power sets:

$$\sum_i x_i - \sum_i y_i = \Delta \tag{2}$$

where $x_i = 1$ and $y_i = 1$ for non-zero (non-empty) elements of the compared sets M, N and respectively $m_i = 0$ and $n_i = 0$ for zero (empty) elements of the compared sets, and Δ is a numerical estimate of the proximity

– evaluation of the standard deviation of the power sets:

$$\sqrt{(\sum_i x_i)^2 - (\sum_i y_i)^2} = \delta, \tag{3}$$

where δ is a numerical estimate of the proximity

– probabilistic estimate of the maximum deviation of the power sets:

$$\sum_i P(x_i = 1)x_i - \sum_i P(y_i = 1)y_i = \Delta_P, \tag{4}$$

where $P(.)$ is the probability, Δ_P—probabilistic numerical evaluation of the proximity

– probabilistic estimate of the standard deviation of the power sets:

$$\sqrt{(\sum_i P(x_i = 1)x_i)^2 - (\sum_i P(y_i = 1)y_i)^2} = \delta_P, \tag{5}$$

where δ_p is a numeric probability estimation of proximity.

The use of these binary functional relations allows to rank combinations of images by their proximity to the reference image and at the same time allows to enter a numerical estimate of the proximity.

The process of formation of imagers based on information from the senses of the robot is carried out in the following sequence.

Initially searched for the presence of images that are close to the standard imagers of the database in each of the cards K_V, K_R, K_S, K_G and K_T. Found images are the names of the benchmarks. They are recorded in the CNSR observable database together with their coordinates and are excluded from the relevant maps.

At the second stage, in the intersections of sets from K_V, K_R, K_S, K_G and K_T maps, the presence of images close to the standard imagers in databases is searched.

At the same time, in the beginning, successive pairs of cards are superimposed on each other (Fig. 1) with the implementation of the operation of intersection of sets ($K_V \cap K_R$; $K_V \cap K_S$; $K_V \cap K_G$; $K_V \cap K_T$; $K_R \cap K_S$; $K_R \cap K_G$; $K_R \cap K_T$; $K_S \cap K_G$; $K_S \cap K_T$; $K_G \cap K_T$). At each intersection of the images, the presence of images close to the database standard imagers is searched. The found intersections of images are given names of standards. They are recorded in the CNS observable database together with their coordinates and are excluded from the corresponding map intersections. Therefore, in each subsequent intersection participate corrected by the results of the removal of the map.

Then, the three cards are sequentially superimposed on each other with the operation of crossing the sets ($K_V \cap K_R \cap K_S$; $K_V \cap K_R \cap K_G$; $K_V \cap K_R \cap K_T$; $K_R \cap K_S \cap K_G$; $K_R \cap K_S \cap K_T$; $K_S \cap K_G \cap K_T$). At each intersection of the images, the presence of images close to the database standard imagers is searched. The found intersections of images are given names of standards. They are recorded in the CNSR observable database together with their coordinates and are excluded from the corresponding map intersections. Therefore, in each subsequent intersection participate corrected by the results of the removal of the map.

Then, four cards are sequentially superimposed on each other with the operation of crossing the sets ($K_V \cap K_R \cap K_S \cap K_G$; $K_V \cap K_R \cap K_S \cap K_T$; $K_R \cap K_S \cap K_G \cap K_T$). At each intersection of the images, the presence of images close to the database standard imagers is searched. The found intersections of images are given names of standards. They are recorded in the CNSR observable database together with their coordinates and are excluded from the corresponding map intersections. Therefore, in each subsequent intersection participate corrected by the results of the removal of the map.

Next, all five adjusted maps are overlaid on each other ($K_V \cap K_R \cap K_S \cap K_G \cap K_T$). In the intersection of images, the presence of images close to the database standard imagers is sought. The found intersections of images are given names of standards. They are recorded in the CNSR observable database together with their coordinates and are excluded from the corresponding map intersections.

In the third stage, the combinations of sets of maps K_V, K_R, K_S, K_G and K_T are searched for the presence of images close to the standard imagers in the databases. At the same time, it is similar at the beginning of the operation of combining two sets ($K_V \cup K_R$; $K_V \cup K_S$; $K_V \cup K_G$; $K_V \cup K_T$; $K_R \cup K_S$; $K_R \cup K_G$; $K_R \cup K_T$;

$K_S \cup K_G$; $K_S \cup K_T$; $K_G \cup K_T$), then three—($K_V \cup K_R \cup K_S$; $K_V \cup K_R \cup K_G$; $K_V \cup K_R \cup K_T$; $K_R \cup K_S \cup K_G$; $K_R \cup K_S \cup K_T$; $K_S \cup K_G \cup K_T$), четырех—($K_V \cup K_R \cup K_S \cup K_G$; $K_V \cup K_R \cup K_S \cup K_T$; $K_R \cup K_S \cup K_G \cup K_T$) and five—($K_V \cup K_R \cup K_S \cup K_G \cup K_T$). The found image associations are assigned the names of the standards. They are recorded in a database of the observed data CSR together with their coordinates and are excluded from the relevant associations of the cards. Therefore, in each subsequent Association, the maps adjusted according to the results of the deletions participate.

At the fourth stage, the symmetric difference sets from K_V, K_R, K_S, K_G and K_T maps are searched for the presence of images close to the standard imagers in the databases. In this case, it is similar to the first implementation of the operation of the symmetric difference of two sets ($K_V \Delta K_R$; $K_V \Delta K_S$; $K_V \Delta K_G$; $K_V \Delta K_T$; $K_R \Delta K_S$; $K_R \Delta K_G$; $K_R \Delta K_T$; $K_S \Delta K_G$; $K_S \Delta K_T$; $K_G \Delta K_T$), then three—($K_V \Delta K_R \Delta K_S$; $K_V \Delta K_R \Delta K_G$; $K_V \Delta K_R \Delta K_T$; $K_R \Delta K_S \Delta K_G$; $K_R \Delta K_S \Delta K_T$; $K_S \Delta K_G \Delta K_T$), four—($K_V \Delta K_R \Delta K_S \Delta K_G$; $K_V \Delta K_R \Delta K_S \Delta K_T$; $K_R \Delta K_S \Delta K_G \Delta K_T$) and five—($K_V \Delta K_R \Delta K_S \Delta K_G \Delta K_T$). The found symmetric image differences are assigned the names of the standards. They are recorded in a database of the observed data CNSR together with their coordinates and are excluded from the relevant associations of the cards. Therefore, in each subsequent association, the maps adjusted according to the results of the deletions participate.

If any more intersections, joins, or symmetric differences are corrected, they are given new names in the form of a new word of English letters. They are also recorded with their coordinates in the database of the observed data of the CNSR and the corresponding message is transmitted to the community of robots to legitimize the new reference word and its corresponding image.

Thus, in the database CNSR formed semantic data about the robot's space, based on which a robot accepts behavioural solutions [4] using the stored knowledge base of the robot behavioral model algorithms. These algorithms are recorded in the knowledge base robot at the stage of its creation, based on its purpose. Therefore, such algorithms will be called genetic. However, after the formation of a database of semantic data about the surrounding space of the robot may be that in the same place of space partially or completely there are two or more imagers. Therefore, it is necessary to adjust the semantic database in order to eliminate the detected collisions. This adjustment is closely related to the formation from semantic data—pragmatic, relevant to the problem being solved by the robot at the moment.

For example, if a robot-car is faced with the problem of accident-free passage of the intersection, then it requires the formation of images for displaying the surrounding space only from the sense organs of vision $C_V \subset C$ and hearing $C_R \subset C$. Accordingly, the process of obtaining pragmatic information to solve the problem is sharply reduced.

4 Genetic Robot Algorithms

In general, we can distinguish five large interacting modules of robot behavior algorithms.

4.1 Module 1 Saving Lives

This module contains algorithms that, based on the analysis of imagers in the surrounding space and their own state, form the movements of robots in such a way as to avoid collisions and hits in hazardous areas of the environment, as well as algorithms for moving to favorable zones with dynamic changes in the environment. In addition, there may be algorithms of environmental impact in order to change it for the most comfortable of its own functioning.

4.2 Module 2 Self-improvement

This module contains algorithms for adapting the robot's own subsystems to changes in the environment, as well as algorithms for optimizing the subsystems and the robot as a whole for the purposes of performing certain tasks coming from higher-level systems (coordinators) or from the systems of planning their own behavior.

4.3 Module 3 Curiosity

This module contains algorithms for finding new imagers in the environment by improving the sensor system of the Central nervous system, adding new sensors and improving the methods of processing sensory, syntactic and semantic data.

4.4 Module 4 Training

This module contains search algorithms for obtaining new knowledge about the properties of the environment by trial and error, as well as the synthesis of new behavioral algorithms for previous modules based on the analysis of options for responding to changes in knowledge about the environment.

4.5 Module 5 Creating Your Own Kind (Continuation of the Genus)

This module contains algorithms for the synthesis of similar robots with the search and use of appropriate objects in the environment.

5 Methods of Decision-Making Based on Robot Reflections in the Environment of Choice

A robot's decision about its own behavior can be reflexive and conscious. In reflexive decision-making robot based on semantic data selects the behavioral algorithm from the modules of genetic algorithms described above the one with the highest priority. Setting priorities is carried out at the design stage by solving an optimization problem using mathematical programming methods [17]. For example, if there is an object in the environment that is dangerous and which moves in the direction of the robot, then the robot takes to the implementation of the algorithm of leaving the trajectory of this object, since the preservation of life for it has the highest priority.

When making a conscious decision, the robot analyzes the coordinator's goal of functioning, as well as the behavior and intentions of neighboring robots, selects from the semantic data on the environment pragmatic data related to the purpose of functioning, and then selects from the genetic algorithms those that most optimally lead to the goal of functioning. In this case, different approaches to decision-making can be used.

In deductive decision-making process of thinking in the CNSR begins with the global level and then goes down to the local. The technical analogue of this type of thinking can be the process of optimization, when first, on the basis of available information, the best solution is sought out of all possible, and then, by checking on the basis of available information, all restrictions, the solution is corrected.

The choice of the optimal solution from all ui solutions can be carried out in different ways. The most simple case, to use methods of mathematical programming [17], when required to compute an n-dimensional vector U that optimizes (paying in a minimum or a maximum depending on the substantive problem definition) the criterion for the quality of the solutions $f_0(U)$ subject to constraints $f_j(U) \leq c_j,\ j \in 1, 2, \ldots r;\ u \in W$, where $f_j(U)$ are known scalar functions, c_j—assigned number, W is a predetermined multiple n-dimensional space.

In the case where the constraints describing the selection environment are given in the form of logical-probabilistic relations [16], i.e. when the attributes of logical variables in the equations describing the constraints are probabilities $P\{u_i = 1\}$, the quality criterion can be expressed as follows:

$$f_0(U) = P\{u_i = 1\} \to \max. \tag{6}$$

The probabilities $P\{u_i = 1\}$ can be calculated approximately according to the algorithm described in [18].

If, as a result of the analysis of data in any CNSR, it is revealed that the influence of certain components of y_i on the behavior of the robot is different, then the quality criterion (6) is advisable to lead to the form:

$$f_0(U) = \beta_i P\{u_i = 1\} \to \max, \tag{7}$$

where β_i—assigned weights.

In the case where the constraints describing the selection environment are given in the form of logical-linguistic relations [16], i.e. when the attributes of logical variables in the equations describing the constraints are the membership functions $\mu(y_i)$, the quality criterion can be expressed as follows:

$$f_0(U) = \mu\{u_i\} \to \max. \tag{8}$$

The values of the membership functions $\mu(y_i)$ can be calculated using the algorithms described in [16].

If, as a result of the analysis of data in any CNSR, it is revealed that the influence of certain components of y_i on the behavior of the robot is different, then the quality criterion (8) is advisable to lead to the form:

$$f_0(U) = \beta_i \mu\{u_i\} \to \max \tag{9}$$

In the case when the constraints describing the selection environment are given in the form of logical-interval relations [16], i.e. when the attributes of logical variables in the equations describing the constraints are intervals $\Delta_{ij} = \left[a_{ji}, b_{ji}\right]$, the quality criterion can be expressed as the following expressions:

$$f_0(U) = k_{jl}(b_{ji} - a_{ji}) \to \min, \tag{10}$$

$$f_0(U) = [k_{ji}(b_{ji} - a_{ji}) - c_{ji}]^2 \to \min, \tag{11}$$

$$f_0(U) == [k^b_{ji}(b_{ji} - b^0_{ji})^2 + (a_{ji} - a^0_{ji})]^2 \to \min, \tag{12}$$

$$f_0(U) = [k^b_{ji}(b_{ji} - b^0_{ji})^2 + k^a_{ji}(a_{ji} - a^0_{ji})]^2 \to \min, \tag{13}$$

where $k_{ji}, k^b_{ji}, k^a_{ji}$—are the coefficients of the preferences of decision makers (PDM) on the optimality, c_{ji}—the requested PDM width of the interval, b^0_{ji}, a^0_{ji}—the desired decision maker the boundaries of the intervals.

After calculating the quality criteria of all possible solutions in accordance with Eqs. (6), (7)—with the logical-probabilistic description of the uncertainties, or (8), (9)—with the logical-linguistic description of the uncertainties, or (10)–(13)—with the logical-interval description of the uncertainties, all found solutions are ranked. The solutions are then tested for the feasibility of other constraints describing the eigen state of the robot, starting with the first one having the highest quality criterion. In this case, the first of the tested solutions satisfying these constraints is considered optimal.

When inductive decision-making process of thinking in the CNSR begins with the analysis of individual decisions and then goes to find a common, global conclusion. The technical analogue of this type of thinking can be the process of optimization, when first, on the basis of available information, all solutions are checked for the feasibility of restrictions describing their own state, and then the best solution is sought out of all possible under the conditions of restrictions of solutions according to the criteria of type (6)–(13).

According to Pearce, cognitive activity in the Central nervous system is the interaction of abduction, induction and deduction [19]. In this case, the abduction carries out the adoption of plausible hypotheses by explaining the facts, by induction testing of hypotheses is realized, and by deduction from the accepted hypotheses the consequences are deduced. The technical analogue of this type of thinking can be the process of finding the optimal solution by analogy, when from all possible solutions are first selected by methods of pattern recognition [15] those solutions that are closest to the existing solutions stored in the database of the CNSR and gave good results in the past. Then you can use deductive and/or inductive decision-making methods on the quality criteria (6)–(13) to choose the best.

Comparing the described methods of decision-making, it can be concluded that the abduction method is the fastest by analogy with intuition, but its reliability depends on the completeness of the database of good decisions from past experience, i.e. strongly depends on the time of operation of such robots in similar environmental conditions. The deductive method is faster than the inductive method with a large number of constraints, since it does not require constraint checking for all solutions. With complex quality criteria and a small number of restrictions, the inductive method can give a faster result, as it will reject the search for a solution by complex quality criteria for solutions that are unacceptable by restrictions.

However, for intelligent systems, as a rule, it is not possible to form a scalar quality criterion. Then the choice of the optimal solution from all u_i solutions can be carried out by methods of mathematical programming in ordinal scales, generalized mathematical programming or multistep generalized mathematical programming [16, 20]. In this case, the problem of finding the optimal solution is the problem of choosing the "best" (in the sense of the binary ratio q_0) among the permissible solutions, i.e. as close as possible to the reference solution $u_э$. It is required to find one solution u_0—this being the binary correlation subject to the binary relations that describe constraints:

$$u_0 \, q_0 \, u_э \tag{14}$$

$$u_э \, q_j \, c_j, \tag{15}$$

$$u_0 \, q_j \, c_j, \tag{16}$$

where $j = 1, 2, \ldots, n$, $(u_э, u_0) \in W$.

In the simplest case, the binary relations q_0 and q_j can be numbers 0 and 1. Then the optimal solution is $q_0 = 1$ and $q_j = 1$. Naturally, the search for such a solution will be unlikely. Therefore, when determining the optimal solution of q_o and, accordingly, q_j should be expressed as logical equations in the Zhegalkin algebra [21] reduced to the matrix form [22]:

$$A_0 U = Q_o, \tag{17}$$

$$A_j C = Q_j, \tag{18}$$

where A_0 and A_j—identification strings containing 0 and 1 in a given order, U—vector of logical variables characterizing the behavior of the Autonomous robot, C—vector of logical variables characterizing the environment, Q_o—vector of logical variables characterizing the proximity of the robot's behavior to the standard, Q_j—vector of logical variables characterizing the limitations of the environment.

Dimensions A_0 and U, as well as A_j and C coincide. The attributes of the logical variables in Eqs. (6) and (7) can be probabilities, membership functions, and score estimates set by the decision maker.

6 Conclusion

Wherever there is a group of complex intelligent technical objects that must work together to perform some work or solve some problem, there is a problem of finding the optimal algorithm for decision-making for each control object, for example autonomous robot. At the same time, the formation of the behavior of an autonomous robot is based on pragmatic information obtained by successive transformation of the measuring information from the robot sensors into syntactic, then semantic and finally pragmatic information into the CNSR. As a result, the situational control system forms the behavior of the robot in the group, i.e. determine the sequence of its actions, which in an ever-changing environment are necessary to achieve this goal.

A robot's decision about its own behavior can be reflexive and conscious. In reflexive decision-making robot based on semantic data selects the behavioral algorithm from the available genetic algorithms the one with the highest priority. When making a conscious decision, the robot analyzes the coordinator's goal of functioning, as well as the behavior and intentions of neighboring robots, selects from the semantic data on the environment pragmatic data related to the purpose of functioning, and then selects from the genetic algorithms those that most optimally lead to the goal of functioning. Different approaches to decision-making can be used: deductive, inductive and abductive decision-making. The abduction method is the fastest by analogy with intuition, but its reliability depends on the completeness of the database of good solutions from past experience, i.e. it strongly depends on the time of operation of such robots in similar environmental conditions.

In determining the optimal solution under conditions of not complete certainty based on binary relations, the latter can be expressed as logical equations in the Zhegalkin algebra reduced to a matrix form, which makes it easy to parallelize the process of finding the optimal solution.

Acknowledgements This work was financially supported by Russian Foundation for Basic Research Grants 18-01-00076 and 19-08-00079.

References

1. Dobrynin, D.A.: Intelligent robots yesterday, today, tomorrow. In: X natsional'naia konferentsiia po iskusstvennomu intellektu s mezhdunarodnym uchastiem KII-2006 (25–28 sentiabria 2006 g., Obninsk) [X National Conference on Artificial Intelligence with International Participation (25–28 September 2006, Obninsk)]: Conference Proceedings, vol. 2. FIZMATLIT Publ., Moscow (2006). (in Russian)
2. Gorodetskiy, A.E., Tarasova, I.L., Kurbanov, V.G.: Behavioral decisions of a robot based on solving of systems of logical equations. In: Gorodetskiy, A.E., Kurbanov, V.G. (eds.) Smart Electromechanical Systems: The Central Nervous System, 270 p. Springer International Publishing AG (2017). https://doi.org/10.1007/978-3-319-53327-8
3. Gorodetskiy, A.E., Kurbanov, V.G.: Smart Electromechanical Systems: The Central Nervous Systems, 270 p. Springer International Publishing (2017). ISBN 978-3-319-53326-1. https://doi.org/10.1007/978-3-319-53327-8
4. Gorodetskiy, A.E., Kurbanov, V.G., Tarasova, I.L.: Decision making in the central nervous system of the robot. Informatsionno-upravliaiushchie sistemy **1**, 21–30 (2018). (in Russian). https://doi.org/10.15217/issnl684-8853.2018.1.21
5. Gorodetskiy, A., Kurbanov, V., Tarasova, I.: Formation of images based on sensory data of robots. In: Proceedings of the 14th International Conference on Pattern Recognition and Information Processing PRIPT 2019, Minsk, Belarus, 21–23 May 2019
6. Davydov, O.I., Platonov, A.K.: Robot and Artificial Intelligence. Technocratic Approach, no. 112, 24 p. Preprint IPM im. M. V. Keldysh (2017). (in Russian). https://doi.org/10.20948/prepr-2017-112
7. Krysin, L.P.: Types of pragmatic information in the "explanatory dictionary". Izvestiya RAN seriya literatury I yazyka **74**(2), 3–11 (2015). (in Russian)
8. Meshcheryakov, B.G., Zinchenko, V.P. (eds.): Great Psychological Dictionary. Praym Publ., Moscow (2003)
9. Babich, A.V.: Industrial Robotics, 263 p. Book on Demand Publ., Moscow (2012). (in Russian)
10. Polivtsev, S.A., Khashan, T.S.: The study of geometric and acoustic properties of the sensors for the technical hearing system. Problemy bioniki [Probl. Robots Bionics] **6**, 63–69 (2003). (in Russian)
11. Ying, M., Bonifas, A.P., Lu, N., Su, Y., Li, R., Cheng, H., Ameen, A., Huang, Y., Rogers, J.A.: Silicon Nanomembranes for Fingertip Electronics. IOP Publishing Ltd, 10 Aug 2012
12. Gorodetskiy, A.E., Kurbanov, V.G., Tarasova, I.L.: Methods of synthesis of optimal intelligent control systems SEMS. In: Smart Electromechanical Systems, pp. 25–45. http://dx.doi/org/10.1007/978-3-319-27547-5_4
13. Rachkov, M.Yu.: Tekhnicheskie sredstva avtomatizacii [Technical Means of Automation], 185 p. MGIU Publ., Moscow (2009). (in Russian)

14. Zheltikov, M.O.: Osnovy teorii upravleniya [The Basics of Control Theory]. Lecture Notes. SGTU Publ., Samara (2008). (in Russian)
15. Gorodetsky, A.E., Tarasova, I.L.: Nechetkoe matematicheskoe modelirovanie ploho formalizuemyh processov i sistem [Fuzzy Mathematical Modeling Difficult to Formalize Processes and Systems], 336 p. Politekhnicheskii Universitet Publ., Saint-Petersburg (2010). (in Russian)
16. Gorodetskiy, A.: Osnovy teorii intellektual'nykh sistem upravleniia [Foundations of the Theory of Intelligent Control Systems], 313 p. LAP LAMBERT Academic Publishing GmbH @ Co. KG (2011). (in Russian)
17. Karmanov, V.G.: Matematicheskoe programmirovanie [Mathematical Programming], 263 p. Fiz.-Mat. Literature Publ., 2004. (in Russian)
18. Gorodetsky, A.E., Dubarenko, V.V.: Combinatorial method for calculating the probability of complex logic functions. Zhurnal vychislitel'noi matematiki i matematichskoi fiziki **39**(7), 1201–1203 (1999). (in Russian)
19. Svetlov, V.A.: Methodological concept of Charles Pearce's scientific knowledge: unity of abduction, deduction and induction. Logiko-Filosofskie shtudii **5**, 165–187 (2008). (in Russian). ISSN 2071-9183
20. Iudin, D.B.: Vychislitel'nye metody teorii priniatiia reshenii [Computational Methods of Decision Theory], 320 p. Nauka Publ., Moscow (1989). (in Russian)
21. Zhegalkin, I.I.: Arifmetizatsiia simvolicheskoi logiki [Arithmetization Symbolic Logic]. Matematicheskii sbornik [Math. Collect.] **35**(3–4) (1928). (in Russian)
22. Gorodetskiy, A.E., Dubarenco, V.V., Erofeev, A.A.: Algebraic approach to the solution of logical control problems. Avtomatika i telemekhanika [Autom. Remote Control] **2**, 127–138 (2000). (in Russian)

A Situation Control of Robotized Space Module as Multimode Dynamic Object

Pavel P. Belonozhko

Abstract *Problem statement:* assembly and service robotized space module (ASRSM). An important feature of the system that determines the specifics of the controlled motion modes is the presence of a free (movable) base in the inertial space. The authors consider the problem of situation control over the movement of large cargo in relation to a moving platform with the help of a manipulator. *Purpose of research*: Under specific conditions for the mode with the absence of superposed forces some reduced system can be considered as a control object. In this case the use of the reduced system proper motions corresponding to the initial system inertial motions in terms of their internal degree of freedom is efficient at control synthesis in the number of situations. In particular, the control over cargo movement in relation to the moving platform built with the use of "ballistic" motions of the reduced system is optimal for energy input minimization. Results: It is demonstrated that for the plane motion of a robotized module with a single-level manipulator the reduced system is a non-linear oscillatory system. In addition to mass-inertia and geometrical properties of the initial system (mechanical design model of a robotized space module), the reduced system proper motions are also defined by the initial system kinetic momentum. The initial system kinetic momentum depends on the initial motion conditions for some current operation mode defined by the previous operation mode of the robotized space module in the situation control problem. The problem of bringing the system from some initial position to some required one is considered. It is shown how the nature of control changes depending on the changes of the phase portrait determined by the conditions of situational control.

Keywords Assembly and service robotic space modules · Proper inertial motion by degree of freedom of manipulator · Reduced system · Phase portrait

P. P. Belonozhko (✉)
Bauman Moscow State Technical University, Moscow, Russia
e-mail: byelonozhko@mail.ru

© Springer Nature Switzerland AG 2020

A. E. Gorodetskiy and I. L. Tarasova (eds.), *Smart Electromechanical Systems*, Studies in Systems, Decision and Control 261, https://doi.org/10.1007/978-3-030-32710-1_21

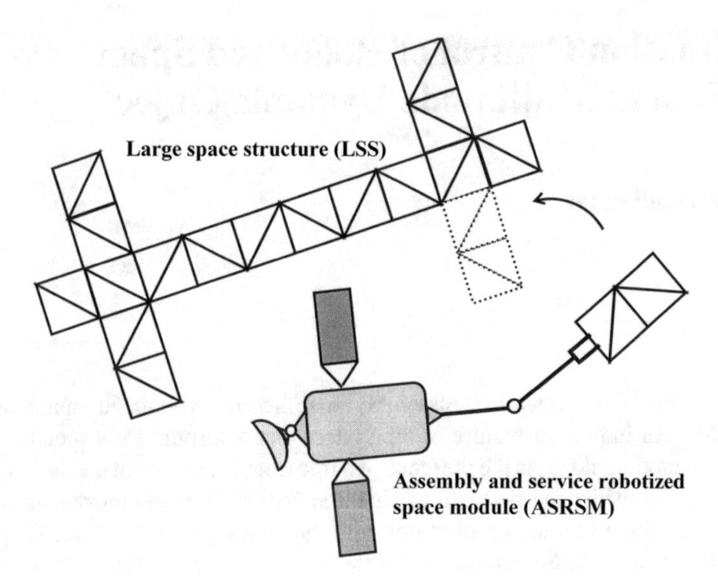

Fig. 1 Robotic assembly of large space structure

1 Introduction

Analysis of the current state of space robotics [1–13] shows the prospects of the concept of automated assembly of large space objects with the help of assembly and service robotized space modules (ASRSM). The problem of situational control of the movement of massive cargo relative to the movable base in the performance of installation and service operations with the use of ASRSM (Fig. 1).

In [10, 13–19] the urgency of the special modes of controlled motion of the MRCM, an important feature of which is the lack of external forces and moments in relation to the system, is noted. Shown [14–19], that introduction in consideration of the reduced system provides an opportunity of visual qualitative research of a question of use of own movements of system on degrees of mobility of the manipulative mechanism at construction of control.

2 The Proper Motion of the Reduced System

Let us consider the case when a space module (SM) with a massive payload (PL) in the capture of a single-stage massless manipulator makes a flat motion in the absence of forces external to the «SM-manipulator-PL» system.

Masses of SM and PL—m_1 and m_2 respectively, J_1 and J_2—moments of body inertia SM and PL in respect to the centers of mass C_1 and C_2, l_1 and l_2—distance between centers of mass and a joint A. The motion is viewed in respect to the

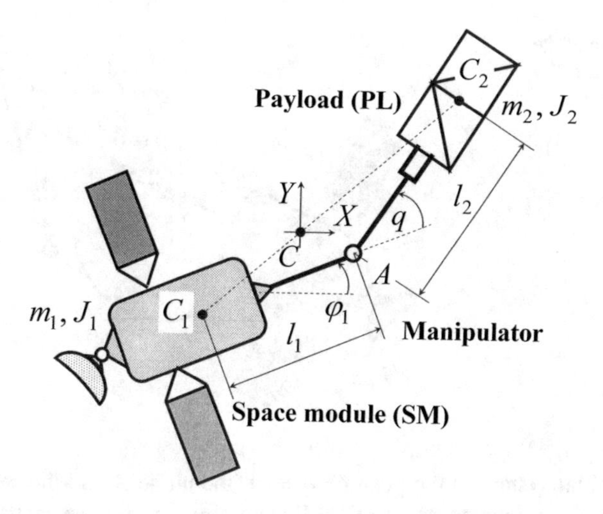

Fig. 2 «SM-manipulator-PL» system

nonrotating inertial coordinate system XCY originating in the system's center of mass C. Position relative to XCY is defined by angle φ_1, which describes absolute motion of the platform, and joint angle q, which describes motion of the load in respect to the platform (Fig. 2).

In [16, 18, 19] the author shows that the considered initial system can be put in accordance with the reduced nonlinear oscillatory system with one degree of freedom, the proper motion of which is described by the equation

$$\begin{aligned}&\ddot{q}\frac{(J_1+\tilde{m}l_1^2)(J_2+\tilde{m}l_2^2)-\tilde{m}^2l_1^2l_2^2\cos^2 q}{(J_1+J_2+\tilde{m}l_1^2+\tilde{m}l_2^2+2\tilde{m}l_1l_2\cos q)}\\&+\dot{q}^2\frac{\tilde{m}l_1l_2\sin q(J_1+\tilde{m}l_1^2+\tilde{m}l_1l_2\cos q)(J_2+\tilde{m}l_2^2+\tilde{m}l_1l_2\cos q)}{(J_1+J_2+\tilde{m}l_1^2+\tilde{m}l_2^2+2\tilde{m}l_1l_2\cos q)^2}\\&+L^2\frac{\tilde{m}l_1l_2\sin q}{(J_1+J_2+\tilde{m}l_1^2+\tilde{m}l_2^2+2\tilde{m}l_1l_2\cos q)^2}=0,\end{aligned}\tag{1}$$

where L—the kinetic moment of the initial system, which remains unchanged by virtue of the assumptions

$$\begin{aligned}L=\;&\dot{\varphi}_1(J_1+J_2+\tilde{m}l_1^2+\tilde{m}l_2^2+2\tilde{m}l_1l_2\cos q)\\&+\dot{q}(J_2+\tilde{m}l_2^2+\tilde{m}l_1l_2\cos q)=const.\end{aligned}\tag{2}$$

With the aim of qualitative analysis of the dependence of the eigenmovements of the reduced system on the quantitative characteristics of the investigated mode of motion, we give a visual interpretation of the kinetic moment L, which is a parameter of Eq. (1).

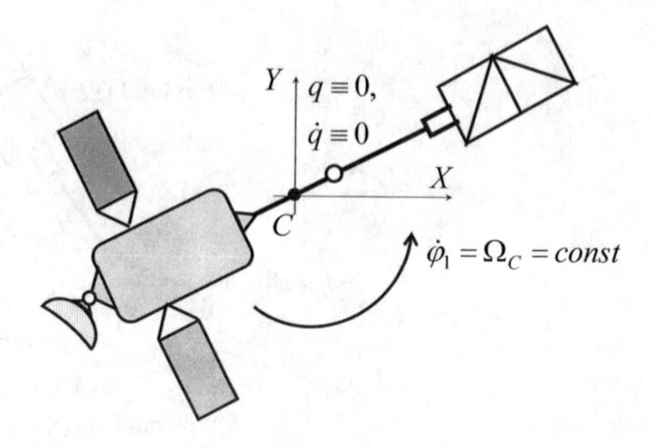

Fig. 3 Rotation of the «SM-manipulator-PL» system as a solid

Let the initial system in the position $q = 0$ (the hinge A and the center of mass of the system C are located on a straight line passing through the centers of mass of bodies C_1 and C_2) rotates relative to the center of mass C with a constant angular velocity Ω_C (Fig. 3).

The reduced system is in the «lower» position of stable equilibrium $q \equiv 0,\ \dot{q} \equiv 0$, and the kinetic moment of the initial system L and the kinetic energy of the initial system T_L are determined by the relations

$$L = J_C \Omega_C, \tag{3}$$

$$T_L = \frac{1}{2} J_C \Omega_C^2 = \frac{1}{2} \frac{L^2}{J_C}, \tag{4}$$

where J_C—the moment of inertia of the system «SM—manipulator—PL» in position $q = 0$ relative to the axis passing through the center of mass C

$$J_C = J_1 + J_2 + \tilde{m} l_1^2 + \tilde{m} l_2^2 + 2\tilde{m} l_1 l_2. \tag{5}$$

Thus, in accordance with (3), the kinetic moment L is a linear function of the angular velocity Ω_C, uniquely determined by the kinematic parameters of the motion of the initial system at the time of termination of the external forces (for example, at the time of switching off the engines SM). In the problem of situational control of any set of initial conditions, $\varphi_1^{(0)}, \dot{q}^{(0)}, q^{(0)}$ corresponds to a certain angular velocity Ω_C, which is thus a visual characteristic of the considered mode of motion of the original system.

In case

$$\dot{\varphi}_1^{(0)} = \Omega_C,\ \dot{q}^{(0)} = 0,\ q^{(0)} = 0. \tag{6}$$

the kinetic energy of the reduced system and the potential energy of the reduced system [16, 18, 19] are zero

$$E_k^{(0)} = \frac{1}{2}\frac{(J_1 + \tilde{m}l_1^2)(J_2 + \tilde{m}l_2^2) - \tilde{m}^2 l_1^2 l_2^2 \cos^2 q^{(0)}}{(J_1 + J_2 + \tilde{m}l_1^2 + \tilde{m}l_2^2 + 2\tilde{m}l_1 l_2 \cos q^{(0)})}\dot{q}^{(0)2} = 0, \tag{7}$$

$$\begin{aligned} E_p^{(0)} &= \frac{1}{2}\frac{1}{(J_1 + J_2 + \tilde{m}l_1^2 + \tilde{m}l_2^2 + 2\tilde{m}l_1 l_2 \cos q^{(0)})}L^2 \\ &- \frac{1}{2}\frac{1}{(J_1 + J_2 + \tilde{m}l_1^2 + \tilde{m}l_2^2 + 2\tilde{m}l_1 l_2)}L^2 = 0. \end{aligned} \tag{8}$$

Suppose now that the internal forces (for example, the control moment M in the hinge) act on the system for some time. Let the angular velocity SM, the hinge coordinate and the hinge velocity have some nonzero values $\dot{\varphi}_{10}$, q_0 and $\dot{q}_0$ at the end of the action of the internal forces. At the same time, it is obvious that in general

$$\dot{\varphi}_{10} \neq \Omega_C. \tag{9}$$

The kinetic moment L of the initial system does not change the value. Thus, $\dot{\varphi}_{10}$, q_0 and $\dot{q}_0$ are related by the ratio

$$\begin{aligned} L = J_C \Omega_C &= \dot{\varphi}_{10}(J_1 + J_2 + \tilde{m}l_1^2 + \tilde{m}l_2^2 + 2\tilde{m}l_1 l_2 \cos q_0) \\ &+ \dot{q}_0(J_2 + \tilde{m}l_2^2 + \tilde{m}l_1 l_2 \cos q_0). \end{aligned} \tag{10}$$

The kinetic energy of the system will increase, and will be equal to

$$T = T_L + E_0, \tag{11}$$

where T_L is still determined by the ratio (4), and the term E_0 is the total energy of the reduced system, equal to the sum of the kinetic energy of the reduced system and the potential energy of the reduced system

$$E_0 = E_{k0} + E_{p0}, \tag{12}$$

$$\begin{aligned} E_{k0} = E_k|_{q=q_0,\dot{q}=\dot{q}_0} &= \frac{1}{2}\frac{(J_1 + \tilde{m}l_1^2)(J_2 + \tilde{m}l_2^2) - \tilde{m}^2 l_1^2 l_2^2 \cos^2 q}{(J_1 + J_2 + \tilde{m}l_1^2 + \tilde{m}l_2^2 + 2\tilde{m}l_1 l_2 \cos q)}\dot{q}^2\bigg|_{q=q_0,\dot{q}=\dot{q}_0}, \\ &= \frac{1}{2}\frac{(J_1 + \tilde{m}l_1^2)(J_2 + \tilde{m}l_2^2) - \tilde{m}^2 l_1^2 l_2^2 \cos^2 q_0}{(J_1 + J_2 + \tilde{m}l_1^2 + \tilde{m}l_2^2 + 2\tilde{m}l_1 l_2 \cos q_0)}\dot{q}_0^2 \end{aligned} \tag{13}$$

$$\begin{aligned} E_{p0} &= E_p|_{q=q_0,\ \dot{q}=\dot{q}_0} \\ &= \frac{1}{2}L^2\left(\frac{1}{(J_1 + J_2 + \tilde{m}l_1^2 + \tilde{m}l_2^2 + 2\tilde{m}l_1 l_2 \cos q)} - \frac{1}{(J_1 + J_2 + \tilde{m}l_1^2 + \tilde{m}l_2^2 + 2\tilde{m}l_1 l_2)}\right)\bigg|_{q=q_0,\ \dot{q}=\dot{q}_0} \\ &= \frac{1}{2}\frac{1}{(J_1 + J_2 + \tilde{m}l_1^2 + \tilde{m}l_2^2 + 2\tilde{m}l_1 l_2 \cos q_0)}L^2 - \frac{1}{2}\frac{1}{(J_1 + J_2 + \tilde{m}l_1^2 + \tilde{m}l_2^2 + 2\tilde{m}l_1 l_2)}L^2. \end{aligned} \tag{14}$$

Next, the system makes its own motion, determined by Eq. (1) with the initial conditions q_0 and $\dot{q}_0$. In this case, there is an integral of energy

$$E_k + E_p = E_0 = const, \tag{15}$$

where

$$E_k = \frac{1}{2}\frac{(J_1 + \tilde{m}l_1^2)(J_2 + \tilde{m}l_2^2) - \tilde{m}^2 l_1^2 l_2^2 \cos^2 q}{(J_1 + J_2 + \tilde{m}l_1^2 + \tilde{m}l_2^2 + 2\tilde{m}l_1 l_2 \cos q)}\dot{q}^2, \tag{16}$$

$$E_p = \frac{1}{2}\frac{1}{(J_1 + J_2 + \tilde{m}l_1^2 + \tilde{m}l_2^2 + 2\tilde{m}l_1 l_2 \cos q)}L^2 - \frac{1}{2}\frac{1}{(J_1 + J_2 + \tilde{m}l_1^2 + \tilde{m}l_2^2 + 2\tilde{m}l_1 l_2)}L^2, \tag{17}$$

which is an equation of the family of phase trajectories of the reduced system. The total energy E_0 of the reduced system determined in accordance with (12)–(14) initial conditions q_0 and $\dot{q}_0$ is a parameter of the family of phase trajectories.

At Fig. 4 the phase portrait of the reduced system corresponding to the initial system with the following mass-inertia and geometric parameters is shown

$$m_1 = m_2 = 2500 \text{ kg}, \quad J_1 = J_2 = 7000 \text{ kg} \cdot \text{m}^2, \quad l_1 = l_2 = 1.5 \text{ m}. \tag{18}$$

The parameters (18) approximately correspond to the parameters of the SM and the manipulator of the ETS-VII experiment [5, 6]. It is assumed that the mass and moment of inertia of PL are equal to the mass and moment of inertia SM.

The angular velocity Ω_C of the system in the configuration shown in Fig. 3, determining the kinetic moment L, is

$$\Omega_C = 0.05\ \frac{rad}{s}. \tag{19}$$

As shown in [16, 18, 19], at a given kinetic moment L, depending on the initial conditions q_0 and $\dot{q}_0$, two types of eigenmovements of the reduced system are possible—oscillations relative to the lower position of stable equilibrium $q \equiv 0,\ \dot{q} \equiv 0$ and rotation. The corresponding groups of phase trajectories are separated by a separatrix. At Fig. 4 it is easy to see that the phase portrait of the proper motions of the reduced system is qualitatively similar to the phase portrait of the proper motions of the mathematical pendulum. From (15)–(17) it follows that the movement of the separatrix corresponds to the value of the total energy

$$E_s = \frac{1}{2}\frac{1}{(J_1 + J_2 + \tilde{m}l_1^2 + \tilde{m}l_2^2 + 2\tilde{m}l_1 l_2)}L^2 - \frac{1}{2}\frac{1}{(J_1 + J_2 + \tilde{m}l_1^2 + \tilde{m}l_2^2 + 2\tilde{m}l_1 l_2)}L^2, \tag{20}$$

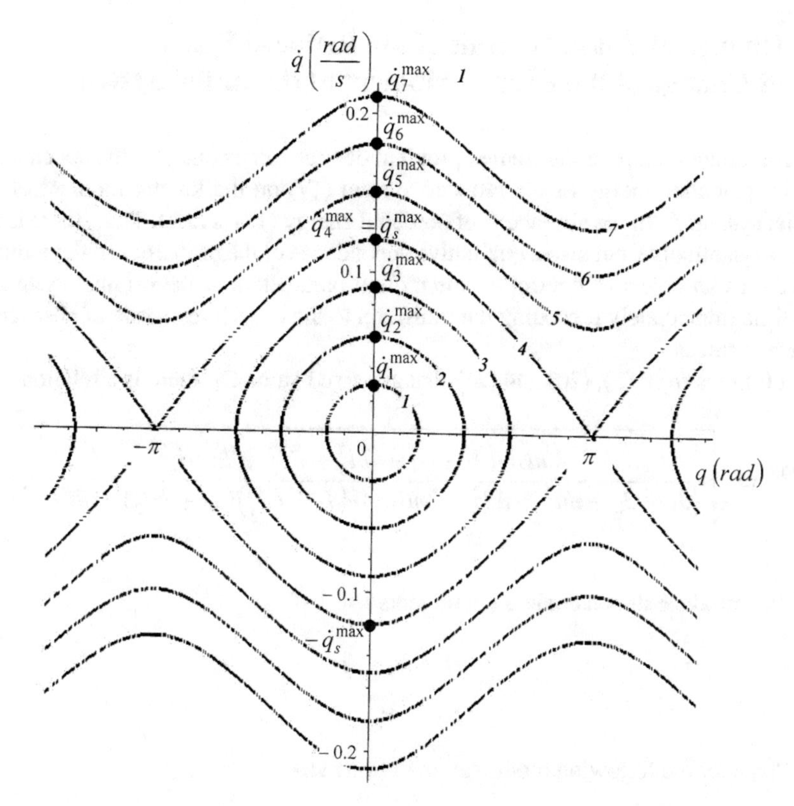

Fig. 4 Phase portrait of own movements the reduced system at $\Omega_C = 0.05\ \frac{rad}{s}$

which is equal to the potential energy of the system in the "upper" position of unstable equilibrium $q = +\pi\ rad$. Therefore, since the kinetic energy of the system is equal to the total energy (20) when passing the "lower" position of the stable equilibrium, we have

$$E_{ks} = \frac{1}{2}\frac{\left(J_1 + \tilde{m}l_1^2\right)\left(J_2 + \tilde{m}l_2^2\right) - \tilde{m}^2 l_1^2 l_2^2}{\left(J_1 + J_2 + \tilde{m}l_1^2 + \tilde{m}l_2^2 - 2\tilde{m}l_1 l_2\right)}\dot{q}_s^{\max^2} = E_s. \quad (21)$$

From (20) and (21), taking into account (3) and (5), the velocity $\dot{q}_s^{\max}$ of the passage of the reduced system moving along the separatrix, the "lower" position of the stable equilibrium, which is thus a visual characteristic of the motion of the reduced system, can be determined.

3 Change of Phase Portrait of the Reduced System at Change of the Kinetic Moment of the Initial System

An important feature of the studied problem of situational control is the dependence of the potential energy of the reduced system (17) on the kinetic moment of the initial system L. In the absence L of potential energy (17) is reset. Thus, there is not only a quantitative, but also a qualitative dependence of the properties of the reduced system as an object of control on the motion parameters of the original system at the time immediately preceding the transition to the considered mode of absence of external forces.

It follows from (3), (20) and (21) that at a fixed value Ω_C there is a relation

$$\dot{q}_s^{\max} = \sqrt{\frac{4\tilde{m}l_1l_2\left(J_1 + J_2 + \tilde{m}l_1^2 + \tilde{m}l_2^2 + 2\tilde{m}l_1l_2\right)^2}{\left(J_1 + J_2 + \tilde{m}l_1^2 + \tilde{m}l_2^2 - 2\tilde{m}l_1l_2\right)\left(\left(J_1 + \tilde{m}l_1^2\right)\left(J_2 + \tilde{m}l_2^2\right) - \tilde{m}^2l_1^2l_2^2\right)}}\,\Omega_C. \tag{22}$$

We introduce dimensionless coefficients

$$\begin{aligned} &0 < \alpha_1 < 1, \\ &\alpha_2 > 1. \end{aligned} \tag{23}$$

Consider the following modification options Ω_C

$$\Omega_{C1} = \alpha_1\Omega_C < \Omega_C, \tag{24}$$

$$\Omega_{C2} = \alpha_2\Omega_C > \Omega_C. \tag{25}$$

It follows from (22) that

$$\dot{q}_{s1}^{\max} = \alpha_1\dot{q}_s^{\max} < \dot{q}_s^{\max}, \tag{26}$$

$$\dot{q}_{s2}^{\max} = \alpha_2\dot{q}_s^{\max} > \dot{q}_s^{\max}. \tag{27}$$

At Fig. 5 shows the phase portrait of the proper motion of the system corresponding to the original system (18) with $\Omega_{C1} = 0.75\ \Omega_C = 0.0375\ \frac{rad}{s}$.

At Fig. 6 shows the phase portrait of the proper motion of the system corresponding to the original system (18) with $\Omega_{C2} = 1.25\ \Omega_C = 0.0625\ \frac{rad}{s}$.

At Fig. 7 shows the phase portrait of the proper motion of the system corresponding to the original system (18) with $\Omega_{C3} = 0\ \frac{rad}{s}$.

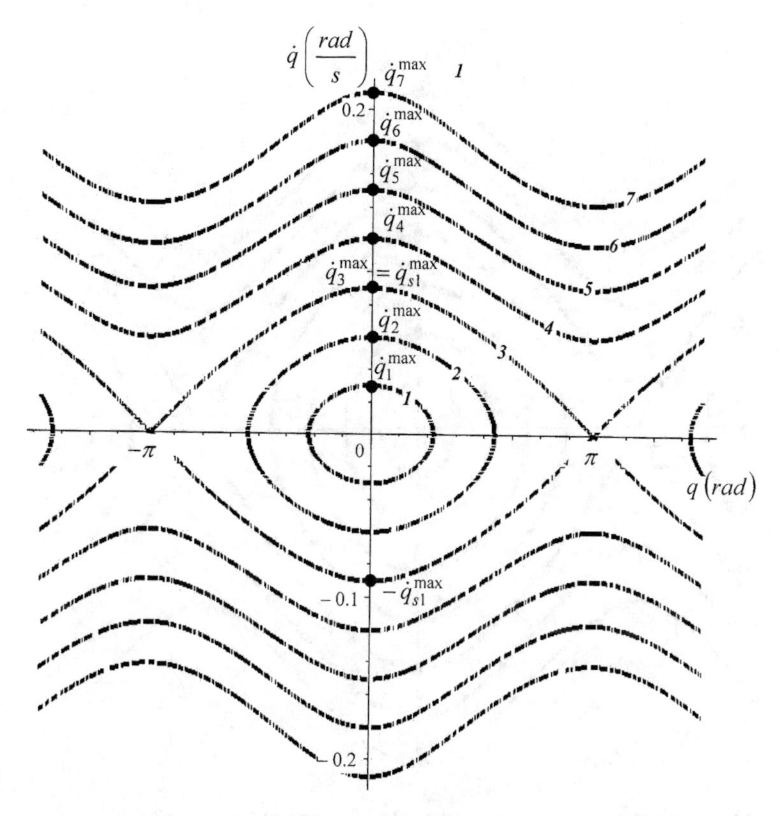

Fig. 5 Phase portrait of own movements the reduced system at $\Omega_C = 0.0375\ \frac{rad}{s}$

Each figure shows seven phase trajectories. Energy levels of phase trajectories having the same number are the same in all cases. Accordingly, the points of intersection of phase trajectories with the ordinate axis coincide, since in the position $q = 0$ the kinetic energy of the reduced system is equal to its total energy.

From a comparison of Figs. 4 and 5 it is seen that as the kinetic moment of the initial system L decreases, the energy level of the separatrix decreases, and the separatrix for the reduced system in Fig. 5 is the phase trajectory of 3. At Fig. 4 phase trajectory 3, having the same energy level, corresponds to the oscillatory motion of the reduced system. Phase trajectory 4 in Fig. 4 is a separatrix, and in Fig. 5 corresponds to the rotations.

From a comparison of Figs. 4 and 6 it is seen that as the kinetic moment of the initial system L increases, the energy level of the separatrix increases, and the separatrix for the reduced system in Fig. 5 is the phase trajectory of 5. In Fig. 4 phase trajectory 5, having the same energy level, corresponds to the oscillatory motion of the reduced system. Phase trajectory 4 in Fig. 4 is a separatrix, and in Fig. 5 corresponds to oscillations.

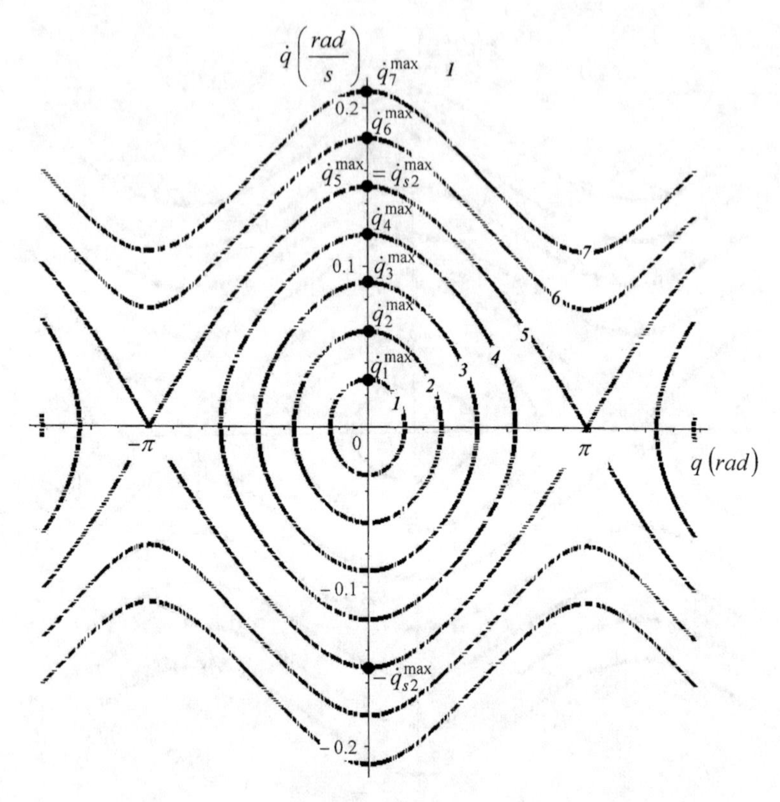

Fig. 6 Phase portrait of own movements the reduced system at $\Omega_C = 0.0625\ \frac{rad}{s}$

From a comparison of Figs. 4 and 7 it is seen that when the kinetic moment of the initial system L is zeroed, there is a qualitative change in the phase portrait of the reduced system. In the absence of the kinetic moment of the initial system, all the proper motions of the reduced system represent a rotation.

4 The Use of Their Own Movements of the Reduced System in the Control

The question of using the proper motions of the reduced system in the construction of the control is considered, in particular, in [17–19]. The controlled motion of the reduced system is described by the equation

$$\ddot{q}\,\frac{\left(J_1+\tilde{m}l_1^2\right)\left(J_2+\tilde{m}l_2^2\right)-\tilde{m}^2l_1^2l_2^2\cos^2 q}{\left(J_1+J_2+\tilde{m}l_1^2+\tilde{m}l_2^2+2\tilde{m}l_1l_2\cos q\right)}$$

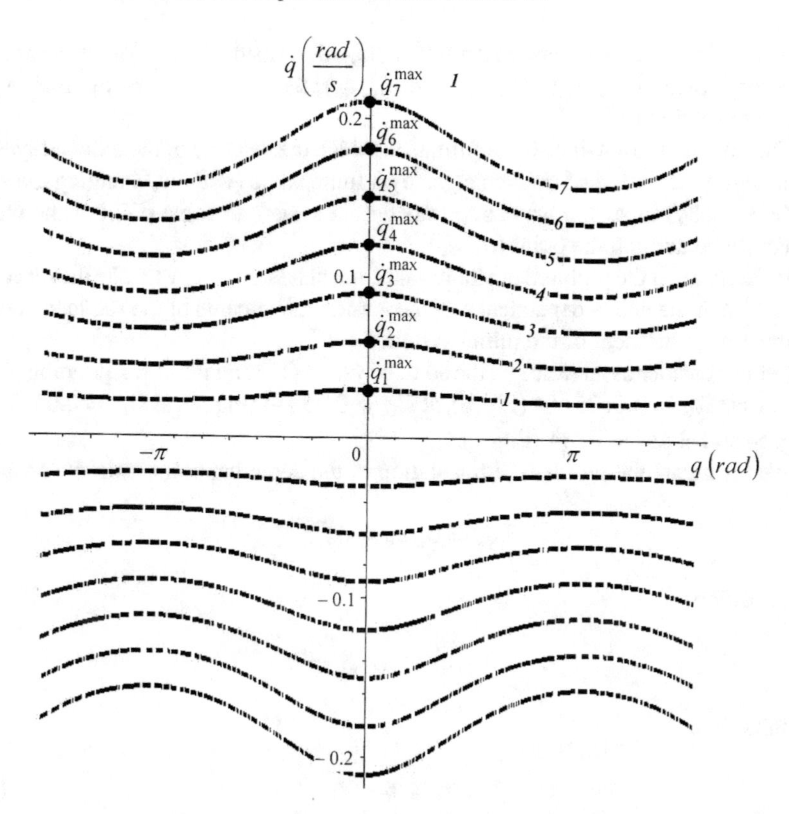

Fig. 7 Phase portrait of own movements the reduced system at $\Omega_C = 0\ \frac{rad}{s}$

$$+\dot{q}^2\frac{\tilde{m}l_1l_2\sin q\left(J_1+\tilde{m}l_1^2+\tilde{m}l_1l_2\cos q\right)\left(J_2+\tilde{m}l_2^2+\tilde{m}l_1l_2\cos q\right)}{\left(J_1+J_2+\tilde{m}l_1^2+\tilde{m}l_2^2+2\tilde{m}l_1l_2\cos q\right)^2}$$
$$+L^2\frac{\tilde{m}l_1l_2\sin q}{\left(J_1+J_2+\tilde{m}l_1^2+\tilde{m}l_2^2+2\tilde{m}l_1l_2\cos q\right)^2}=M, \tag{28}$$

where M—the control moment at the joint of the manipulator.

We will consider the problem of control as the problem of translation of the depicting point from position $(q_0, \dot{q}_0)$ to position $(q_k, \dot{q}_k)$. In accordance with the criterion proposed in [20], the optimal control from the point of view of energy costs will be that provides a minimum of functional

$$J(t_0, t_k) = \int_{t_0}^{t_k} |M\dot{q}|dt. \tag{29}$$

It is obvious that in the case of the self-motion described by Eq. (1) along a ballistic trajectory containing points $(q_0, \dot{q}_0)$ and $(q_k, \dot{q}_k)$, the control will be optimal in the sense of criterion (29).

The pulse control will also be optimal, at which the velocity of the reduced system for a negligible period of time changes by a finite value (without changing the sign of the velocity) in such a way as to transfer the system to some point of the phase trajectory containing the point $(q_k, \dot{q}_k)$.

In this case, in the problem of situational control it is necessary to take into account the nature of the above dependence of the natural movements of the reduced system on the kinetic moment of the initial system L.

Let us consider as an example the source system (18) and the corresponding given systems at $\Omega_C = 0.0625\ \frac{rad}{s}$ (Fig. 6), at $\Omega_C = 0.05\ \frac{rad}{s}$ (Fig. 4), at $\Omega_C = 0.0375\ \frac{rad}{s}$ (Fig. 5) and at $\Omega_C = 0\ \frac{rad}{s}$ (Fig. 7).

We formulate the problem of translation of the depicting point from the position

$$q_0 = 0,\ \dot{q}_0 = \dot{q}_3^{\max} \tag{30}$$

into position

$$0 \le q_k \le \pi,\ \dot{q}_0 = 0 \tag{31}$$

in finite time

$$\tau_k = t_k - t_0 \tag{32}$$

no size limit $\tau_k = t_k - t_0$.

In case

$$q_k = q_{31}^{\max}, \tag{33}$$

where $q_{31}^{\max}$—the maximum deviation from the equilibrium position of the trajectory 3 of the reduced system at $\Omega_C = 0.0625\ \frac{rad}{s}$, the system comes to the required position as a result of its own uncontrolled movement along the ballistic trajectory (Fig. 8).

In case

$$0 \le q_k < q_{31}^{\max} \tag{34}$$

the system can be brought to the desired position by means of a single pulse braking, at which the speed decreases abruptly (Fig. 8). The total energy of the reduced system is also reduced. This control will be optimal in terms of minimizing the functional (29).

In case

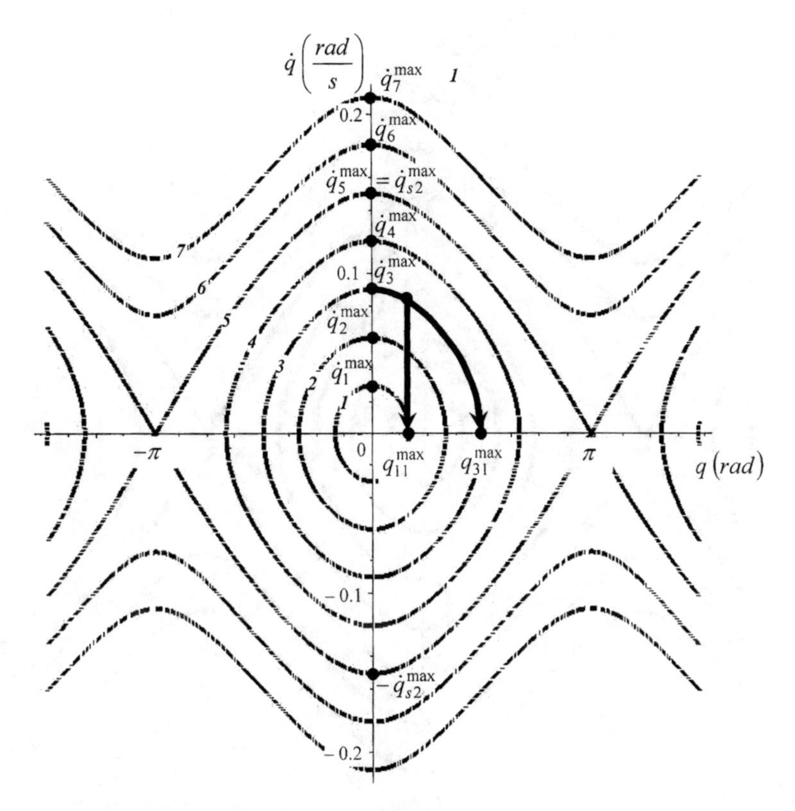

Fig. 8 Transfer the reduced system at $\Omega_C = 0.0625\ \frac{rad}{s}$ into position $0 \le q_k \le q_{31}^{\max}$, $\dot{q}_0 = 0$ with the preservation or reduction of its total energy

$$q_{31}^{\max} < q_k \le \pi \tag{35}$$

the system can be brought to the desired position by means of a single pulse acceleration, at which the speed increases abruptly (Fig. 9). The total energy of the reduced system is increased. This control is also optimal in the sense of minimization of functional (29).

It should be noted that in order to bring the system to a position in a finite time with the help of pulse control, it will be necessary to transfer the system to a trajectory with an energy level above the energy level of the separatrix, followed by braking. This control will not be optimal from the point of view of minimization of functional (29).

In the case of the reduced system at $\Omega_C = 0.05\ \frac{rad}{s}$ to bring it into position $q_{31}^{\max} < q_k < q_{32}^{\max}$ you will not need a pulse acceleration, as in the case $\Omega_C = 0.0625\ \frac{rad}{s}$, a impulse braking (Fig. 10).

Since for the reduced system at $\Omega_C = 0.0375\ \frac{rad}{s}$ trajectory 3 is a separatrix (Fig. 5), ballistic movement followed by pulse braking for a finite time can be reached any position

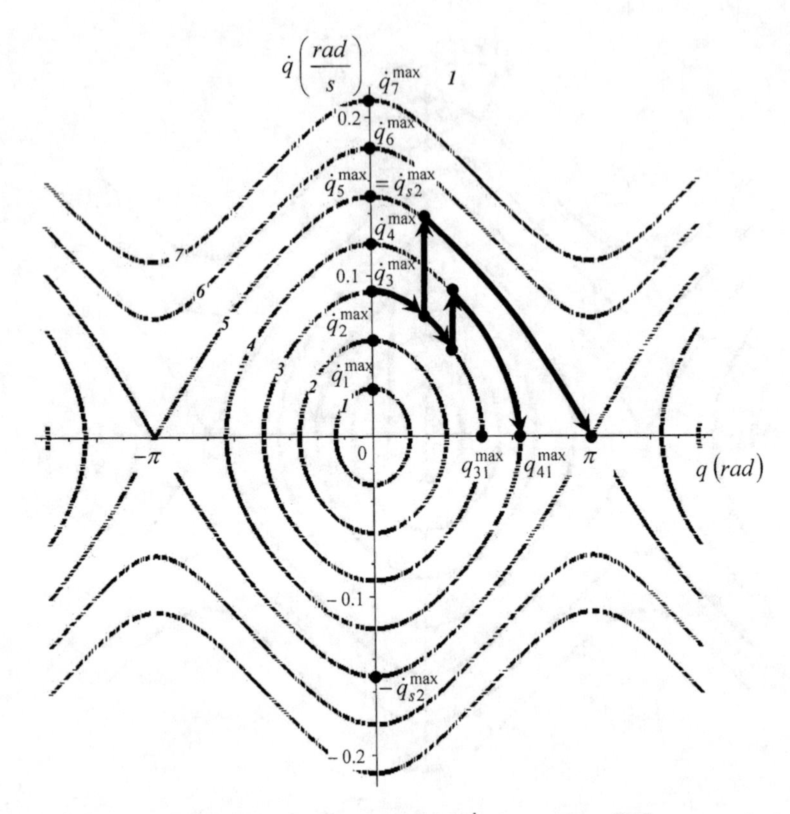

Fig. 9 Transfer the reduced system at $\Omega_C = 0.0625\ \frac{rad}{s}$ into position $q_{31}^{\max} < q_k \leq \pi,\ \dot{q}_0 = 0$ with an increase in its total energy

$$0 \leq q_k < \pi,\ \dot{q}_0 = 0. \tag{36}$$

At $\Omega_C = 0\ \frac{rad}{s}$ all their movements are Rugbymania, and ballistic movement followed by pulse braking for the final time can be achieved any position (31) for an arbitrary value of the speed $\dot{q}_0$ of the system when $q_0 = 0$.

This example illustrates the features of the use of own movements of the system in the problem of situational management.

5 Conclusion

The article considers the use as the control object of the reduced system, put in accordance with the original system «SM–manipulator–PL», functioning in the mode of controlled movement of PL relative to SM in the absence of external forces. Investigated the question of using the proper motions given system when building control. The qualitative and quantitative changes of the phase portrait of the reduced

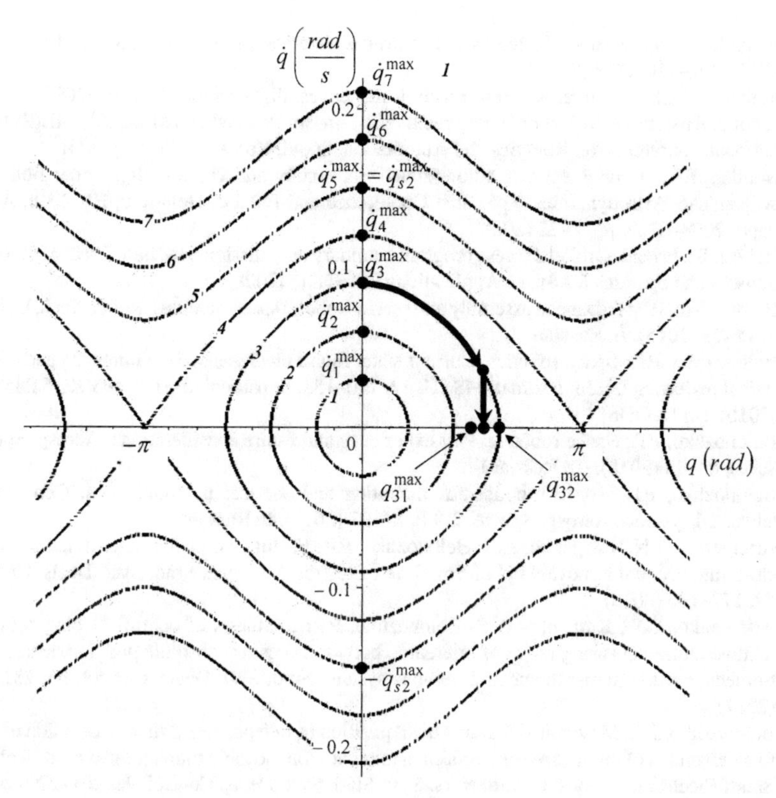

Fig. 10 Transfer the reduced system at $\Omega_C = 0.05\ \frac{rad}{s}$ into position $q_{31}^{\max} < q_k < q_{32}^{\max}$ with the preservation or reduction of its total energy

system depending on the mode of motion of the initial system are analyzed. Examples illustrating the features of situational management in relation to the reduced system.

References

1. Thronson, H.A., Akin, D., Lester, J.: The evolution and primise of robotic in-space servising. In: AIAA SPACE 2009 conference and exposition, Pasadena, California, 14–17 (2009)
2. Flores-Abad, A., Ma, O., Pham, K., Ulrich, S.: A review of space robotics technologies for on-orbit servicing. Prog. Aerosp. Sci. **68**, 1–26 (2014)
3. NASA's Robotic Refueling Mission Practices New Satellite-Servicing Tasks//NASA. http://www.nasa.gov/mission_pages/station/research/news/rrm_practice.html. Accessed 05 May 2019
4. Robotic refueling mission (RRM). https://sspd.gsfc.nasa.gov/robotic_refueling_mission.html. Accessed 04 May 2019
5. Yoshida, K.: Space robot dynamics and control: to orbit, from orbit, and future. In: Hollerbach J.M., Koditschek D.E. (eds.). Robotics Research, pp. 449–456, Springer, London (2000)

6. Yoshida, K.: Space robot dynamics and control: a historical perspective. J. Robot. Mechatron. **12**(4), 402–410 (2000)
7. Yoshida, K., Hashizume, K.: Zero reaction maneuver: flight validation with ETS-VII space robot and extension to kinematically redundant arm. In: Proceedings of the 2001 IEEE International Conference on Robotics and Automation, Seoul, Korea, 21–26 May 2001
8. Mulder, T.A.: Orbital express autonomous rendezvous and capture flight operations. In: AIAA/AAS Astrodynamics. Specialist Conference and Exhibit, Honolulu, HI, 2008, AIAA paper 2008–6768, pp. 1–22 (2008)
9. Robert, B.: Friend. Orbital express program summary and mission overview. Proc. SPIE 6958, Sensors and Systems for Space Applications II, 695803 (2008)
10. Belonozhko, P.P.: Advanced assembly and service robotic space modules. Robot. Tech. Cybern. **2**, 18–23 (2015). In Russian
11. Belonozhko, P.P.: Space robotics. Current state, future challenges, development trends. Analytical review. Sci. Edu. Bauman MSTU. **12**, 110–153. https://doi.org/10.7463/1216.0853919 (2016). (In Russian)
12. Belonozhko, P.P.: Space robotics: Past experience and future considerations. Aerosp. Sphere. **1**(94), 84–93 (2018). (In Russian)
13. Belonozhko, P.P.: Space robotics for mounting and service: potential aims, Concepts of advanced systems. Aerosp. Sphere. **2**(99), 84–97 (2019). In Russian
14. Artemenko, Y.N., Karpenko, A.P., Belonozhko, P.P.: Features of manipulator dynamics modeling into account a movable platform. Smart Electromech. Syst. Stud. Syst. Decis. Control. **49**, 177–190 (2016)
15. Artemenko, Y.N., Karpenko, A.P., Belonozhko, P.P.: Synthesis of control of hinged bodies relative motion ensuring move of orientable body to necessary absolute position. smart electromechanical systems: the central nervous system. Stud. Syst. Decis. Control. **95**, 231–239 (2017)
16. Belonozhko, P.P.: Methodical features of acquisition of independent dynamic equation of relative movement of one-degree of freedom manipulator on movable foundation as control object. Smart Electromech. Syst. Central Nerv. Syst. Stud. Syst. Decis. Control. **95**, 261–270 (2017)
17. Artemenko, Y.N., Karpenko, A.P., Belonozhko, P.P.: Synthesis of the program motion of a robotic space module acting as the element of an assembly and servicing system for emerging orbital facilities. In: Smart Electromechanical Systems: Group Interaction. Studies in Systems, Decision and Control, vol. 174, pp. 217–227, Springer International Publishing, Switzerland (2019)
18. Belonozhko, P.P.: Robotic assembly and servicing space module peculiarities of dynamic study of given system. In: Smart Electromechanical Systems: Group Interaction. Studies in Systems, Decision and Control, vol. 174, pp. 287–296, Springer International Publishing Switzerland (2019)
19. Belonozhko, P.P.: Proper inertial motion of robotic space module. Reduced system dynamics. In: Proceedings—2018 International Conference on Industrial Engineering, Applications and Manufacturing, ICIEAM 2018. May 2018, Article number 8728701. ISBN: 978-153864307-5. https://doi.org/10.1109/icieam.2018.8728701
20. Formal'sky, A.M.: Motion control of unstable objects, 232p. Fizmatlit, Moscow. ISBN 978-5-92221-1460-8 (2014)

Printed in the United States
By Bookmasters